Innovation in Electric Arc Furnaces

Yuri N. Toulouevski · Ilyaz Y. Zinurov

Innovation in Electric Arc Furnaces

Scientific Basis for Selection

The Second Edition
Revised and Supplemented

Springer

Yuri N. Toulouevski
Holland Landing, ON
Canada

Ilyaz Y. Zinurov
Chelyabinsk
Russia

ISBN 978-3-642-36272-9 ISBN 978-3-642-36273-6 (eBook)
DOI 10.1007/978-3-642-36273-6
Springer Heidelberg New York Dordrecht London

Library of Congress Control Number: 2013933295

Printed on acid-free paper

Springer is part of Springer Science+Business Media (www.springer.com)

Preface to the Second Edition

The efforts of the developers of innovations for EAFs have intensified significantly in the last several years. This was caused by tightening of economic and environmental requirements to steelmaking. Development of new processes and new types of the furnaces was a way to meet this challenge. 400 ton capacity furnaces with the 240–300 MVA transformers have been developed and implemented. Productivity of these furnaces exceeds 360 t/h. Even on 120-t EAFs productivity level has reached over 200 t/h. A new promising process namely continuous melting of scrap in liquid metal has been implemented not only in the conveyor furnaces (Consteel furnaces), but in the shaft furnaces as well. New methods of scrap charging in the shaft furnaces have been developed which have great technological and environmental advantages. Consteel furnaces started using not only the heat of off-gases for scrap preheating but the burners as well.

The absence of the unbiased comparative analysis of new furnaces and technological processes makes difficult choosing between them, since the advantages of innovations are advertised, whereas the deficiencies are concealed. Potential users of the abovementioned innovations really need such analysis. The authors' works regarding these important issues were not yet finished by the time of the first edition of this book. That is why the second edition became a necessity.

The promising process of continuous melting of scrap in liquid metal, its advantages, and limiting factors are reviewed in detail in the Chaps. 1, 6, and 7 of the second edition. The performance indices of the conveyor and shaft furnaces using this process as well as heating of scrap by off-gases are compared to those of the modern EAFs operating without scrap preheating. The root causes of low energy effectiveness of heating of scrap by off-gases are discussed, and rationale for turning away from such heating is given.

Developed by the authors design concepts of fuel-arc furnaces (FAF) with continuous scrap melting and its preheating up to 800 °C by powerful oxy-gas burners either on a conveyor or in a shaft are suggested. In FAF, electric energy consumption is reduced to 200 kWh/t with gas flow rate of 20 m^3/t. Such furnaces can successfully compete with the most advanced modern EAFs and replace them. In the new edition, the latest innovations and other up-to-date information are added to the Chaps. 1, 10, and 14, etc.

The authors express their gratitude to the readers who gave their feedback. All their observations and considerations are taken into account in the second edition of the book. The authors thank Ch. Baumann and I. Falkovich for their help and cooperation. Our special gratitude goes to G. Toulouevskaia for the great work she did for preparing of this publication.

The Authors

Preface to the First Edition

Selection of innovations for each plant as well as selection of directions of further development is one of the crucial problems both for the developers and for the producers of steel in EAF. Ineffective selection leads to heavy financial losses and waste of time. In practice, this happens quite frequently.

The main objective of this book is to help the readers avoid mistakes in selecting innovations and facilitate successful implementation of the selected innovations. The entire content of the book is aimed at achieving this objective. This book contains the critical analysis of the main issues related to the most widespread innovations in EAF. The simplified methods of calculations are used for quantitative assessment of innovations. These methods are explained by numerous examples. Considerable attention is given to the new directions of development which the authors consider to be the most promising.

In the process of writing of the book, its content was discussed with many specialists working at metallurgical plants and for scientific research and development organizations. The authors express deep gratitude for their valuable observations and considerations.

A number of the important issues covered in the book are debatable. The authors would like to thank in advance those readers who will consider it possible to take the time to share their observations. Their input will be really appreciated and taken into account in further work.

Our heartfelt thanks go to G. Toulouevskaia for her extensive work on preparation of the manuscript for publication.

Ontario, Canada　　　　　Yuri N. Toulouevski

Contents

Introduction

Electric Arc Furnaces (EAF) are being greatly improved at a fast pace. Only 20–30 years ago, today's EAF performance would be impossible to imagine. Owing to the impressive number of innovations, the tap-to-tap time has been shortened to 30–35 min for the best 100–130 t furnaces operating with scrap. Hourly productivity increased by six times, from 40 up to 240 t/h. Electrical energy consumption got reduced approximately 1.8 times, from 630 to 340 kWh/t. Electrical energy share in overall energy consumption per heat dropped to 50 %. Electrode consumption was reduced by about six times, Fig. 1. One might expect such performances should be normal for most of steelmaking shops in the immediate future.

The technological function of EAF was drastically changed. All the technological processes providing both steel qualities required and its special properties have been moved out the furnaces to secondary ladle metallurgy equipment.[1] The necessary increase in furnace productivity could not be achieved without this revolutionary change in EAF steelmaking. The main technological processes in the modern furnaces are melting of solid charge materials and heating of liquid bath. It is precisely these substantial thermal-energy processes that now define furnace productivity. To get these processes going, it is necessary to obtain heat from other kinds of energy (electrical or chemical) and transfer it to zones of solid charge or liquid bath. This is why the electric arc furnaces themselves and the processes in them are reviewed in this book mainly from the unified thermal-energy point of view.

These furnaces turned to be very flexible in terms of charge materials selection. They could readily accommodate to melting in various combinations steel scrap, pig iron and hot metal, and reduced iron as pellets or briquettes. In the majority of furnaces, metal charge consists of scrap with small additions of pig iron. Traditionally, scrap is charged into the furnace from above as a single charge or in two-three portions. Only at the so-called Consteel furnaces and at certain shaft furnaces, scrap is practically continuously charged by means of a conveyer or a hydraulic pusher via a furnace sidewall door. The wide variety of innovations being

[1] These processes and equipment are not considered in the book.

offered by the developers for each particular case corresponds to the various furnace operation conditions.

Changes in heat techniques, furnace designs, and equipment are taking place at a fast pace. Every year, new technical solutions are offered and widely advertised. Steel manufacturers have difficulty navigating through the flood of innovations. Under steep competition, advertisement information is somewhat biased and incomplete. This makes selection of innovations for solving particular problems even harder. It is not easy to decide which information is trustworthy enough. But it is much more difficult to decide what to select: an innovation which was already proved by practice or it is better to take a risk of the first realization which in the case of success promises maximum economical effect. Frequently, the cost of new technologies exerts the decisive influence on this selection. Certainly, the price is one of key criteria. But the other not less important factors such as, for instance, a new equipment reliability have to be taken into consideration. Therefore, when being based for the most part on a price, a serious error can be made.

What could help to carry out unbiased analysis of innovations and select those which could yield the best results for particular circumstances of a given plant? First of all, comprehensive understanding of mechanisms and basic laws defining the main processes of the EAF heat is required. The modern concepts of these processes are presented in numerous magazine papers and reports from technical conferences which are held worldwide on a regular basis. For a practical steelmaker, it is hard to get reliable general information necessary to solve specific practical problems. Meanwhile, the knowledge of general simplified yet correct in principle concepts is sufficient for decision making. These general concepts are currently commonly accepted. Without compromising scientific strictness, these principles are discussed in this book at the level easy to understand for the readers who do not have an adequate background in this field.

Data on effectiveness of any proposed innovation must not contradict proven principles of the processes of the heat. If such a contradiction takes place, a proposal should be excluded from further consideration. Regretfully, experience proves that innovations which do contradict to the basic principles are proposed rather frequently.

A typical case can be shown. Up to this point, various methods of bath oxygen blowing are proposed in order to provide carbon oxidation inside the bath not to carbon monoxide CO as it takes place in reality but to carbon dioxide CO_2. If this would be possible, both heat amounts released in the bath and bath heating rate could be increased several times. However, according to the basics of physical chemistry of steelmaking processes, formation of CO_2 in presence of liquid iron is possible only in practically insignificant amounts. This basic principle was responsible for the failure of all attempts to oxidize carbon in steel bath to CO_2. These attempts were repeatedly undertaken in the past in both open hearth furnaces and oxygen converters.

This example demonstrates that historical approach to the analysis of innovations proposed is very helpful. Such an approach is widely used in various chapters of this book. In certain cases, data, obtained not only in the modern

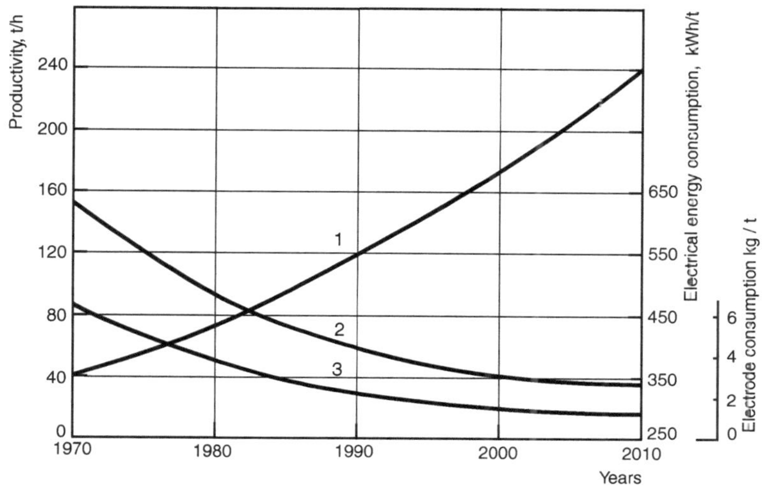

Fig. 1 Improvement in the 120-t EAF performances. *1* productivity, t/h, *2* electrical energy consumption, kWh/t, *3* electrode consumption, kg/t

steelmaking units but also in the obsolete open hearth furnaces, are used. When evaluating innovations for electric arc furnaces, the experience from open hearth furnaces as well as from converters proves to be highly useful. This is particularly relevant for the results of scientific and industrial studies of oxygen blowing in open hearth bath since the studies similar in scale, accuracy of experimental procedure, and resultant effectiveness have not been conducted in EAF.

Simplified calculations should be used for the preliminary comparative evaluation of innovations. Such calculations can be done manually by using regular calculators. Their accuracy is quite sufficient for the purpose pointed out. In many cases, the accuracy is not inferior to the accuracy of calculations which use complex methods of mathematical simulation. It can be explained due to the fact that often the input parameters for calculations are known quite approximately, and the accuracy of final results cannot exceed the accuracy of the input data regardless of calculation technique applied. In this regard, the mathematical calculations are similar to millstones: whatever you pour in that what you will get.

It should be emphasized, that carrying out even the very simple calculations greatly promotes comprehensive understanding of physical basics of processes and effects produced by various factors. Using "off the shelf" programs developed by means of the mathematical simulation of processes does not provide such possibilities. For the consumer, these programs are similar to a "black box" which does not reveal the mechanism of the process. The "black box" produces the final result but does not allow judging the conformity of the calculation to all the conditions of the specific case. Therefore, common "off the shelf" programs must be used with a great caution for evaluation of specific innovations.

When evaluating innovations which require heat balances of EAF, it is necessary to calculate thermal effects of exothermic reactions of oxidation of carbon, iron and its alloys. These thermal effects strongly depend on temperature of the initial substances and chemical reaction products. In a series of important cases, an effect of temperatures is not taken into account or it is not completely considered in the tables available to the readers. This leads to significant errors in calculations. In this book, an accurate and universal method, which is appropriate in all cases, is offered to determine influx of chemical heat. It is based on so-called method of full enthalpies and is very convenient for practical use.

Currently, most of innovations for EAF are aimed at the development of means and methods providing further intensification of processes of solid charge melting and liquid bath heating. Calculations in this field require knowledge of processes of heat transfer as well as hydro- and aerodynamics. To help readers mastering such calculations, several chapters containing required minimum of information in these fields of science are included in the book. This information is presented in a rudimentary form yet not compromising strict scientific meaning. Formulae for calculations are given in simplified form convenient for practical computing. Nevertheless, in doing so, the accuracy of calculations is maintained. Application of these formulae is illustrated by a large number of examples for analysis of innovations. Getting familiar with material in this book will allow the reader to perform required calculations on his own. In order to facilitate still further the performance of calculations, reference data needed for calculations are given in the book. This permits the readers to do away with the problem of searching for such data in various handbooks.

The book covers a wide variety of topics ranging from scientific concepts to state-of-the-art improvement practice of steelmaking in EAF. The book also contains new, progressive, in authors' opinion, ideas on key issues regarding intensification of the heat such as scrap heating using high power oxy-fuel burners, deep bath blowing with oxygen and carbon using high-durable tuyeres, etc.

Significant attention is given to analysis of various directions of automation of the energy modes control of the heat. The descriptions of different automated control systems are drawn up by their developers according to the same principle and in essence differ only slightly from each other. Usually, the system functions are enumerated in detail. For example, the system controls the consumptions of electrical energy, oxygen, and fuel ensuring their savings and the increase in furnace productivity. But there is no information on how this is being done or on a specific algorithm (mechanism) of the system operation. Therefore, both estimation and selection of innovations in this field present great difficulties for metallurgists. The method for comparing the automation systems based on analysis of information used for controlling the heat is outlined in this book. This method provides a means for easy understanding of real and alleged advantages of a particular system as well as for making a justified decision.

The last chapter of the book deals with environment protection from gas and dust emissions of arc furnaces. A problem of reduction in energy gas evacuation costs is reviewed with consideration for current tendencies.

You can assess this book based on its contents. It is addressed to a wide range of EAF-steelmakers and all other metallurgists related to this industry. This range includes, among other, three categories of specialists: those who have to effectively use innovations in day-to-day practical work, those responsible for selection of innovations for their factories, and the developers of new processes and equipment for EAF. The book can also be used as a textbook for students of all levels studying metallurgy.

Chapter 1
Modern Steelmaking in Electric Arc Furnaces: History and Development

1.1 General Requirements to Steelmaking Units

The structure of modern steelmaking has been formed gradually during the last 100 years. In this period, due to many different reasons, the requirements to steelmaking units have changed substantially. Some production methods have appeared and developed, while other ones have become noncompetitive and have been rejected. All these changes were interrelated and influenced each other. The understanding of electric steel production development and its prospects can not be complete if this process is studied separately setting aside the development of steelmaking in general. Therefore, it is necessary, even if briefly, to review the history of not only electric arc furnaces but also other steelmaking units competing with each other.

Steelmaking units should meet a number of requirements that could be classified into four groups in the following way:

1. Process requirements ensure the necessity to produce various steel grades of required quality.
2. Economic requirements call for reduction of manufacturing costs so as to increase profitability and competitiveness of products.
3. Environmental requirements do not permit any excessive environment pollution, the level thereof being governed by state regulations.
4. Health and safety requirements exclude the use of physically and psychologically straining labor which, at a certain stage of social development of society, becomes unacceptable for the population of a given country.

In any case, all innovations introduced in steelmaking have always been aimed at fulfilling some or all of the above mentioned requirements. However, the influence of these requirements has been changing greatly in the course of time.

Y. N. Toulouevski and I. Y. Zinurov, *Innovation in Electric Arc Furnaces*, DOI: 10.1007/978-3-642-36273-6_1,

1.1.1 Process Requirements

Up to the middle of the twentieth century, the most important changes in steelmaking were instigated by these very requirements. At the very beginning of the century, they led to development and wide spread of the electric arc furnaces (EAFs), since these units made it possible to easily achieve highest temperatures and ensured the best conditions for producing of high-quality alloyed steel grades and alloys. Previously, such metal could be produced by the crucible method only. Due to its inefficiency and too high requirements to the purity of raw materials, this method could not compete with the EAF process. A demand for special expensive steels and alloys with particular properties was quickly increasing. Electric arc furnace became the main supplier of such metals, though it was also used for production of relatively small quantities of common steel.

The process requirements were also a reason for replacement of acid and basic Bessemer converters with open-hearth furnaces. Due to increased nitrogen content, the quality of steel produced in the air-blast converters was greatly inferior to that of the open-hearth steel. As a result, the open-hearth method has become prevailing method of steel mass production, right up to the development of oxygen converters and even somewhat later.

The process requirements ceased to have a substantial effect on the relative competitiveness of basic steelmaking units when the ladle furnaces were introduced and became widespread as molten metal treatment units. At present, both oxygen-blown converters and EAFs usually produce semi-products of preset temperature and carbon concentration. This metal is treated to reaching the final chemical composition, refined by removing dissolved gases and non-metallic inclusions therein, and heated up to optimal temperature in ladle furnaces and other secondary metallurgy units.

Practically every steel grade can be produced by this way. The only obstacle encountered when producing some specific steel grades in EAFs is the contamination of scrap with copper, nickel, chrome and other residual contaminants which can not be removed in the course of processing of the finished steel. Permissible content of these contaminants is strictly limited in quality steel grades. This obstacle is overcome by means of more careful scrap preparation as well as by partial substitution of scrap with hot metal or products of direct iron reduction. Recently, such products are used in electric steelmaking rather widely.

1.1.2 Economic Requirements

The cost of scrap and ferroalloys amounts to approximately 70% of the general costs in EAFs operating on scrap. The so-called costs of operating constitute the rest 30%; the cost of electrical energy, fuel and electrodes account for about 40% of the latter. There are three possible ways of reducing the costs:

1. By cutting down specific consumption of charge materials, energy-carriers, refractory materials etc. per ton of steel.
2. By increasing output and thus reducing specific manufacturing costs, such as maintenance staff costs etc.
3. By replacing expensive charge materials and energy-carriers with cheaper ones.

Innovations developed in the first direction are always justified as well as those in the second one. For more than half a century, the main direction of development of electric arc furnaces is increasing of their productivity. Almost all innovations, implemented in this period of time, were aimed at this problem. Without solving this problem the EAF could have never become the very steelmaking unit which along with oxygen converter is a determinant of world steelmaking.

Excluding the cost of metal charge the productivity is a parameter on which the entire economics of steelmaking process depends to the greatest degree. As a rule, when productivity is increased, manpower and maintenance costs are reduced, as well as costs of electrical energy, electrodes, fuel, refractories and other so-called costs of operating, including overall plant expenditures.

Electric arc furnaces are mostly intended to be installed at mini-mills where they determine productivity of the entire plant. Increasing output of mini-mills to one-two million of tons per year or even more had decisive effect on the maximum productivity level of EAF. It is most reasonable to equip steelmaking shops at such plants with one furnace, two at the maximum. Such organization of production allows minimizing manpower and operating costs in general.

If the shops are equipped with a number of furnaces then under conditions of extremely high pace of operation it is impossible to avoid some organizational delays. Any disruption of the preset production pace at one of the furnaces adversely affects other furnaces thus reducing significantly the shop productivity and that of the plant as a whole. Therefore, preference is given to the shops equipped with one furnace, even in the cases when required output exceeds 2.0–2.5 million ton per year.

Innovations developed in the third group are not always justified. Prices on materials and energy-carriers are subjected to rather abrupt fluctuations so that they are difficult forecast. In different countries, they can change dissimilarly and even in the opposite directions. That is why the innovations, determined only by a price difference, are associated with relatively high risks, especially when they are aimed for long-term and wide spread.

Let us discuss a number of examples. Scrap was substantially cheaper than hot metal for a long time nearly everywhere. Under such conditions, increasing amount of scrap re-melted in oxygen converters aiming at reducing hot metal consumption, could promote a significant increase in converter steel profitability. To achieve this various methods were developed to introduce additional heat into converters, such as scrap preheating by powerful oxy-fuel burners, introduction of coal and other carbon-containing additives into the charge, post-combustion of carbon monoxide evolved in the converter etc. Developing out and mastering these innovations was associated with significant difficulties. To overcome these difficulties long-term extensive industrial research accompanied by vast spending was conducted in a number of countries.

However, the interest in all these innovations was gradually declining as scrap price was increasing and approaching the price of hot metal. Therefore, replacing hot metal with scrap in converters was stopped. On the contrary, in the recent years, hot metal started to be used in EAFs in increasing amounts. This assured significant reduction of tap-to-tap time and electrical energy consumption, and also promoted production of such steel grades which require charge rather free from foreign contamination.

At present, a situation similar to that of replacing hot metal with scrap is developing with regard to innovations aiming at substituting electrical energy with the natural gas energy in EAFs. Just recently, this aim was justified by low cost of gas compared to electrical energy. At first glance, such price ratio could not change substantially, since a significant share of electrical energy is produced at thermal power plants using gas. However, in reality in most of countries, the price of natural gas was rising many times more quickly than the price of electrical energy. For example, in the USA the price of natural gas has increased by more than four times since 1990 to 2008. This limited use of oxy-gas burners in EAFs. Presently, in the North America, the prices for natural gas have sharply dropped down to about the price level that has taken place in the early twenty-first century due to commercial development of a shale gas deposit. A new economical situation contributed to the fact that replacing electrical energy with energy from natural gas in EAFs has again become an urgent problem.

Attention has to be given to the fact that increase of price of scrap and natural gas cannot be explained by alleged rising shortage of these resources. On the contrary, the supplies of unused dormant scrap are constantly growing in the majority of the developed countries. For instance, presently in the US supplies of steelmaking-worth scrap exceed 800 million tons. Natural gas price is growing even in the countries where gas reserves are practically unlimited. The increase in the mentioned prices is not linked directly to either scrap preparation costs or gas production and transportation costs.

Along with long-term rising trend, world scrap prices are subject to very sharp fluctuations, depending on the demand. Scrap price is growing during the years of industrial development and steelmaking increase, while it is falling in the period of industrial stagnation. In some years, scrap prices have changed repeatedly by 1.5–2 times. Absence of these fluctuations cannot be guaranteed in the future; they obstruct investments in scrap-processing industry.

The technical advantages of using scrap and natural gas in steelmaking are beyond any doubt. It was calculated that each ton of steel produced out of scrap instead of hot metal provides saving of approximately 1100 kg of iron ore, 640 kg of coal and 2.9 MWh of energy. However, purely technical considerations are not prevalent in this case. The analysis of the actual situation leads to the conclusion that a very contradictory and unpredictable practice of formation of prices on scrap, hot metal and energy-carriers is not caused by objective technical reasons but is dictated by transient short-term considerations both of political and purely commercial nature. These considerations are formed under the influence of many factors.

The world scrap market is rather sensitive to situation with sharp fluctuations in some countries with highly-developed steelmaking, especially in those of them where the government exerts substantial influence and even strictly controls scrap import and export. It must be taken into account that scrap could be considered not only as waste to be disposed, but also as raw material resources of great strategic significance. All this influences the prices and rather negatively affects general technological progress in steelmaking.

Prices of various types of metal charge and energy-carriers are subject to sharp fluctuations with time and country to country; therefore the price factor should not be used as a unique criterion that would allegedly divide all new developments into promising and prospectless. The very processes, which are unprofitable now, could prove, in the nearest future, not only economically sound but also the most efficient in some countries or maybe everywhere. Hence, it is quite necessary to pay considerable attention to developing innovations aimed at radical improving such objective performance parameters of EAFs as output, energy efficiency, environmental protection level, operational reliability etc., even if today's prices of resources required for EAFs seem to be unacceptable. Entire history of technological development proves this point of view.

1.1.3 Environmental and Health and Safety Requirements

In the first half of the last century, the effect of these factors was practically insignificant. Afterwards, it started to rise gradually. In the recent decades, steelmakers, especially in developed countries, encounter, increasingly stringent restrictions regarding emissions of CO_2, CO, NO_x, dioxins and other harmful gases and dust. There is a point of view rather worthy of notice that environmental requirements are observed to become more and more stringent to a degree that in some cases is beyond any reasonable limits. Further even rather insignificant reduction of emissions requires multiple cost increase from steelmakers, although steelmaking is not by far the main air pollution source on a national scale.

According to the data provided by Thyssen Stahl AG, dust emissions were reduced at the company's plants in the 1960s from 18–19 to 2 kg per ton of steel while the costs amounted to 100 DM per ton of dust. In the 1970s, dust emissions were further reduced by 0.45 kg per ton of steel while the costs amounted to 3,000 DM per ton of dust. In the early 1980s, emissions were further reduced by 0.5 kg per ton of steel which had required an increase in cost up to 10,000 DM, and to satisfy the new requirements it would take more than 100,000 DM per ton of dust [1].

At present, cost of environmental protection amounts to 15% of a new plant's price. Extremely stringent environmental requirements cause moving steelmaking plants from Europe and the USA to Third World countries where these requirements are much less strict. Mini-mills are being built equipped with EAFs, since, along with other decisive advantages compared to the integrated plants, emissions into atmosphere are significantly lower when producing steel from scrap. At integrated plants CO_2 emissions amount to 1825 kg/t of liquid steel. For EAFs where the charge consists of 50% scrap and 50% hot metal these emissions are 1717 kg/t, and at furnaces operating on 100% scrap those are only 582 kg/t. Besides, water pollution decreases by 40% approximately when producing steel at mini-mills.

It should be emphasized that using hot metal in EAFs to a great degree strips them of their environmental advantages. Some rather efficient innovations in EAFs such as slag foaming by carbon injection or use of oxy-gas burners provide significant reduction of electrical energy consumption but increase emissions of CO_2. Such innovations should be evaluated, from the environmental requirements point of view, not only within the bounds of steel melting plants but also on a national scale, taking into consideration the reduced demand for electrical energy for steelmaking and, consequently, reduced emissions at the thermal power plants.

The significance of health and safety requirements was growing concurrently with increased effect of environmental factors. Presently, in the US and Europe mostly immigrants from the Third World countries are employed at the positions requiring hard physical and mentally demanding labor in steelmaking shops. This, in turn, leads to complex social problems. Therefore, more attention is paid to the innovations aimed at elimination of hard physical labor by changing the processes, introducing mechanization and automation thereof.

The health and safety requirements affected significantly the very structure of steelmaking. Contrary to the well-known opinion, open-hearth steelmaking was replaced by oxygen converter method not so much for economic reasons as for social ones. There was only a slight difference between the cost of steel produced by open-hearth method and oxygen converter method. According to various estimates, the difference amounted to approximately one dollar per ton of steel. However, labor conditions in open-hearth shops compared to those in oxygen converter shops were extremely gruelling, especially during furnace repairs.

Shrinking and later complete stopping of open-hearth steelmaking caused, in most countries, sharp increase of available scrap resources, since a scrap share in open-hearth charge was approximately 40–45% whereas in the converter charge it was about 20–25%. That was one of the key factors which caused the appearance and fast spread of mini-mills equipped with electric arc furnaces operating on scrap.

The examples presented above show the close connection of all changes in the structure of steelmaking. Under the pressure of the requirements discussed above and within the evolving economic conditions these changes have led eventually to the modern steelmaking in EAFs.

1.2 High-Power Furnaces: Issues of Power Engineering

1.2.1 Increasing Power of EAF Transformers

This innovation plays a decisive role in sharp shortening tap-to-tap time and increasing EAF productivity per hour. The first so-called ultrahigh-power (UHP) furnaces appeared in the US in 1963. These 135-t furnaces were equipped with 70–80 MVA transformers, specific power amounting to 520–600 kVA/t. Previously, specific power of 50 to 100-t furnaces did not exceed 200–250 kVA/t. Due to their successful operation UHP furnaces became widespread rather quickly, and their specific power was increased to 1000 kVA/t.[1]

In recent years, furnaces of 300 to 400-t capacity have gained some spread. In such furnaces, the specific transformer power is reduced despite a sharp increase in its absolute power. It is mostly explained by a limited current carrying capacity of electrodes, Sect. 1.2.3. The very powerful transformers in the world have been installed at a 420-t EAF with a power of 240 MVA in Gebze [2], and at a 300-t EAF with that of 300 MVA [3] in Iskenderun, Turkey.

At the first stage of UHP furnaces, increase of their average monthly and annual productivity was limited by to sharp deterioration of durability of sidewalls and roof refractory lining and subsequent increase of downtime during repairs. This obstacle has been eliminated by replacing up to 85% of total lining area with water-cooled panels.

Hourly productivity of the furnace at the given capacity is inversely proportional to overall tap-to-tap time τ. The value of τ represents a sum of two components: power-on furnace operation time when arcs are on (τ_{on} period) and so-called power-off time of operations requiring electric arc switching-off (τ_{off} period) $\tau = \tau_{on} + \tau_{off}$. The τ_{off}-operations include tapping, closing of taphole after tapping, scrap charging by one or several baskets etc. By increasing electric arc power, only the duration of scrap melting and liquid bath heating (τ_{on} period) is reduced. Therefore, if duration of power-off period τ_{off} is too long and τ_{on}/τ ratio drops below 0.7, using UHP furnaces become economically inexpedient.

The higher the power is the greater part of the tap-to-tap time should be taken up by the power-on furnace operation time, and the closer the average power value within τ_{on} time to the maximum value should be. Any power reduction, occurring within that period, decreases EAF's productivity. When UHP furnaces were implemented, these requirements significantly promoted reducing the duration of power-off operations and transferring process operations, which required reduced power, from furnaces to secondary ladle metallurgy units. Such process operations which increase the tap-to-tap time significantly are desulphurization of steel and refining it to the required chemical composition. As it has been already mentioned, it would be impossible to achieve the current parameters of UHP furnaces without

[1] See Sect. 1.2.2 on difference in measuring EAF power in VA (volt-ampere) and watts, W.

converting them over to producing semi-product. Further below, these furnaces will be referred to as "high power".

1.2.2 Specifics of Furnace Electrical Circuit

An increase in electrical power aggravates not only water-cooled elements and durability problems of refractory lining associated with the increase of thermal radiation from electric arcs. Problems of electrical nature caused by the specifics of the furnace electrical circuit arise as well. This circuit includes electric arcs, electrodes and so-called secondary circuit connecting electrodes with the furnace transformer. The secondary circuit consists of busbars, flexible cables and current-conducting arms with electrode clamp. Without going into details let us discuss only the basic electrical specifics of the circuit.

Overwhelming majority of EAFs operates on alternating current (AC). As it is well known, alternating current I, measured in amperes (A) and voltage U, measured in volts (V), change in a sinusoidal manner. If the current passes through an active resistance[2] both values reach their maximum and pass zero simultaneously, i.e., they coincide in phase, Fig. 1.1. In this case, consumed actual electrical power P, measured in watts, W, is converted entirely into heat, $P = U \cdot I$.

If the AC circuit includes not only active resistance but also inductive impedance, e.g., a conductor wound around an iron core then the maximum and minimum values of current and voltage will not coincide in phase, Fig. 1.1, curve a. In that case, electrical energy is not entirely converted into heat; a part thereof is consumed to form an alternating electromagnetic field in the space surrounding the electrical circuit. Phase shift in such circuit is characterized by the so-called power factor $\cos \varphi < 1$. The actual power consumed by the circuit and converted into heat is calculated by formula $P = U \cdot I \cdot \cos \varphi$. Electric furnace circuit is characterized by certain inductance. Therefore, the EAF power is usually expressed both in actual power units, MW, and in total power units, megavolt amperes, MVA. P, MW = P, MVA. $\cos \varphi$.

The electric circuit specifics are determined first of all by the specifics of the arcs themselves. For simplicity sake, let us discuss first the processes which occur in a direct current (DC) arc; similar processes take place in high power AC arcs. DC arc column between a graphitized electrode (cathode) and either scrap lumps or a furnace bath, is high-temperature plasma consisting of neutral molecules and atoms of various gases and vapors present in the furnace freeboard, as well as of electrically charged particles, i.e., of electrons and ions. Current transfer in the arc is conducted mainly by electrons emitted from the cathode heated up to a high temperature. According to various estimates, in high power furnaces, the

[2] For alternating current, active resistance is, for instance, the resistance of straight wire the electromagnetic field energy of which could be ignored.

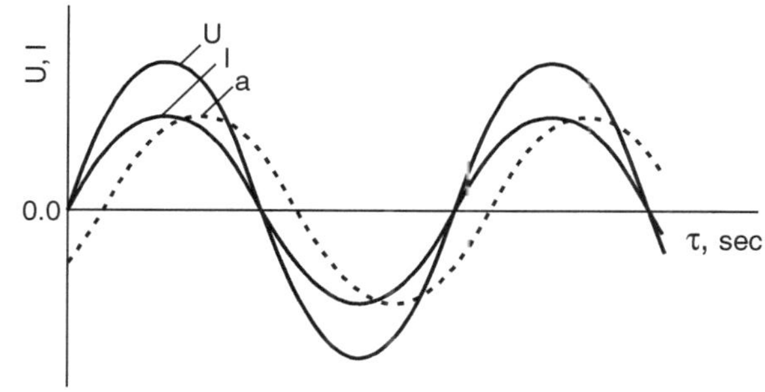

Fig. 1.1 Curves of voltage *U* and current *I*. *a* phase shift

temperatures in the arc column range within 6000 to 7000°C and current density reaches several thousands of A/cm^2.

Arc column is compressed by electro-dynamic forces resulting from the interaction of the arc current with its own electromagnetic field surrounding the arc. The resulting pressure affects the liquid bath surface causing the arc to submerge into the melt to a certain degree. If the current increases electro-dynamic forces compressing the arc rise as well as heat energy concentration within the arc space and depth of arc submersion into the liquid bath.

Similarly to the regular conductors, e.g. metallic ones, arc voltage rises as its length increases. However, contrary to these conductors which obey Ohm's law stating linear dependence between voltage and current, active resistance of the arc decreases as current increases. Therefore, an increase in current does not require voltage rise. Such a nonlinear volt-ampere characteristic of the arc does not provide conditions required to stabilize arc discharge. The secondary circuit should have a certain resistance for stable arcing, active resistance in DC EAF, as well as inductance for the AC EAF. All the above stated with respect to the DC arc can be to a considerable degree applied to arcs in AC EAFs, considering values of current and voltage within each half-cycle, with taking into account inductance in the secondary circuit.

In the AC arc cathode and anode alternate at each voltage direction change. Either electrode or surface of scrap lumps or liquid bath serves as the cathode, by turns. In modern furnaces, the arc discharge does not cease as voltage approaches zero at the end of each half-cycle, since high power arcs have significant inertia regarding both conductivity and temperature condition of the arc column. However, the shape of arc voltage curves can significantly differ from sinusoidal. That difference is getting smoothened as power and current grow.

Arcing stability could vary significantly during the course of the heat. Immediately after charging a new basket of scrap, arcing takes place on the surface of separate scrap lumps which are continuously moving while the charge settles down. During this period arcs are not stable. Despite operation of automatic controllers the arcs break rather often as a result of sudden sharp increase of arc length. Arc discharge breaks occur also during short circuit when electrodes get in contact with scrap pieces. After the initial bore-in period, arcing is observed between the electrodes and the surface of molten metal collected at the bottom, and later, between the electrodes and the surface of formed bath. In this case, arc stability increases significantly. As arc current and power are increased their stability grows during all periods of the heat. The same is observed when scrap is preheated to high-temperature.

Since the arc volt-ampere characteristic is nonlinear, the entire electrical circuit of EAF becomes nonlinear as well. The processes taking place in this circuit as well as dependences between electrical parameters of the circuit are rather complex. These dependences are shown schematically in Fig. 1.2. With an increase of current I actual furnace power P_{EAF} and arc power P_{ARC} (P_{ARC} is lower than P_{EAF} due to power losses in the secondary circuit) change according to extremum curves, i.e. curves with a maximum. At the beginning, the P_{EAF} and P_{ARC} grow to certain maximal values and then fall quickly as current grows further. During short circuits, when currents are the highest the arc power drops to zero. Arc voltage U_{ARC} and power factor cos φ decrease steadily as current increases, Fig. 1.2. The power factor cos φ is close to 0.8 in the modern high power furnaces.

The dependence between current value I and electrical energy consumption E_{EL} is also of extremum nature. If only this kind of energy were used in the furnace, then to minimum energy consumption a certain current value I_1 would correspond. Current I_1 is considerably lower than current I_2, at which maximal arc power P_{ARC} is reached. It could be assumed approximately that maximum rates of scrap melting and liquid bath heating, i.e., maximal furnace productivity correspond to

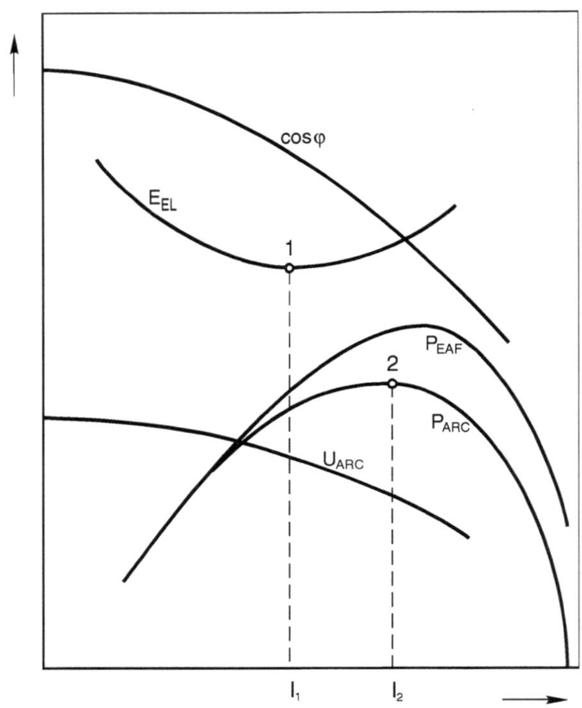

Fig. 1.2 Electrical characteristics of EAF

maximal arc power. Thus, most economical mode of the heat as far as electrical energy consumption is concerned, does not match the mode of the maximum productivity as the former requires operation at lower current and lower arc power. It is basically impossible to combine these two modes.

The above mentioned principle is based upon the fundamentals of thermodynamics and is valid not only for EAF but also for all other types of furnaces. As the power of external energy source increases from low values corresponding to the furnace idling (in this case, actual power of electric arcs is meant), the thermal efficiency of the source first rises rapidly from zero to its maximum level and then starts dropping. This could be explained by the fact that further increase of useful, i.e. assimilated power falls more and more behind the total rise of the power source due to of the advanced growth of heat losses. These problems are discussed in detail in Chap. 5.

1.2.3 Optimum Electrical Mode of the Heat

Electrical mode is the program of changing, in the course of the heat, of such electrical parameters of the furnace circuit as current and voltage, arc power etc. The design of the furnace transformers allows changing these parameters stepwise within a wide range, whereas current and voltage may vary at the constant maximum actual power as well. Switching over the transformer voltage steps "on-load" is performed either automatically or by operator's command.

Since the introduction of high power furnaces, their electrical modes were developed based on the following general principle. At the period of melting solid charge, when scrap still shielded sidewalls against arc direct radiation, the maximum transformer power and long arc were used, i.e., high voltages at reduced currents. As liquid bath formed arcs were shortened gradually by reducing voltage and raising current. At final stages of the heat the furnace operated at reduced power with maximum currents which assured maximum submersion of the short arcs into the bath, the highest heat absorption by metal, and the lowest heat losses through water cooling the sidewall panels. Earlier it was assumed that the basic principles of such electrical mode are not subject to any revision [4].

However, affected by foamy slag technology introduced ubiquitously, Sect. 1.3.2, these principles have undergone fundamental changes. Instead of decreasing length of arcs in order to achieve their immersion in the melt through the increased current and pressure force of arcs onto the liquid bath surface, the level of the melt is raised and long arcs are covered with the foamed slag. This possibility has lead to the development of new principles of the optimum electrical mode of the heat.

Electrical power can be increased by increasing either current or voltage because the power is proportional to the product of these values. Both these ways are tightly associated with the problem of graphitized electrodes which is one of bottlenecks limiting further increase of electrical power EAFs. As current increases, electrode current density grows, the latter being strictly limited to avoid the sharp drop in electrode durability. Presently, high power furnace electrodes operate at the current densities reaching to 35 A/cm^2. Further increase of the current density requires significant improvement of such quality characteristics electrode as electric conductivity, density, strength etc., which greatly complicates electrodes manufacturing technology and increases their cost sharply.

In order to increase electrode current carrying capacity, usually electrode diameter is increased. Alongside with widely spread electrodes of diameter 610 mm, those 710 and 750 mm are used; reportedly, 800 mm electrodes have been produced. However, increasing electrode diameter is associated with overcoming obstacles no less significant than the growth of permissible current density, thus leading to their sharp cost increase. At present, the possibilities of both ways are basically exhausted. On the contrary, it is rather promising to raise electrical power of EAF by increasing maximum secondary voltage of transformer. Previously, this voltage did not exceed 1000 V. Today, furnaces operate successfully at voltage up to 1600 V.

Voltage increase has another advantage compared with current increase. The former is not accompanied by the increase in electrode consumption which is approximately half-determined by current density and rises proportionally to the growth of that value. The opportunities of further voltage increase are also limited by a number of factors including the danger of spark-over in dust-laden gas gap between electrode holders over the roof as well as between the electrodes and the furnace roof.

Method of foamed slag allows to outline the optimum electrical mode of a modern EAF as operation at constant electrical parameters during the entire heat, i.e., at maximum active arc power at maximum voltage and minimum current for the given power value. Such mode enables to obtain the maximum furnace productivity at minimum steelmaking costs. The closest to the implementation of the optimum mode were furnaces operating with continuous scrap charging into the liquid bath, Sects. 1.3.8 and 6.2.4. However, this mode can be also implemented successfully on the furnaces operating with a single scrap charge. Only at the very beginning of the heat, while arcing takes place above the scrap pile, in order to the avoid furnace roof damage, slight arc shortening, achieved by minor voltage lowering and current increase, could be required.

1.2.4 Direct Current Furnaces

These steelmaking units were created and widespread during the last decades of the past century due to the development of reliable high power sources of direct current. First such furnaces, identified as DC furnaces were equipped with graphitized electrodes which were inserted into the freeboard through the roof, as usually, and with several bottom electrodes installed in the bottom lining.

The bottom electrodes were cooled with water or compressed air. Zone of water cooling was as usual located outside of the bottom lining.

Later on, DC furnaces equipped with one graphitized electrode located at the axis of the freeboard became wide spread. In these furnaces, the central current conducting bottom section of the lining is used as bottom electrode whereas multiple steel rods or plates penetrate through the lining. Bottom electrodes of this type reached durability of 5,000 heats and more.

At the initial stage of the operation, DC furnaces differed advantageously from alternating current furnaces (identified as AC furnaces) due to significantly reduced electrode consumption and noise level, as well as some lower electrical energy consumption. These very advantages, undoubted at that time, made for DC furnaces certain spread despite the fact that the cost of electrical equipment for these furnaces was significantly higher compared with that of AC furnaces.

Later on, performances of AC furnaces improved to the extent that DC furnaces lost advantages mentioned above. This occurred due to mainly more efficient usage of slag foaming technology. Arc immersion in the slag to a great degree levels out the difference between direct current and alternating current arcs.

However, in comparison with AC furnaces DC ones are distinguished by a sharp reduction in the level of electric interferences generated into supply grids. In certain cases, such an advantage makes have a preference for direct current because of insufficient electric power supply grids. For instance, in 2010 a 420-t DC furnace with a power of 175 MW has been put into operation at Tokyo Steel plant. The furnace is equipped with two graphitized electrodes 760 mm in diameter and four water-cooled bottom electrodes. In this case, choosing DC is explained by the fact that AC furnace of required power could not be installed since it would generate unallowable interferences in insufficient electric power grids in a region of the plant [5].

1.2.5 Problems of Energy Supply

Modern EAFs are heavy users of electrical energy. They pose severe requirements to the power of electric supply grids. The level of required power is determined by grid capability of withstanding electrical loads, created by furnaces, without noticeable deteriorating of quality of electrical energy. Power of grids significantly limits selection of furnaces which can be installed in a given region. This selection is affected by the necessity to provide so-called furnace electromagnetic compatibility with other electrical energy consumers supplied from the same grids. The electromagnetic compatibility is described by the furnace capability of functioning normally without exerting unacceptable effect on other consumers. This effect is caused by the fact that EAFs deteriorate the quality of electrical energy of the grid and the more so the less the difference between the power of the grid and the total power of the furnaces supplied by it is. This is associated with the nonlinearity of the furnace circuit as well as with short circuiting caused by electrodes touching scrap.

Electric energy quality deterioration is expressed in the distorting the shape of sinusoidal voltage curve as well as in voltage fluctuations and depression, and frequency fluctuations. All these disturbances cause lighting flickering (flicker), variations of TV sets brightness and color, noise in operation of electronics particularly electric clocks, and other unacceptable phenomena. Compensation of detrimental effects of furnace operation on quality of electrical energy in the grids requires significant costs on installation of additional expensive electrical equipment such as transformers with built-in reactors, compensating devices etc. The cost of this equipment amounts to 25–30% of a high power furnace transformer. These costs should be taken into consideration while selecting a furnace and a process, as well as costs of solving environmental problems.

1.3 The Most Important Energy and Technology Innovations

1.3.1 Intensive Use of Oxygen, Carbon and Chemical Heat

Alongside with electrical power increase, this innovation has played an exceptionally great part in increasing EAF productivity and reducing electrical energy consumption due to the fact that the bath absorbs a large quantity of chemical heat released when oxidizing carbon, iron and its alloys as Si, Mn, etc., by oxygen. The commercial use of oxygen in steelmaking began only after development and implementation of new and relatively cheap methods of its production. Methods of injecting oxygen into the freeboard and liquid bath were improved continuously.

At first, oxygen was used in EAFs in rather limited amounts, mainly for cutting scrap and bath decarburization. Oxygen was injected into the furnace manually through a slag door using steel pipes. Later on, this operation was mechanized completely. To introduce oxygen various manipulators started to be applied, and not only consumable pipes but also water-cooled lances were used which were inserted through the openings in the roof and sidewalls of the furnace. Sidewall oxy-gas burners of 3–3.5 MW were also widely used in electric arc furnaces. All that contributed to increase of oxygen consumption. However, only 15–20 year ago, oxygen consumption in the bath did not exceed 10–15 m^3/t of steel.

Further sharp increase of oxygen consumption, especially for the bath blowing, is inseparably linked to use of carbon powder injected into the bath concurrently with oxygen. The impressive results achieved by expanded oxygen use would have never been obtained without carbon injection. As the intensity of oxygen blowing of the bath increased, the amount of oxidized iron increases inevitably. Carbon reduces iron oxides thus preventing unacceptable drop of yield. Besides, injected carbon causes slag foaming. Immersion of arc in foamy slag assures sharp increase in efficiency of electrical energy usage.

In modern high power furnaces operating on scrap, an average oxygen consumption is approximately 40 m^3/t, not infrequently, it is as high as 50 m^3/t. Oxygen consumption increases, in some cases, to 70 m^3/t in furnaces where oxygen is used also for post-combustion of carbon monoxide (CO) in the freeboard. Consumption of carbon powder injected into the bath reaches as high as 15–17 kg/t.

Since the heat duration is very short, such specific oxygen and carbon consumption values require quite high-intensity injection. In modern furnaces, the specific intensity of oxygen blowing is usually 0.9–1.0 m^3/t per minute and it may also reach 2.5 m^3/t per minute if hot metal and reduced iron are used in large amounts; the latter value approaches the blowing intensity observed in oxygen converters. Increased consumption of oxygen and carbon was promoted by the development of new methods of blowing oxygen and carbon into the bath. All these issues are discussed in detail in the respective chapters of this book.

Wide application of oxy-gas burners and oxygen in modern EAFs has sharply increased a share of chemical energy in total amount of energy consumed per heat. As it has been already mentioned, the share of electrical energy was reduced to approximately 50% of total energy consumption, whereas in furnaces where hot metal and reduced iron are used it dropped even significantly lower. As far as energy aspect is concerned, such furnaces have very little in common with the furnaces of the past, where the role of other energy sources was quite insignificant compared with electric arcs. It could be stated today that the energetics of electric arc furnaces is quite close to that of oxygen converter processes based exclusively on the use of internal sources of chemical heat.

1.3.2 Foamed Slag Method

One of the most significant results of implementation of new engineering devices for joint bath blowing with oxygen and carbon powder was the possibility of reliable slag foaming and maintaining slag foam thickness at a layer assuring complete immersion of electric arcs in slag. Such technology increased heat transfer from arcs to the melt up to the maximum possible level and provided a number of other advantages already mentioned previously.

Slag foaming mechanism, during concurrent blowing oxygen and carbon into the bath, is as follows. Oxygen oxidizes carbon contained in the bath and dissolved in it, by reaction $C + 0.5\ O_2 \rightarrow CO$. A certain portion of oxygen is consumed for iron oxidation with formation of FeO. Carbon injected into the bath is dissolved there and reduces iron oxides by reaction $FeO + C \rightarrow Fe + CO$. Thus, both reactions running concurrently generate small bubbles of CO gas which float upwards foaming the slag. If proper correlation of oxygen and carbon consumptions is assured FeO reduction by carbon allows to use oxygen in large volumes without any fear of yield drop.

Carbon consumption for creating a foamed slag layer of required thickness depends on the arc length and amounts to approximately 6–10 kg/t in modern furnaces. However, in order to increase furnace productivity in practice this consumption is approximately doubled which also allows increasing oxygen consumption significantly. It should be stressed that excessive thickness of foamed slag layer reduces the productivity and other basic parameters of furnace performance. Therefore, it is very important to develop reliable means of controlling the slag foam level.

Consumption of carbon powder depends on its quality as well as on injection method and slag composition. The quality of this expensive material is mainly determined by the content of carbon in it and by the ability to dissolve quickly in the melt. Coke powder contains on average only as much as 80% of carbon. Carbon content in graphite powder is considerably higher. Besides, graphite dissolves much quicker.

To achieve as complete absorption of injected carbon as possible, it is quite important to distribute points of injection around the bath perimeter; the distance from the injectors to liquid metal level is also of great significance. Of all practiced injection methods, the best carbon assimilation, close to 100%, is achieved when carbon is injected directly into the slag near slag-metal boundary. The worst option is to inject carbon powder from above onto the surface of slag. In that case, a significant portion of the material is removed out of the furnace by off-gas flow.

The selection of injector locations is of great importance for the efficiency of carbon injection. It is not recommended to combine carbon and oxygen injection points. When such combining occurs, oxygen and carbon jets come into direct contact with each other, so carbon burns out partly prior to being dissolved in the melt. This portion of carbon is uselessly lost both for slag foaming process and for iron oxides reduction. Therefore, the oxygen and carbon injectors should be located at a certain distance from one another. Carbon injection into the bath zone in front of the slag door is inexpedient since foamed slag tends to flow over the furnace sill, and carbon is partly lost for the process. In this regard, the best results are provided by carbon injection into oriel zone and the bath zone adjacent to it since foamed slag moves towards the arcs and covers them.

Foaming ability and stability of formed foam depend greatly on the physical properties of slag, such as its viscosity, density, surface tension, and the concentration of undissolved solid particles. All these properties are determined by slag composition and its temperature. In basic slags, foaming ability increases as SiO_2 concentration grows. In the modern EAFs the duration of liquid bath existence is rather short-term, and there is not enough time for slag to be completely formed. The slag is rather non-homogeneous and contains great quantities of undissolved particles of lime and other small-size particles. That contributes to better and easier foaming and excludes any necessity of increasing SiO_2 concentration in the slag significantly since it reduces basicity of slag compared with the ordinary level. Foaming is also facilitated by injecting dolomite and lime powder into the slag, as well as by adding coke to slag at an early melting stage.

In order to keep a thick layer of foamed slag without it flowing over a sill, a slag door is shut. At the same time, air infiltration into a freeboard is eliminated. The mode of operation with a shut slag door is in common use on the modern furnaces. This mode of operation, however, leads to forming of thick metal-slag build ups in front of a shut slag door. These build ups hinder monitoring of temperature and metal composition during a heat and cause difficulties in furnace maintenance. During the heat it is required to open the door in order to purge a sill for slagging-off. For this purpose forklift hoists and other special mechanisms which occupy a lot of room at operating platform are used.

To eliminate these shortcomings, V. Shver (PTI Inc. Company, the USA) has developed an original design of a slag door, so called SwingDoor™, Fig. 1.3. The key feature of this door is that it moves around its upper horizontal axis rather than up and down as it typically takes place. In a shut position, this door becomes an extension of wall panels, which precludes freezing of slag on a sill. The whole length of a sill is free from buildup, thus all the complications associated with buildup removal are eliminated. For slagging-off the slag door is turned through an angle of about 40°. In so doing, no air infiltration into a freeboard is allowed. The slag door is opened in full solely in case of shutdowns of the furnace in order to examine and repair refractory of bottom and water-cooled panels.

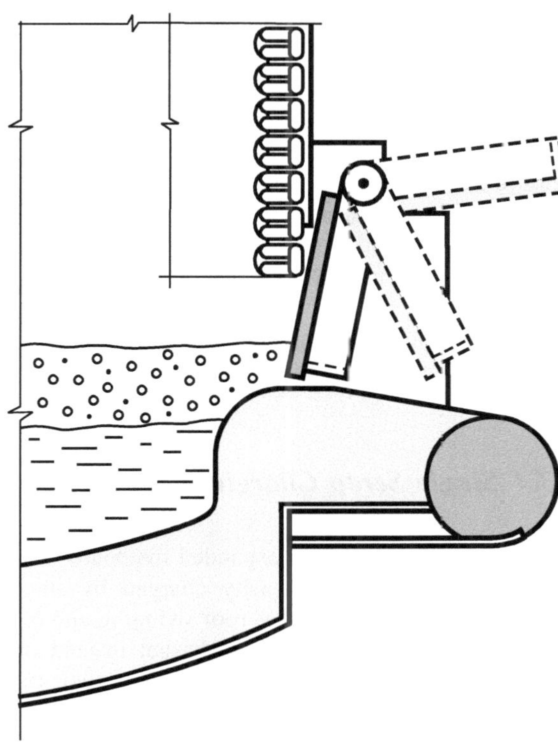

Fig. 1.3 SwingDoor™

1.3.3 Furnace Operation with Hot Heel

At present, a steelmaking method is widely used where about 15–20% of metal and certain amount of slag are left at the furnace bottom after each tapping. The rest of slag is removed from the furnace over the sill. If steel-tapping hole is made according to the modern requirements such method allows to pour practically slag-free metal into the ladle. That provides ferroalloys savings and facilitates performing subsequent operations of secondary metallurgy. Yet, the main advantage of EAF operation with hot heel is energy efficiency.

In high power furnaces, boring-in scrap pile occurs so quickly that melt layer is not deep enough when electrodes reach closely to the bottom. There is a danger of damaging bottom refractory by powerful arcs. This factor restricts increasing electric power of the furnaces. Presence of hot heel eliminates the said limitation and allows increasing electrical power with the aim of further productivity increase.

Operation with hot heel extends the capabilities of effective use of oxygen for blowing the bath which also promotes growth of productivity. Presence of hot heel allows starting oxygen blowing almost immediately after scrap charging. When blowing hot heel in the presence of carbon charged with scrap, slag is foamed and the arcs immerse into the melt, thus increasing their efficiency. Carbon monoxide CO, escaping from the hot heel, combusts and heats the scrap layer when passing through the scrap thus accelerating settling and melting down of charged metal-charge. Oxygen blowing of relatively cold scrap pile, when started too early without hot heel, is ineffective. Although such blowing accelerates the heat but this is achieved due to intense iron oxidation which leads to unjustified yield drop. At the same time, electrodes oxidation and consumption increase as well.

Maintaining the mass of metal, left in the furnace, at a relatively constant level close to the optimum is a necessary condition of substantially complete and stable using the advantages provided for by operation with hot heel. Significant fluctuations of hot heel size from one heat to another significantly reduce the efficiency of this important process element. Visual control is inefficient. It is necessary to use engineering tools to enable easier operator's control of hot heel. For this purpose, furnaces are equipped by sensors which allow controlling a furnace weight varying during the heat and consequently a hot heel weight.

1.3.4 Single Scrap Charging

In recent years, EAFs with expanded freeboard size capable of receiving all scrap of about 0.7 t/m^3 bulk density charged by single basket are getting spread. Charging each basket requires roof swinging and current switching off. With short tap-to-tap time, using one scrap basket instead of two leads to a considerable increase in EAF's hourly production. However, the advantages of furnaces with single scrap charging are not limited to that.

Freeboard volume is expanded in such furnaces mainly by means of increasing its height. Greater height of scrap pile in the furnace provides for better scrap absorbing the heat of hot gases, obtained when post-combusting of CO, passing upwards through scrap layer from below. The same can be said about absorbing heat from flames of oxy-gas burners installed in the lower parts of furnace sidewalls. Increasing depth of pits bored-in by arcs in scrap also increases the degree of arc heat assimilation. All this increases scrap heating temperature prior to its immersion into the melt, and accelerates melting. At the same time, electric energy consumption is decreased due to the reduction in the time when the furnace is open and loses a lot of heat. Dust-gas emission into shop atmosphere is reduced also while scrap charging. In furnaces of 300 to 400-t capacity, due to the freeboard height extend, the number of charges is decreased to two per heat.

However, considering the effect of increasing the furnace freeboard height on the utilization of heat in it, it should be taken into account that sidewall area is increased and, consequently, heat losses with cooling water are increased as well. To reduce these losses, measures are taken to increase the thickness of skull layer on the sidewall panels. For instance, Danieli Company uses panels consisting of two layers of pipes. The pipes of the internal (with respect to freeboard) layer are spaced apart much wider than in the external one. That facilitates formation of thicker skull layer and its better retention on the pipes [5]. As freeboard height is increased considerably, electrode stroke and their length are respectively increased as well, thus increasing the probability of electrode breaking. To prevent breaking the rigidity of arms and of the entire electrode motion system should be increased.

1.3.5 Use of Hot Metal and Reduced Iron

Charging certain quantity of hot metal into the furnace to replace a portion of scrap provides quite substantial increase in productivity and sharp reduction of electrical energy consumption. At integrated plants where hot metal is in available in excess and scrap is in short supply this method of intensification of electrical steelmaking has recently become rather widespread. However, even with this group of furnaces the use of hot metal is rather a temporary opportunistic approach than promising direction. An EAF is far worse suited for hot metal processing compared with oxygen converters. Besides, as it was stated previously, EAFs lose their environmental advantages over converters when using hot metal. Scrap is the main raw material used in electric arc furnaces both at present and in the very long run and the reserves of scrap are constantly growing all over the world. Reduced iron and hot metal will be likely used only in certain regions and in limited amounts, particularly, in order to reduce a content of copper and nickel in a charge when producing certain steel grades.

1.3.6 Post-Combustion of CO Above the Bath

The problem of post-combustion of CO has two aspects. Firstly, post-combustion of CO, as well as H_2, is necessary due to considerations of explosion-proof operation of off-gas evacuation system of the furnace. Explosive mixtures of gases containing CO, H_2 and O_2, are necessary to burn in the freeboard and within the limits of off-gas evacuation system with sufficiently high temperatures. To prevent explosions, under any circumstances such mixtures must not penetrate into the low-temperature zones of the gas evacuation systems where their burning can stop. Besides, the CO emissions into atmosphere are unacceptable due to the environmental considerations and are limited by the relevant standards. This first aspect of the post-combustion of CO is discussed in Chap. 14 in relation to problems of gas evacuation and environmental protection.

This problem has the energy aspect as well. At the first sight, post-combustion of CO to CO_2 promises a sharp increase in the input of chemical energy in the thermal balance of the furnace. Let us remind that when carbon is oxidized to CO, only 28% of the energy of the full combustion of carbon to CO_2 is released. The remaining 72% of chemical energy of carbon can be obtained by post-combustion of CO, Table 4.2. However, efficient utilization of this energy in the steelmaking process encounters significant difficulties.

The first of them is the fact that CO evolves relatively uniformly throughout the entire bath surface. To return the chemical energy into the bath, the post-combustion must occur near its surface. If CO burns in the space under the roof, then the emitted heat is absorbed by the wall and roof panels rather than the bath, which only increases heat losses with water.

However, in order to post-combust CO near the bath surface, it is necessary to cover this entire large surface with oxygen jets. This requires quite significant oxygen resources which are not quite often available at the plants. On those furnaces where complete post-combustion of CO in the freeboard was attempted, the oxygen consumption reached 60–70 $м^3$/m. Yet, a considerable share of CO escaped into the off-gas evacuation system.

Usually, the oxygen flow rates for post-combustion of CO were maintained at the constant level or adjusted according to preset program. However, the intensity of CO evolution from the bath is very non-uniform in time and fluctuates from heat to heat. Therefore, a significant portion of oxygen supplied for post-combustion of CO at the constant intensity was wasted.

The oxygen consumption for post-combustion can be sharply reduced if it is supplied according to the actual fluctuation of intensity of CO evolution. In order to achieve this, the oxygen flow rate must be automatically adjusted according to the content of CO, CO_2 and O_2 in the gases leaving the freeboard. Presently, this opportunity is available, since the systems of continuous off-gas analysis are implemented and operate successfully at a number of arc furnaces in various countries.

The second and the main difficulty is that the return of energy of post-combustion of CO back to the steelmaking process is hampered by quite unfavorable conditions of heat exchange. High-temperature layer of gases over the bath obtained by post-combustion of CO by oxygen jets is separated from the liquid metal by a quite thick layer of foamed slag with low thermal conductivity. As a result, the slag surface rather than the metal is mostly heated, and the former transfers heat to the water-cooled panels by radiation. Judging by the published data on operation of CO post-combustion systems in the EAFs, the efficiency of utilization of heat obtained from post-combustion hardly exceeds 20–25%. Due to relatively low efficiency and very large oxygen consumptions, the systems of post-combustion of CO in the freeboard so far have not become spread.

The results of persistent attempts of post-combustion of CO in the oxygen converters in order to increase the scrap share in the metal charge are of interest. For long period of time, these attempts were made in Russia and other countries. The research was conducted in two different directions. For post-combustion of CO inside the bath, the so-called two-tier tuyeres were used. In these tuyeres, the additional nozzles intended for post-combustion of CO above the reaction zone were located above the first row of the main blowing oxygen nozzles. For post-combustion of CO above the bath, the additional water-cooled oxygen tuyeres were installed in the lining of the converter throat. Both of these directions have not yielded positive results. The development of oxidation reaction of CO to CO_2 in the liquid bath is hampered by the laws of thermodynamics. The beneficial absorption of heat from post-combustion of CO in the converter throat was obstructed by the unsatisfactory conditions of heat transfer. At the same time, in the case of post-combustion, the durability of throat lining reduced sharply.

1.3.7 Increase in Capacity of Furnaces

The productivity of oxygen converters has approached their limit whereas the performance potential of EAFs is still quite significant. This is confirmed particularly by the fact that furnaces of 300–400 t capacity have been put into operation in recent years. Hourly productivity of these furnaces exceeded of 350 t/h. Such a furnace can provide an annual steel production of mini-mill up to 3 million tons. The prefix “mini” slightly agrees with the plant of such productivity.

Even in the early 1980s, it was considered that an increase in EAFs’ capacity over 200-t was inexpedient [4]. To a considerable extent, such an opinion was based on relatively low productivity of a 360-t EAF at Sterling Steel plant, the USA. Scrap was charged into the furnace with three baskets; tap-to-tap time was nearly two hours.

However, this experience cannot serve as a proof of inexpediency of an increase in EAF’s capacity. The low productivity of the furnace under consideration was mainly explained by low transformer power which was equal to160 MVA or merely 440 kVA/t. The furnace equipped with such a transformer, which does not absolutely correspond to its capacity, cannot be considered as a unit of high power category.

Innovations such as an increase in transformer power, reduction in the number of charges to two, an increase in hot heel weight, etc., resulted in sharp improvement of performances of furnaces of high capacity. At present, when annual productivity of mini-mills exceeds 2 million tons, it is preferable to install one furnace of 300–400 t capacity instead of two furnaces of lower capacity. Let us review the following instances.

As it was mentioned in Sect. 1.2.1, at Gebze mini-mill, Turkey, the furnace of 420-t capacity with the tapping weight of 323-t is in operation. The transformer power is 240 MVA with an option to increase by 20%. During the melting period, active arc power reaches 205 MW at a voltage of 1600 V. The metal charge is loaded using two baskets. The productivity exceeds 370 t/h at tap-to-tap time of 52 min. Electrical energy consumption is 326 kWh/t [2].

An EAF of 300-t capacity must be installed at a new Iskenderun mini-mill, Turkey. Tapping weight is 250-t, transformer power is 300 MVA (1200 kVA/t); metal charge is loaded using two baskets. The productivity is 320 t/h, or 2.4 million tons per year [3].

An increase in furnace capacity up to 300–400 tons takes place in arc furnaces operating with continuous scrap charging into the liquid bath. Energy features of this technological process are considered below.

1.3.8 Continuous Charging and Melting of Scrap in the Liquid Bath

This innovation is used in Consteel conveyor furnaces and shaft furnaces of a new generation. In Consteel furnaces, scrap is transported by a conveyor to the furnace at a preset rate and discharged into the liquid bath through the door in the furnace sidewall.

In shaft furnaces, scrap is charged into a shaft placed on a car next to the furnace. From the shaft scrap is charged into the bath through a door in the furnace sidewall by means of a hydraulic pusher. The charging of scrap is performed by separate portions following one by one continuously. In both cases, before charging into the bath the scrap is preheated with off-gases leaving the furnace. Only main energy features of the continuous scrap melting process which are common for conveyor and shaft furnaces are analyzed bellow. Design and working data of these furnaces are discussed in Chap. 6.

The scrap charging is carried out at a rate which is equal to the scrap melting rate. At any instant, amount of heat obtained by the bath from the electric arcs is equal to that consumed for scrap meltdown and heating the forming melt up to the bath temperature. Thus the temperature of the bath is kept at a constant level. The scrap does not protrude from the melt surface, and arcs are burning under conditions of the flat bath and immersed into foamed slag during the entire heat. Stability of arcing and electrical energy efficiency are considerably increased.

Acoustic noise level is reduced, so is the level of electrical noise generated by powerful arcs in external grids. Concentration of FeO in the slag is reduced as well, while the yield grows. Durability of water-cooled panels increases significantly due to virtually complete elimination of the heat stages when the furnace walls and roof are exposed to direct radiation of open arcs.

Electrode breaking is eliminated as well as the necessity to open the roof for scrap charging with baskets. The roof is opened at the first heat of series only in order to charge one basket of scrap after tapping all the metal and slag as well as to inspect and maintain the furnace bottom. In the rest of the heats the furnaces operate with a hot heel. Under the roof of furnaces a small negative pressure is maintained which eliminates uncontrolled dust-gas emissions into the shop atmosphere through electrode ports. The latter factor considerably reduces the costs on off-gas evacuation system. All these and some other advantages contributed to spreading of the furnaces with continuous charging and melting of scrap. However, the new process has certain disadvantages.

The scrap melting rate in liquid metal is determined by an intensity of convective heat transfer from metal to scrap (amount of heat transferred to scrap per unit of time). This value is determined by the difference of temperatures between metal and scrap as well as by the bath stirring intensity. The liquid metal temperature is maintained at a constant level of 1560–1580°C during the scrap charging process. This level exceeds the temperature at which the scrap begins to melt just by 50°C what considerably reduces the intensity of heat transfer by convection from metal to scrap; and scrap melting process is slowed down.

Exceeding a metal temperature of 1560–1580°C during the charging and melting of scrap is not allowed to avoid reduction in durability of bottom refractory. The area of scrap charging zone is very limited. In this area, the temperature of metal is always lower than that of the bath on the average because of insufficiently intensive stirring of the melt. Upon completion of charging little extra time is required to completely melt the rest of the scrap and heat the melt up to the preset final tapping temperature.

In such a process, an insufficient intensity of heat transfer in the liquid metal is a bottleneck. Selection of methods for increasing the melting rate is highly limited. It is impossible to increase the difference between temperatures of liquid metal and scrap in the course of melting. Unlike EAFs, the duration of melting cannot be shortened due to increasing the power of arcs either. The electric arcs do not contact the scrap and do not affect the intensity of the heat transfer to the scrap inside the bath. They only indirectly contribute to the process of scrap melting by keeping the temperature of the bath at the required constant level. At the given values of the furnace capacity and scrap preheating temperature, such a mechanism of melting excludes the possibility of increasing the rate of melting by increasing the power of arcs. The power of the arcs must strictly correspond to the rate of melting of scrap, which at the specified conditions does not depend on the power of the arcs. On the contrary, the scrap melting rate limits the power of the arcs. In EAFs, such a limitation of the power does not exist: with increasing the power of arcs the melting rate increases.

The intensity of the heat transfer to the scrap inside the bath can be enhanced due to an increase in the bath stirring rate only by using, for this purpose, the same equipment which is utilized in EAFs[3] as well. The melting down process can also be accelerated due to increasing a scrap preheating temperature. An increase in this temperature diminishes the quantity of heat needed to be transferred to the scrap for its melting down and shortens the duration of the process. In EAFs, in order to melt scrap oxy-gas burners and process gases emitted by the bath are used. Neither burners nor process gases can be used in the process of scrap melting in the liquid metal. If a temperature of scrap preheating is insufficiently high, aforesaid draws to conclusion that the furnaces with a continuous scrap charging into the liquid bath must lag behind EAFs in the scrap melting rate and therefore in productivity. This assumption is confirmed by the actual data, Chap. 6.

References

1. Progress or wasteful erroneous development? Ind—Anz (1985) 107(103–104):58–59
2. Abel M, Hein M (2009) The SIMETAL ultimate at colakoglu, Turkey. Iron and Steel Technology 56–64
3. Sellan R, Fabbro M, Burin P (2008) The 300-t EAF meltshop at the new Iskenderun minimill complex. MPT Int 2:52–58
4. Morozov AN (1983) Modern steelmaking in arc furnaces. Metallurgiya, Moscow
5. Adachi T, Sellan R (2012) The jumbo size 420 t EAF at Tokyo steel, Japan. MPT Int 2:54–62

[3] See Chap. 11.

Chapter 2
Electric Arc Furnace as Thermoenergetical Unit

2.1 Thermal Performance of Furnace: Terminology and Designations

There are different forms of energy. Heat is one of them. Heat is a form of energy used to realize the furnace thermal performance in a steel melting process carried out at high temperatures. Therefore, the term “thermal performance of a furnace” has a rather profound meaning [1]. According to the energy conservation law, heat does not appear out of nothing and does not disappear. All other forms of energy can be transformed into heat, for instance, electrical or chemical energy, in strictly equivalent amounts.

During heating of any body, a certain quantity of heat transfers to it from the heat source. This process is called heat transfer. The heat assimilated by the body increases its internal energy. The body temperature thus rises. When cooling down, the body gives a part of its internal energy (in a form of heat) to objects surrounding it. Heat transfer processes proceed at a practically constant pressure in EAFs and in other furnaces. In all such cases a change in internal energy of the body is equivalent to a change in what is called enthalpy. This thermodynamic parameter is widely applied in thermo-technical calculations. Enthalpy, like other kinds of energy, can be measured in Joules (J) or in kW-hours (kWh). As a Joule is a very small quantity (3600 kJ = 1 kWh), energy unit of kWh will be used further in most cases.

While solving heat engineering problems, knowing absolute body enthalpy values is not necessary. It is sufficient to determine the enthalpy changes ΔE, which are calculated per 1 kg of substance or per 1 m^3 (for gases). In processes taking place at constant pressure, these changes are calculated by the following formula:

$$\Delta E = c_p x \,\Delta t \quad \mathrm{kWh/kg\,(or\,kWh/m^3)}, \tag{2.1}$$

Δt – temperature change,°C

Y. N. Toulouevski and I. Y. Zinurov, *Innovation in Electric Arc Furnaces*,
DOI: 10.1007/978-3-642-36273-6_2, © Springer-Verlag Berlin Heidelberg 2013

c_P – mean heat capacity of the body at a constant pressure (for gases) within temperature range Δt, kWh/kg°C (or kWh/m^3°C)

Heat capacity c_P is measured by a quantity of heat transferred to a body with 1 kg mass or with 1 m^3 volume (for gases) to raise its temperature by 1°C. Correspondingly, we can distinguish between specific mass or volumetric heat capacities. Since only enthalpy differences are determined, they are counted off the initial temperature equal to 0°C according to formula (2.1). Mean heat capacity values presented in the tables correspond to temperature differences between 0°C and t°C.

So far we did not pay attention to the fact that the same mass of gases can occupy a different volume depending on their temperature and pressure. It should be made clear what specific gas volume is meant in the determination of heat capacity given above, formula (2.1). The same is also true for all the other values related per 1 m^3 of gas as well as for all data and results of calculations where gas volumes appear.

In order to avoid this uncertainty and have a chance of comparing various values to each other it is assumed to reduce gas volumes to standard conditions, namely to temperature of t = 0°C and pressure of 760 mmHg (1.01 bar). The standard volumes of gases are designated as m^3 (*s. t. p.*).

From the ideal gas laws any volumes could be reduced to the standard conditions. According to these laws at a constant pressure the volume of a gas increases in direct proportion with an increase in its absolute temperature T, which is measured in Kelvin's degrees, K. On Kelvin's scale, degrees K are counted out from the absolute temperature zero equal to −273°C on Celsius' scale. The values of temperature differences Δt are identical in both scales. Conversion from K to °C is determined by the following expression: T, K = t°C + 273. Thus, the water boiling point 100°C amounts to 373 K as well as 0°C = 273 K on Kelvin's scale.

At a constant temperature the volume of a gas decreases in direct proportion with an increase in its absolute pressure p_{ABS}, which is equal to the sum of excessive pressure of a gas p (from a manometer) and atmospheric pressure 1.01 bar: $p_{ABS} = p + 1.01$ bar. In regular calculations the atmospheric pressure might be assumed as 1.0 bar (10^5 Pascal, Pa).

To reduce a gas volume V, m^3 with a temperature of t°C > 0°C and pressure p bar > 0.0 bar (from a manometer) to standard volume V m^3 (*s. t. p.*), the volume V m^3 should be divided by the value of $(t + 273)/273$ and multiplied by the value of $(p + 1.0)/1.0$. For instance, if with t = 500°C and p = 3 bar (from a manometer), V = 300 m^3 then V m^3 (*s. t. p.*) = 300 × 4 × 273/(500 + 273) = 423.8 m^3 (*s. t. p.*).

Further, as well as in Chap. 1, volumes of gases are assumed to be given in m^3(*s. t. p.*) unless otherwise specified but designation of (*s. t. p.*) is left out.

It is necessary to note some specifics of designations of different kinds of energy accepted in this chapter as well as in the following ones. Enthalpy is usually designated as I in thermodynamics. In this book, a common designation E is accepted for enthalpy and other kinds of energy, which are added quite often. This simplifies both presentation and learning of this material. In sections dealing exclusively with heat transfer processes, heat quantity is designated as Q, kWh and q, kWh/m^2.

Chemical energy of reactions of iron and other elements' oxidation and reduction plays a great part in thermal performance of EAFs. Oxidation reactions release heat and are called exothermic. Reduction reactions are accompanied by the absorption of heat; they are called endothermic. An amount of heat released or absorbed per unit of an element involved in a reaction is called thermal effect or enthalpy of chemical reaction; in thermodynamics it is usually designated as H. According to energy conservation law, thermal effect of exothermic reaction is equal to decrease in total enthalpy of the reactants. Therefore, by convention a minus sign is ascribed to it ($-H$). Respectively, a plus sign is ascribed to thermal effect of endothermic reaction ($+H$). It is rather convenient to measure enthalpies of chemical reactions in kWh per 1 kg of an element (e.g., iron Fe), or 1 m^3 of gas (e.g., oxygen O_2).

In modern steelmaking shops, the required steel qualitative characteristics are achieved in ladle furnaces and other units of secondary metallurgy. The EAF itself is designed purely for thermo-technical task, namely to melt the materials charged into the furnace and to heat the melt up to the tapping temperature. At the same time, carbon percentage in the melt is reduced to required final value. To solve this problem, it is necessary, first, to release (generate) certain amount of heat in the furnace by means of electric arcs, oxygen-fuel burners, and other energy sources; second, to transfer the heat to materials melted and to the liquid bath. Thus, heat generation and heat transfer are the basic processes determining the production and other basic parameters of modern EAFs. All this gives grounds to classify electric arc steel melting furnaces as thermoenergetical units.

2.2 External and Internal Sources of Thermal Energy: Useful Heat

All heat energy sources could be divided into two groups, namely the external and internal sources, depending on their position relative to the materials melted and to the liquid bath heated. The external sources include electric arcs (not completely immersed into slag), oxy-fuel burners, CO (carbon monoxide) post-combusted above the bath, and other sources. Only a portion of the heat generated by external sources can be transferred to the solid charge or to the liquid bath. This portion is useful heat. The rest of the released heat energy is inevitably lost in the process of heat transfer. The better the heat transfer process is organized, the lower the level of the heat loss, the greater the portion of the useful heat absorbed by the charge or the bath, and, consequently, the lower the consumption of electrical energy and fuel.

The internal energy sources include oxidation reactions of carbon, iron, silicon, and other elements contained in the melt. All these exothermic reactions release great amounts of heat. The heat generated in the bath itself is practically fully absorbed by the bath, because in this case the heat transfer process is not involved. Thus, almost all the heat generated by internal sources is useful.

In the course of the heat, the useful heat transferred to the solid charge and the liquid bath is accumulated in the melt and increases its enthalpy. At the same time, the temperature of the melt also rises. The final enthalpy of the bath just prior to tapping is equal to the sum of enthalpy of the metal at the tapping temperature E_{MET} and enthalpy of the slag E_{SL}. The value of enthalpy E_{MET} is equal to the quantity of useful heat accumulated by metal over the course of the heat. This quantity is called useful heat of the entire heat. It depends on the tapping temperature and is close to 400 kWh/ton of liquid steel.

If scrap and other metallic components of the charge, for instance, pig iron, are charged into the EAF in a cold state (at ambient temperatures), their enthalpy at the beginning of the heat can be disregarded. In that case, the enthalpy E_{MET} consists only of contributions of useful heat obtained from all the abovementioned external and internal energy sources. If preheated scrap or hot metal is charged into the furnace, then their enthalpy E_{SCR} contributes greatly to the useful heat of the heat E_{MET}. In that case, the share of all the external energy sources is reduced accordingly, as well as the specific consumption of electrical energy and fuel.

Sometimes, the enthalpy of final slag E_{SL} is also included into the useful heat of the heat. However, there is no substantial ground for that. The purpose of the heat is to produce metal, but not slag. The enthalpy of the slag tapped out of the furnace together with the metal is, in essence, lost heat, as well as that of the slag tapped out prior to metal tapping. Along with other heat losses, heat losses with the slag should be reduced as much as possible by reducing the amount of slag.

2.3 Factors Limiting the Power of External Sources

In order to shorten tap-to-tap time and to increase EAF productivity, it is necessary to increase the power of electric arcs, oxy-gas burners, and other external heat energy sources. Increasing the power of these sources in itself is not a problem. Thus, for instance, the power of a burner, used in oxygen converters for preheating scrap prior to charging of hot metal, is usually equal to 150–170 MW. The power of the burners used in EAFs is ten times lower than that of the converter burners. Usually it does not exceed 3.5 MW.

A similar situation is observed with regard to the EAF's transformers. Their power is much lower than that of the transformers used in electrical circuits. It is obvious that the problem consists not in the burners and the transformers themselves, but in eliminating the factors which limit the power of heat energy sources under specific conditions of electric arc furnaces. All these limiting factors could be divided into two fundamentally different groups, the first being the factors inherently connected with the physical–chemical processes taking place in the EAF, and the second being the external, sometimes random factors not connected with those processes directly. Let us identify, in brief, the first group as inherent factors and the second group as external factors. The vast group of external factors includes, for example, insufficient power of the external network supplying electricity for the EAF, insufficient fume exhauster draught which restrains the carrying capacity of the furnace gas removal ducts, insufficient power of bag house, etc. The importance of these factors is quite obvious. Eliminating their restrictive effect is not connected with interference into physical–chemical processes of the heat and does not require improvement in carrying out of these processes. The problem is frequently reduced to the additional capital investments required for eliminating the bottlenecks such as replacement of fume exhausters or modernization of gas cleaning system.

The inherent limiting factors are inseparably linked to basic physical–chemical processes such as carbon oxidation during the oxygen blowing of the bath, stirring of the bath with carbon monoxide bubbles and oxygen jets, scrap iron oxidation by products of complete gas combustion in oxy-gas burners, and slag foaming. All these factors limit the reasonable power level of external heat energy sources as they strongly affect the heat transfer process conditions.

2.4 Key Role of Heat Transfer Processes

As a rule, it is the processes of heat transfer from a heat source to charge materials or the liquid bath that finally determine what part of energy generated in the source is converted into useful heat. If this portion of energy decreases substantially while the power of the source increases, then such an increase in heat generation becomes inexpedient. In such cases, it is necessary in the first place to improve the heat transfer conditions rather than to increase the power of the source. This will ensure an increase in both furnace productivity and its energy efficiency.

Let us illustrate the aforesaid by an obvious example. If a deep vessel containing water is heated from above, the upper layer of water boils while the main mass of water remains cold. It could be explained by the fact that stagnant water conducts heat poorly, and that hot water is lighter in weight than cold water and stays on the top. If under such conditions the power of the heat source is increased, then the entire additional power is used on an increase in the boil-off rate of the top layer of water, i.e., an increase in the amount of vapor generated per unit time. The rate of heating of the lower water layers remains practically unchanged.

When the vessel is heated from below, the heated water rises to the top, while the cold water goes down. The whole volume of water is stirred intensively and is quickly heated up to the required temperature, although the power of the energy source is several times lower than that of when heating from above. Correspondingly, the energy consumption for heating is reduced. In this case, the intensity of heat transfer from the top water layers to the lower ones is an inherent factor limiting the power of energy source. With water stirring, the heat transfer intensity increases many-fold, which makes it expedient to sharply increase the power of the energy source for the purpose of acceleration of water heating.

The same situation is observed in the case of heating the bath in electric arc furnaces, open-hearth furnaces, and other hearth furnaces. The baths of these furnaces obtain the greater part of the heat supplied by the external energy sources from above: mainly from the electric arcs in the EAFs and from the burner flames in the open-hearth furnaces. Switching to high-power electric arc furnaces would not have been justified, if, concurrently with increasing the power, the measures were not taken to optimize the conditions of heat transfer from the arcs to the bath by means of submerging the arcs into the metal or in the foamed slag.

In open-hearth furnaces, increasing the power of the burners ensured substantial reduction of tap-to-tap time and increase in productivity, but only after successful intensification of heat transfer from the flames to the liquid bath by means of extensive use of oxygen. Before that, the increase in the power led only to excessive fuel consumption, while the tap-to-tap time was shortened insignificantly. A huge difference between the power level of the wall oxy-gas burners in the EAF and of the converter burners can also be explained by very different conditions of heat transfer from flames to scrap in these units, Chap. 6, Sect. 6.5.2.

All of the raised above issues related to the effect of heat transfer conditions on power level of external heat energy sources in the EAFs will be examined in detail in corresponding chapters. But the following fundamental conclusions can be made right now. Indeed, it is the heat transfer processes that limit the possibility of further increasing of EAF's productivity due to increase in the power of electric arcs and oxy-fuel burners. These same processes, basically, determine the energy efficiency of the EAF. When increasing the power of the heat generation sources, it is necessary to increase correspondingly the intensity of the heat transfer processes. Otherwise, the heat absorption rate can be insufficient, thus leading to unjustified increase in heat losses and energy costs. On the other hand, the very fact of improving the heat transfer conditions ensures an increase in both productivity and energy efficiency of the EAF with the same power of external heat sources.

It should be emphasized that in contrast to eliminating bottlenecks caused by external limiting factors, increasing the intensity of the heat transfer appears to be a far more complicated problem, which requires some new original engineering solutions. The total consumption of electrical and chemical energy in the EAF is still too high as compared to that of theoretically required. Too little attention was paid to this aspect of the process at the early stages of electric steelmaking. However, at present, due to a sharp increase in the price of all basic kinds of energy, as well as due to a shortage of energy in many regions and drastic toughening of environmental requirements, improvement of power engineering of electric arc furnaces becomes a matter of utmost importance. On the other hand, this is the direction that ensures the greatest potentialities for improving the process.

A sufficiently profound understanding of basic principles of heat transfer processes as applied to the processes taking place in the furnace is necessary for the critical analysis of innovations developed in this field. The technical literature examines the relevant problems in detail. However, such literature is not always accessible to those readers who have got no special education. In order to overcome this difficulty, the next chapter contains a brief presentation of basic heat transfer laws outlined in elementary form. Fundamental formulae and reference data required for calculations of heat transfer using the simplified methods are also presented.

Such calculations help to quickly resolve the problems related to estimating the innovations aimed at improving the thermal performance of the EAF. Studying this chapter would help the readers to carry out such calculations on their own.

Reference

1. Glinkov MA (1970) Thermal performance of steelmelting baths. Metallurgiya, Moscow

Chapter 3
The Fundamental Laws and Calculating Formulae of Heat Transfer Processes

3.1 Three Ways of Heat Transfer: General Concepts

According to the second law of thermodynamics spontaneous heat propagation in space (with no additional energy consumption) always occurs in the direction of temperature reduction from an area of a higher temperature to an area of a lower one. The heat transfer occurs in three basic ways, i.e., conduction, convection, and radiation.

Conduction heat transfer takes place in immovable continuous mediums as a result of exchange of energies among elementary particles, i.e., molecules, atoms, and unbounded electrons. In pure form this process occurs in solids only or in very thin stationary layers of fluids or gas.

Convective transfer of heat energy (convection) is inseparably linked with displacement of the medium itself (usually fluxes of liquid or gas) in space. Heat exchange between these fluxes and the surface of a solid during their contact is called convective heat exchange. This process runs with the assistance of thermal conductivity of the thin boundary layer of fluid which is in contact with the solid.

Heat radiation occurs inevitably when bodies are heated. It is a result of intra-atomic processes when the internal energy of the heated body partly turns into the energy of radiation. The carriers of radiant energy are electromagnetic waves which spread through vacuum or any other medium which is transparent for them at the velocity of light. Physical properties of these waves are determined by the wavelength. At the temperatures of bodies observed usually inside furnaces (particularly steelmaking ones) practically the whole amount of energy radiated by heated bodies is concentrated within the range from 0.4 to 80 micron (μm); 1 μm is equal to 10^{-6} m. This type of radiation is usually called thermal or infrared radiation. As it falls onto other bodies thermal radiation is partly absorbed by them and turns into the internal energy of body once again. Such process of heat energy transfer from radiative bodies to absorbent ones is called radiant heat exchange.

Both in engineering objects and in nature, heat can be transferred in many cases by two or even three ways simultaneously which complicates greatly heat transfer calculations. In EAFs electric arcs transfer heat to scrap and the liquid bath both by radiation and by convection. Through the furnace bottom heat is transferred to its

Y. N. Toulouevski and I. Y. Zinurov, *Innovation in Electric Arc Furnaces*,
DOI: 10.1007/978-3-642-36273-6_3,

external surface by thermal conductivity of the refractory and from the external surface to the environment by convection and radiation. From internal walls of pipes of roof and wall panels, heat is transferred to cooling water by convection, and by conduction it is transferred through the skull layer on the pipes and through the pipe walls.

Let us discuss each way of heat transfer in detail. Thereat, we will basically consider stationary processes, i.e., such processes where temperatures of bodies participating in heat exchange and heat fluxes do not vary over time. Heat transfer laws in non-stationary processes are much more complex. The introduction to numerical methods of calculation of such processes is given in Sect. 3.2.8. It should be mentioned that, in most cases, for analysis of innovations in EAFs the knowledge of stationary heat transfer laws is quite enough.

Heat transfer intensity is described by heat flux density, i.e., amount of heat transferred per unit of time through the unit of surface area. This value is measured in W/m^2 and is designated by q. The amount of heat transferred per unit of time through surface F (m^2) is called heat flux power or simply heat flux. This value is measured in W and designated by Q.

3.2 Conduction Heat Transfer

3.2.1 Fourier's Law. Flat Uniform Wall. Electrical–Thermal Analogy

Fourier's law is the basic law of conduction heat transfer. Let us explain it by the example of heat transfer from a flat uniform wall surface at temperature t_1 to another one at temperature t_2 (where $t_1 > t_2$), Fig. 3.1. It is assumed that compared to the thickness of the wall δ its other two dimensions (width and length) are infinitely large. In this case, the temperature varies only in the direction of X-axis perpendicular to the wall surfaces. Such temperature field, i.e., the combination of temperature values at all points of the body, is one dimensional since only coordinate X is sufficient to describe the field.

It has been found experimentally that the density of heat flux in this case is directly proportional to the temperature difference ($t_1 - t_2$) and inversely proportional to the wall thickness δ:

$$q = \lambda \times \frac{t_1 - t_2}{\delta}, \mathrm{W/m^2}. \tag{3.1}$$

Formula (3.1) expresses Fourier's law in the integral form. The coefficient of proportionality λ is called the thermal conductivity coefficient. This value describes ability of wall substance to conduct heat and is measured in W/m × °C. The total quantity of heat Q transferred across the section of the flat wall of area F (m^2) is as follows:

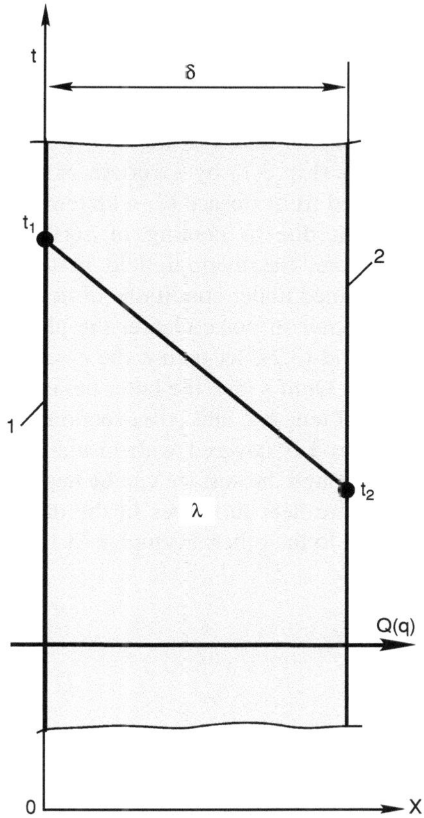

Fig. 3.1 Variations in temperature in a flat uniform wall thickness

$$Q = \lambda F \cdot \frac{t_1 - t_2}{\delta}, \mathrm{W} \tag{3.2}$$

Values

$$R = \delta/\lambda \times F, \, ^\circ\mathrm{C/W} \tag{3.3}$$

$$r = \delta/\lambda, ^\circ\mathrm{C} \times \mathrm{m}^2/\mathrm{W} \tag{3.4}$$

are called thermal resistances of wall surface section F m^2 area and the section equal to 1 m^2 area, respectively. Using values of R and r formulae (3.1) and (3.2) can be expressed as follows:

$$q = (t_1 - t_2)/r, \tag{3.1'}$$

$$Q = (t_1 - t_2)/R. \tag{3.2'}$$

If the coefficient λ does not depend on t and is a constant, the temperature inside the wall varies from t_1 to t_2 linearly. Isothermal surfaces, i.e., surfaces with the equal temperature, are parallel to the wall external surfaces, Fig. 3.1.

Of course, temperature distribution in the wall shown in Fig. 3.1 can exist only on condition that heat flux q is permanently supplied to surface (1) with temperature t_1 (Eq. 3.1) by a certain external energy source and the same heat flux is removed from surface (2) with temperature t_2. The temperature t_2 can be sustained constant due to cooling of surface (2), for example, by the flowing water. Therefore, the thermal field in the wall cannot exist by itself. It is uniquely determined under conditions of heat exchange of wall surfaces with environment.

In order to make clearer the physical meaning of Fourier's law and formulae (3.1) and (3.2), let us use the electro-thermal analogy between the Fourier's law and the Ohm's law, the latter being well known from high school. Let us examine a rod of length L and cross-sectional area S, as shown in Fig. 3.2. The side surface of the rod is covered with insulation of such low thermal conductivity that heat loss through the surface can be neglected. Under this condition, the propagation of the entire heat flux goes in the direction along the rod axis only, from one end surface to the other. Formulae (3.1) and (3.2) for this case look like the following:

$$q = \lambda \times \frac{t_1 - t_2}{L}, \tag{3.5}$$

$$Q = \lambda S \times \frac{t_1 - t_2}{L}. \tag{3.6}$$

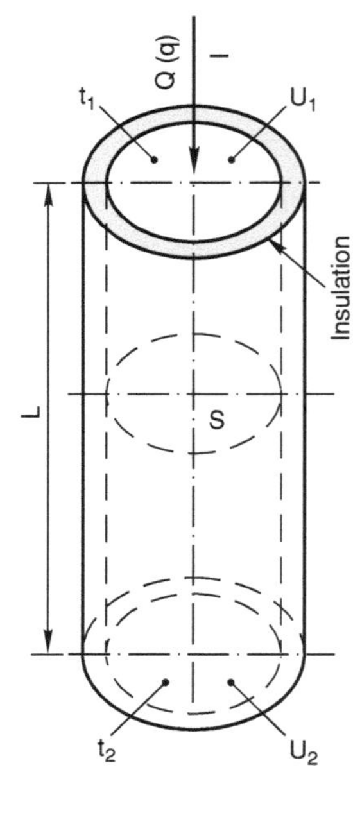

Fig. 3.2 As to electro-thermal analogy

If a current is passed through the rod, the strength of the current *I*, A, according to Ohm's law is as follows:

$$I = (U_1 - U_2)/R_{EL}, \tag{3.7}$$

$(U_1 - U_2)$ – the difference in electric potential between the rod ends.
R_{EL} – the rod electrical resistance, Ω.
$R_{EL}=L/\sigma S$, where σ is the electrical conductivity of the rod material.

If the value $L/\lambda \cdot S$ measured in °C/W, by analogy with formula (3.3), is called as the rod thermal resistance R_λ and the value L/λ as the specific (per unit cross-sectional area) thermal resistance r_λ, and formulae (3.5) and (3.6) are written in a form:

$$q = (t_1 - t_2)/r_\lambda \tag{3.8}$$

$$Q = (t_1 - t_2)/R_\lambda, \tag{3.9}$$

then the total analogy between Fourier's law and Ohm's law will become evident. Each thermal parameter in formulae (3.1′); (3.5) ÷ (3.9) corresponds to a certain electrical analogue: the heat flux Q (W), the strength of the current I (A), the temperature difference Δt (°C), the potential difference ΔU (V), the heat flow density q (W/m^2), the current density $j = I/S$ (A/m^2), the thermal conductivity λ (W/m × °C), the electrical conductivity σ, the thermal resistance $R_\lambda = L/\lambda \times S$ (°C/W), the electrical resistance $R_{EL} = L/\sigma \times S$ (Ω), the specific (per unit cross-sectional area) thermal resistance r_λ, and the specific electrical resistance ρ.

3.2.2 Coefficient of Thermal Conductivity

The coefficients of thermal conductivity are listed in handbooks of thermo-physical properties of substances. Difference between the values of λ for solids and fluids is 1000-fold or even 10,000-fold, Fig. 3.3. They depend a lot on structure, composition, temperature, density, and other parameters of a specific substance.

Among gases, the lightest ones, namely helium and hydrogen, have the highest thermal conductivity. The coefficients of λ for these gases at 0°C are 0.14 and 0.17 W/m × °C, respectively. The thermal conductivity of air and other gases is from 5 to 7 times less. As the temperature increases, the thermal conductivity of gases increases strongly.

As a rule, for non-metallic liquids $\lambda < 0.7$ W/m × °C. Water is one of the best liquid heat conductors. Its $\lambda \approx 0.6$ W/m × °C, and in contrast to the majority of other liquids, the thermal conductivity of water does not fall but grows with an increase in temperature. The thermal conductivity of liquid metals is much higher, e.g., for liquid iron $\lambda \approx 23$ W/m × °C.

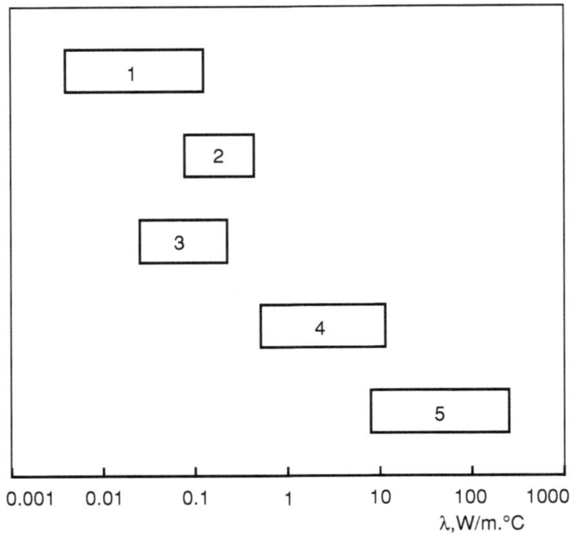

Fig. 3.3 Intervals of thermal conductivity values for different substances 1 – gases; 2 – liquids (non-metals); 3 – heat-insulating materials; 4 – refractories; 5 – metals

For porous and fibrous heat-insulating materials, the values of $\lambda < 0.25$ W/m × °C are typical. The thermal conductivity of cotton wool-like materials at low rammed density approaches the thermal conductivity of the air filling the pores. The values of λ for heat-insulating refractory materials lie within 0.1–0.4 W/m × °C. The thermal conductivity of these materials increases as the temperature rises. It especially increases with the increase in humidity level, which has a great effect on the thermal conductivity. Moreover, the thermal conductivity of a humid material is much higher than that of a dry one and that of water measured separately.

The thermal conductivity of refractory products used in EAFs depends on their composition and varies within the limits of 1.5–3.0 W/m × °C. The thermal conductivity of most of the refractory materials decreases as temperature rises. The thermal conductivity of metals is much higher than that of the other solids. Pure silver and copper have the greatest thermal conductivity. As temperature rises, the thermal conductivity of copper decreases substantially, whereas that of the other metals and Cu–Ni alloys either increases, Table 3.1.

All the data on λ have been obtained from experiments carried out with certain samples and under certain conditions. These data can be used for other products which are not quite identical and which work under different conditions, but only as rather rough data. This warning refers mostly to solids such as metals, refractory, and heat-insulating materials.

Even quite insignificant admixtures to the pure metals reduce their thermal conductivity sharply. Mere traces of arsenic decrease thermal conductivity of copper to about three times. The addition of 0.4% chromium to copper for the purpose of increasing its thermal stability reduces λ by 1.5–2 times at 200–300°C, Table 3.1. The thermal conductivity of copper strongly depends on its density and oxygen content. The thermal conductivity of incompletely deoxidized cast copper

Table 3.1 Thermal conductivity coefficients for certain metals and alloys, λ W/m × °C

Metal or alloy	Temperature, t °C				
	0.0	100	200	300	400
Copper Cu 99.9%	393	385	378	371	365
Aluminum, Al 99.8%	228	226	227	–	–
Nickel Ni 99.9%	92	83	74	68	64
Alloy Cu 99.3%, Cr 0.4%	153	167	202	233	286
Alloy Cu 90%, Ni 10%	57	64	73	79	86
Alloy Cu 68%, Ni 30%	25	32	38	45	–
Low-carbon steel	60	55	53	49	45
Medium-carbon steel	–	49	48	46	43
Stainless steel	–	16	18	19	21
Iron, C 4.0–4.7% (mean values)	57	43	41	39	37

is reduced by 15–20% as compared to that of the rolled electrolytic copper. The thermal conductivity of aluminum, nickel and other metals are also strongly affected by the impurities as well as by their structure. For example, decrease in nickel purity from 99.9 to 99.2% reduces λ by 1.4 times at 0°C.

Thermal conductivity data on refractory and heat-insulating materials given in the handbooks and the original literature differ greatly. For the identical in chemical composition and porosity samples of refractory materials, the scatter in the values of λ, according to different sources, is from 30 to 100%. It is mainly explained by ambiguous characteristics of the samples. Usually, only the chemical composition and the porosity are noted, which is obviously insufficient. Phase composition of refractory materials has a great effect on their thermal conductivity. For the same chemical composition, the phase composition may vary greatly. Samples with the same porosity may differ significantly with regard to the number of micro-cracks and other structural features. The thermal stresses in the products as well as many other factors also have their effects.

Under conditions of the high-temperature furnaces, especially the steelmaking ones, the refractory materials undergo profound physical–chemical transformations. They get saturated with iron oxides and other elements that enter into the composition of the liquid metal and slag, which leads to a sharp change in all the initial characteristics of the refractory materials, including their thermal conductivity. This also refers to the heat-insulating materials which undergo essential structural changes when heated. All of this reduces a possible accuracy in calculations of the heat transfer.

3.2.3 Multi-Layer Flat Wall

Let us examine a furnace wall which consists of three layers adjoining each other tightly. Layer (1) consists of highly durable refractory material; layer (2) consists of refractory material used for lower temperatures; and layer (3) consists of a

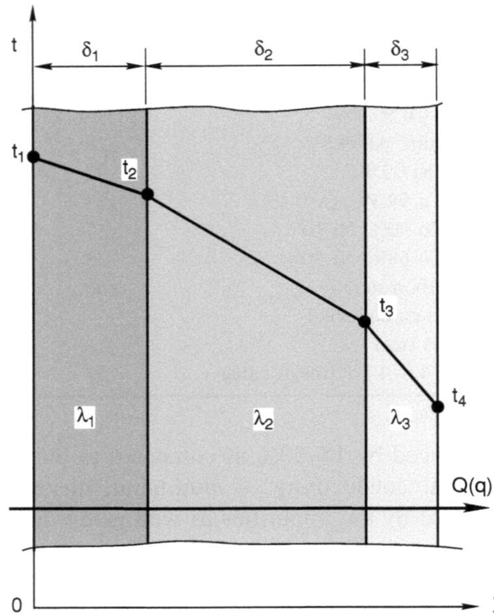

Fig. 3.4 Temperature distribution in a flat three-layer wall

heat-insulating material, Fig. 3.4. The thermal resistance of such a wall is equal to the sum of the thermal resistances of its separate layers. By analogy with formula (3.1′) we have the following:

$$q = \frac{t_1 - t_2}{\delta_1/\lambda_1 + \delta_2/\lambda_2 + \delta_3/\lambda_3}. \tag{3.10}$$

If there is good contact between the layers, their contacting surfaces will have the same temperature, t_2 and t_3, Fig. 3.4. The heat flow in the case of stationary process (t_1 = const and t_2 = const) is constant and is the same for all the layers. Using Eq. (3.1′) for each layer and knowing q, we will find the expressions for determining the unknown temperatures t_2 and t_3:

$$t_2 = t_1 - q \times (\delta_1/\lambda_1), \tag{3.11}$$

$$t_3 = t_2 - q \times (\delta_2/\lambda_2), \tag{3.12}$$

$$\text{or } t_3 = t_4 + q \times (\delta_3/\lambda_3). \tag{3.13}$$

The values $q \times (\delta/\lambda)$ are equal to the temperature differences Δt in each layer. Within each layer, the temperature varies linearly. However, altogether, a temperature curve is a polygonal line. The slopes of the line segments are different for each layer. The temperature of the layers with lower thermal conductivity varies more sharply: $\lambda_1 > \lambda_2 > \lambda_3$, Fig. 3.4. When designing a multi-layer wall, the thickness of separate layers should be selected so as to ensure that the temperatures at the boundaries between the layers inside the wall do not exceed the permissible values for the materials used. In the case under consideration, the temperature t_3 should not exceed the maximum permissible value for the heat-insulating material of layer (3).

In formula (3.1) as well as in the following formulae, it is assumed that λ is constant (λ = const) with regard to both the thickness of the uniform wall and the thickness of each layer of the multi-layer wall. In reality, λ depends on the temperature. Therefore, the temperatures in the wall and in its layers do not vary linearly, but rather curvilinearly. This, however, does not affect the accuracy of calculating the temperatures inside the wall, if the mean values $\lambda = (\lambda_n + \lambda_m) \times 0.5$ are assumed as design values for each layer. λ_n and λ_m are thermal conductivity coefficients at the temperatures of the boundaries between the layers t_n and t_m.

3.2.4 *Contact Thermal Resistance*

Formulae (3.10), (3.11), (3.12) and (3.13) are true on the assumption that the contacts between the single layers of the wall are ideal and the thermal resistances of these contacts can be neglected. Such assumptions, however, are often not met. In many cases, the thermal resistances of the contacts between the layers are significant and should be considered. To do that, the thermal resistances of the contacts (the gaps) between the layers $r_\Delta = \delta_\Delta/\lambda_\Delta$ should be added to the denominator of formula (3.10); δ_Δ is the gap width and λ_Δ is the coefficient of thermal conductivity of the medium filling the gap.

If the surfaces of the layers are rough, their close touch is not possible, and the thin air gaps are left between the layers. As thermal conductivity of air is very low ($\lambda = 0.026$ W/m $\times$ °C at room temperature), even the thin gaps may cause a considerable increase in the total thermal resistance of the wall.

The solid bodies touch each other only with the apexes of the spikes of the rough surface. The contact area of the apexes is negligible, and practically all the heat flow passes through the air gap. After finishing of the rough surfaces (i.e., turning, planing, milling), the average width of the air gap between two contacting metal plates is close to 0.015 mm. The corresponding thermal resistance of the gap filled with air is $r_\Delta = 0.015 \times 10^{-3}/0.026 = 0.58 \times 10^{-3}$ °C $\times$ m^2/W. This is equivalent to the thermal resistance of a steel layer with thickness of 30 mm. The contact thermal resistance between two metal surfaces can be reduced practically to zero, if these surfaces are thoroughly ground and strongly pressed to each other. The same effect can be achieved for the refractory wall as well, if the gaps are filled with fireclay mortar while bricking.

3.2.5 Uniform Cylindrical Wall

The practical tasks often require calculating the heat flow through the cylindrical wall of a pipe. Such are the calculations of the tubular water-cooled panels and the gas duct of the EAFs, for instance. Just like the thermal field of the flat wall, that of the cylindrical wall of the pipe is one dimensional, since in this case the temperature varies along one direction only, namely along the radius, Fig. 3.5. Along the length of the pipe as well as along its perimeter, the temperatures of the external and internal walls remain constant.

In the cylindrical wall, the heat flow density increases when approaching the inner surface, since its area is less than that of the outer one. Therefore, even if λ = const, the temperature in a radial direction does not vary linearly, but rather curvilinearly. As the distance from the outer wall surface increases, the curvature of this line grows. For this reason, the formulae for calculation of the heat flows of q and Q in the cylindrical walls get complicated. For the wall with length L,

$$Q = \frac{2\pi \times \lambda \times L}{\ln(d_2/d_1)} \times (t_1 - t_2)\ ,\ W. \tag{3.14}$$

It is presumed that $t_1 > t_2$, Fig. 3.5. Since $Q = (t_1 – t_2)/R$, the total thermal resistance of the wall with length L is determined by the expression:

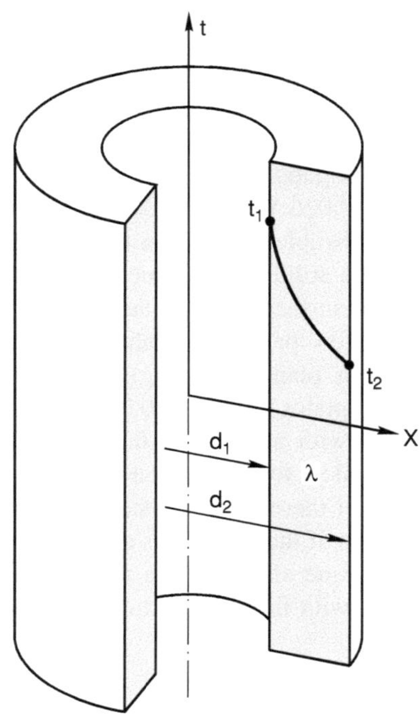

Fig. 3.5 Variations of temperature across a thickness of cylindrical uniform wall

$$R = \frac{1}{2\pi \times \lambda \times L} \times \ln(d_2/d_1) \quad (3.15)$$

Equation (3.14) is also true, if $t_1 < t_2$, i.e., when the heat flow is directed from the inner surface to the outer one. According to formula (3.14), only the heat flow direction changes, whereas its absolute value does not.

The specific heat flow q passing through the pipe wall can be related to one running meter of its length, formula (3.16), or to a unit of the internal surface, formula (3.17), or to a unit of the external surface, formula (3.18):

$$q_L = \frac{2\pi \times \lambda}{\ln(d_2/d_1)} \times (t_1 - t_2), \quad (3.16)$$

$$q_1 = \frac{2\lambda}{d_1 \times \ln(d_2/d_1)} \times (t_1 - t_2), \quad (3.17)$$

$$q_2 = \frac{2\lambda}{d_2 \times \ln(d_2/d_1)} \times (t_1 - t_2). \quad (3.18)$$

The respective specific thermal resistances r in the formulae of $q_L = (t_1 - t_2)/r_L$ type are determined by the expressions:

$$r_L = \frac{1}{2\pi \times \lambda} \times \ln(d_2/d_1), \quad (3.19)$$

$$r_1 = \frac{1}{2\lambda} \times d_1 \times \ln(d_2/d_1), \quad (3.20)$$

$$r_2 = \frac{1}{2\lambda} \times d_2 \times \ln(d_2/d_1). \quad (3.21)$$

3.2.6 Multi-Layer Cylindrical Wall

Let us examine a cylindrical wall which consists of three layers adjoining each other tightly, Fig. 3.6. In the case of stationary process, the quantity of heat Q passing through each layer is equal and constant. Yet the density of the heat flow q increases from the outer layers to the inner ones due to the same reason as in the uniform cylindrical wall. In order to determine the heat flow rate through the multi-layer cylindrical wall, it is necessary to sum up the thermal resistances of the separate layers, just like in the case of a multi-layer flat wall. For the three-layer cylindrical wall q_L is determined by the expression:

$$q_L = 2\pi(t_1 - t_2) \bigg/ \left(\frac{1}{\lambda_1}\ln\frac{d_2}{d_1} + \frac{1}{\lambda_2}\ln\frac{d_3}{d_2} + \frac{1}{\lambda_3}\ln\frac{d_4}{d_3}\right). \quad (3.22)$$

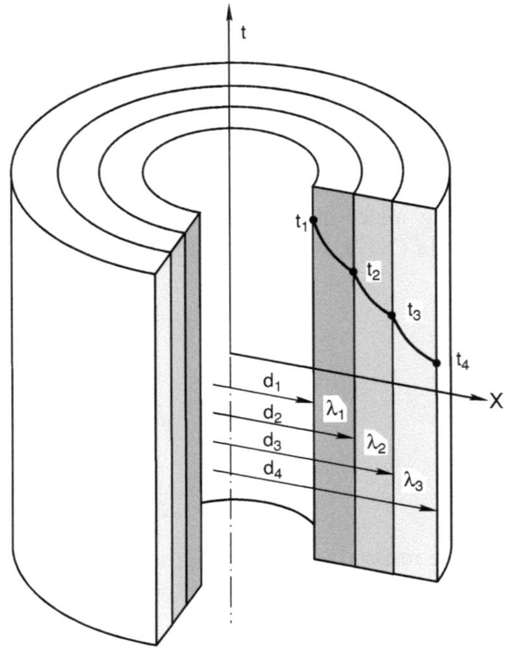

Fig. 3.6 Temperature distribution in the three-layer cylindrical wall

Using formulae (3.20) and (3.21), it is easy to obtain the analogous expressions for q_1 and q_2. Inside each layer the temperature varies logarithmically (Eq. 3.14), but altogether the temperature curve of the multi-layer cylindrical wall is a polygonal line, Fig. 3.6.

3.2.7 *Simplifying of Formulae for Calculation of Cylindrical Walls*

The formulae given above are rather inconvenient because they include natural logarithms. The calculations may be substantially simplified by using the formula analogous to formula (3.2) for the flat wall and introducing into it the coefficient φ to account for the curvature:

$$Q = \frac{1}{\varphi} \times \frac{\lambda}{\delta} \times F_m(t_1 - t_2) \tag{3.23}$$

In this expression, $\delta = 0.5(d_1 - d_2)$ is the thickness of the pipe wall, and $F_m = \pi \times d_m \times L$ is the average area of the pipe surface which is calculated by its arithmetic mean diameter $d_m = 0.5\ (d_1 + d_2)$. Hence, the simplified expressions for q_L are as follows:

Table 3.2 The curvature coefficient of the cylindrical wall, φ

d_1/d_2	1.2	1.4	1.6	1.8	2.0	3.0	4.0	5.0	6.0
φ	1.003	1.010	1.018	1.028	1.040	1.10	1.16	1.21	1.25

$$q_L = \frac{1}{\varphi} \times \frac{\lambda \times \pi \times d_m}{\delta} \times (t_1 - t_2) \tag{3.24}$$

and for the specific (per unit length) thermal resistance r_L:

$$r_L = \frac{\varphi \times \delta}{\pi \times \lambda \times d_m}. \tag{3.25}$$

By adding the thermal resistances of all layers, we obtain a simplified expression for an n-layer of the cylindrical wall which is analogous to formula (3.22):

$$q_L = \frac{\pi(t_1 - t_{n+1})}{\varphi_1 \times \delta_1/\lambda_1 \times d_{m1} + \varphi_2 \times \delta_2/\lambda_2 \times d_{m2} + \ldots. + \varphi_n \times \delta_n/\lambda_n \times d_{mn}}, \tag{3.26}$$

<φ_n, δ_n, and d_{mn} – the coefficient of the curvature, the thickness, and the mean diameter of the n-layer, respectively.

The coefficients of the curvature φ are determined by the ratio of the outer and inner wall diameters (formulae 3.23 and 3.24) or each of the layers of the wall (formula 3.26). As this ratio grows the value of φ increases as well, Table 3.2.

Data in Table 3.2 show that if d_2/d_1 is less than 1.6, the coefficient φ can be assumed to be equal unity and that the cylindrical wall can be calculated as the flat wall. An error of calculations under these assumptions does not exceed 2%. For practical calculations higher accuracy is not required. Except pipes with very small diameters and increased wall thickness the ordinary gauge pipes have the ratio d_2/d_1 far less than 1.6.

3.2.8 *Bodies of Complex Shape: Concept of Numerical Methods of Calculating Stationary and Non-Stationary Conduction Heat Transfer*

Changing of temperature in the bodies of complex shape should be considered not by single coordinate but by two or even all the three coordinates as temperature fields of such objects are two or three dimensional. All formulae shown above for simple bodies of regular symmetrical shapes (plates, cylinders, etc.) have been derived analytically by integration of Fourier's differential equation. As a rule, analytical solutions of heat conduction problems for irregular-shaped bodies which are encountered in practice are not possible. In such cases approximate numerical methods of solutions are applied; there are so-called finite difference methods, etc., in which differential equations are replaced by algebraic ones.

One of such methods is as follows. A body under consideration is divided into a number of small-sized (elementary) volumes. Within each volume generally in its center the nodal point is selected. It is assumed that the entire mass of the substance located in an elementary volume is concentrated at the nodal point. Nodal points are connected with each other by heat conductive rods. Heat resistance values R of the rods are assumed to be similar to the heat resistance of the wall, the thickness of which is equal to the distance between nodal points and the area of which is equal to contact area of the adjacent volumes.

Certain temperatures are assigned to the outermost nodal points at the boundaries of a body. In elementary cases they are constant or determined by heat exchange conditions with the environment. The system consisting of the nodal points and the heat conductive rods is called the computational grid. The more accurate and detailed the thermal field of a body should be determined, the finer the grid should be; however, the amount of calculations increases significantly. If stationary heat conduction problem is being solved, some stationary thermal field should be found within the body under preset conditions at its external boundaries.

The idea of the computation is to trace in time the process of heat exchange between elementary volumes step by step until further changes of temperatures at the nodal points of the grid may be neglected, or the same, temperature field inside the body comes to a practically stationary state. In order to make a smooth transformation of initial arbitrarily chosen thermal field to its finite stationary state, duration of each step of calculation $\Delta\tau$ should be small enough. The value $\Delta\tau$ is chosen according to certain principles depending on size and weight of elementary volumes as well as on heat conduction and heat capacity of the body substance.

The sequence of the calculation will be explained by the example of a long homogeneous rod; its cross-section and computational grid are shown in Fig. 3.7. The nodes of the grid are indicated by numbers. Three external sides of the body *abcd* are heated. The nodes of the grid cells adjoining these sides are indicated by number (1). On assumption, these nodes have equaland constant temperature t_1 (e.g., 300°C). The fourth external side of the rod *ad* is cooled by water. Nodes (2) of the cells adjacent to this side also have constant but lower temperature t_2 (e.g., 20°C). Since the temperatures of the external sides are constant, the two-dimensional thermal field inside the rod, which is determined by temperatures of internal nodes of the grid, should also be constant (stationary). Before the start of calculations these temperatures may be set arbitrarily, e.g., equal to 20°C. As a result of the calculation their real values should be obtained, which would correspond to the permanent temperatures of the external sides of the rod.

Even after the first step of the calculation with duration $\Delta\tau$ the temperatures of all the inner nodes of the grid change. Since the value $\Delta\tau$ is small, these changes are slight too. They can be calculated using equations of heat balance at each nodal point. The physical meaning of these equations is very simple: the quantity of heat received by the given node through all heat conductive rods of the grid during $\Delta\tau$ according to the law of conservation of energy should be equal to the increase in the

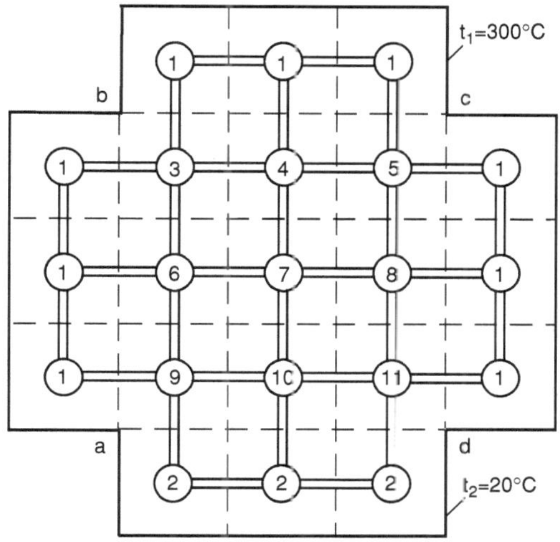

Fig. 3.7 The grid for calculation of two-dimensional thermal conductivity in a long uniform rod

enthalpy of the mass of node ΔE, which in this case is unambiguously linked to its temperature increase Δt per equation $\Delta E = c_P \times m \times \Delta t$, where c_P, kWs/kg × °C, is heat capacity of the substance of the node and m is its mass, kg. For example, if the initial temperature of a node (5) was t_5 and after the first step, i.e., in the period $\Delta\tau$ it rises to t_5', then this new temperature is linked to the initial one by the equation:

$$c_P \times m_5(t_5' - t_5) = \frac{2 \times (t_1 - t_5)}{R_{(1-5)}} + \frac{t_4 - t_5}{R_{(4-5)}} + \frac{t_8 - t_5}{R_{(8-5)}}, \tag{3.27}$$

R – the thermal resistance of respective heat conductive rods.

Similarly, it could be written for nodes (9) and (7) as follows:

$$c_p \times m_9(t_9' - t_9) = \frac{t_1 - t_9}{R_{(1-9)}} + \frac{t_6 - t_9}{R_{(6-9)}} + \frac{t_{10} - t_9}{R_{(10-9)}}, \tag{3.28}$$

$$c_p \times m_7(t_7' - t_7) = \frac{t_6 - t_7}{R_{(6-7)}} + \frac{t_4 - t_7}{R_{(4-7)}} + \frac{t_8 - t_7}{R_{(8-7)}} + \frac{t_{10} - t_7}{R_{(10-7)}}. \tag{3.29}$$

In Eq. (3.28) the term $(t_2 - t_9)/R_{(2-9)}$ is absent since the initial temperatures of nodes (2) and (9) are assumed to be equal (20°C). The total number of equations similar to (3.27), (3.28), and (3.29) will be equal to the number of the nodes with temperatures to be determined. In the given example there are nine such nodes (from No.3 to No.11). Having solved the system of nine equations with nine unknown values we find values of the temperatures t_3', t_4', $t_5' \ldots t_{11}'$ of these nodes after the first step of the calculation. Then the calculated temperature distribution is assumed as the initial and the second step of $\Delta\tau$ duration is made. Multiple iteration of the calculation allows determining, after a certain number of steps, considerable enough, the required virtually stationary distribution of temperatures in the rod under preset constant temperature conditions on its external sides.

Similarly, these and alike methods allow to solve problems of not only stationary but also non-stationary heat conduction, for example, various heating problems of complex-shaped bodies by external heat sources. In this case, heat exchange conditions at the boundaries of the body with the environment can be any conditions encountered in practice and can vary during the heating of the body. Dependence of thermal conductivity and other physical properties of the body on temperature also can be taken into consideration. By using rather fine grid such numerical methods of calculation of stationary and non-stationary heat conduction allow to determine the finite stationary temperature distribution in the body and variable field of temperatures at any preset moment of time with any accuracy required for practice.

Numerical methods of calculation of complex problems of heat conduction were developed yet in the 1930s–1940s, although at that time they were far too labor-intensive. These methods became widespread only after of fast computers with large memory cache. The inconvenience of numerical methods as compared with analytical ones is only in the fact that a new computer program with its own distinctive features is required to solve every new problem.

Example 1. There is a two-layer flat wall. Parameters of the first refractory layer are $\delta_1 = 0.4$ m and $\lambda_1 = 2.0$ W/m × °C and those of the second insulating layer are $\delta_2 = 0.05$ m and $\lambda_2 = 0.2$ W/m × °C. Temperature of the hot wall surface t_1 is 1600°C and that of cold one t_3 is 300°C. Both density of heat flux through the wall q, kW/m^2, and temperature of an interface between layers t_2 should be determined.

By analogy with formula (3.10) and Fig. 3.4, the expression for q could be written down as follows: $q = \frac{t_1 - t_3}{\delta_1/\lambda_1 + \delta_2/\lambda_2} = \frac{1600-300}{0.4/2+0.05/0.2} = 2889$ W/m^2; $q \cong$ 2.9 kW/m^2. For the first layer, it follows from formula (3.1) that $q = \lambda_1\,(t_1 - t_2)/\delta_1$; hence $t_2 = t_1 - q\delta_1/\lambda_1 = 1600 - 2889 \times 0.4/2 = 1022$°C.

Example 2. Parameters for a pipe are given as follows: $d_2 = 300$ mm; $d_1 = 275$ mm; $\delta = 0.0125$ m; $t_1 = 300$°C; $t_2 = 100$°C; $q = 850$ kW/m^2. The thermal conductivity coefficient λ as well as a type of pipe wall material has to be determined.

The ratio d_2/d_1 is 300/275 = 1.09. Such a cylindrical wall can be considered as a flat wall, Table 3.2. The λ-coefficient is calculated with formula (3.1): $\lambda = q\times \delta/(t_1 - t_2) = 850{,}000 \times 0.0125/(300–100) = 53.1$ W/m × °C; $\lambda \cong 53$ W/m × °C. The mean wall temperature is $t_m = (t_1 + t_2)/2 = (300 + 100)/2 = 200$°C. According to Table 3.1, at $t = 200$°C, the value of $\lambda = 53$ W/m × °C corresponds to low-carbon steel.

3.3 Convective Heat Exchange

3.3.1 Newton's Law: Coefficient of Heat Transfer α

Convective heat exchange between a surface of a solid body and liquid or gaseous heat carrier flowing around the surface is a very complex process. It depends on a great number of factors such as the motion nature of heat carrier and its physical properties which vary greatly with temperature, shape, and dimensions of a surface which is flown around, etc. Analytical solutions of convective heat exchange problems are known only for a limited number of elementary cases under a great deal of simplifying assumptions which do not correspond to the real conditions. In fact, all our knowledge of convection which is of any practical value is based upon the results of experimental studies. This situation has led to the fact that for the convenience of generalization of experimental data and simplification of practical calculations, all extremely various convection processes are calculated using the same simple formula. It is derived from Newton's law: quantity of heat Q transferred from a fluid (gas)[1] to a wall or in the reverse direction is proportional to the surface area of heat exchange F and the difference between the temperature of the wall t_W and that of the fluid t_{FL}:

$$Q = \alpha \times F \times (t_W - t_{FL}). \tag{3.30}$$

Coefficient of proportionality α is called coefficient of convection heat transfer. It describes intensity of a heat transfer process; its unit is W/m^2 $\times$ °C. Formula (3.30) seems very simple only at first glance. In fact, the entire complexity of convective heat transfer process and difficulties of its calculation are concealed in the only value, i.e., in coefficient of heat transfer α. Usually, this formula is used in research of convection processes which are very different by their physical nature. Experimental results are processed in such way that allows determining the dependence of α on numerous factors which determine conditions of the process run. All these dependences usually have a sophisticated nature; however, they may be rather simplified for practical calculations, particularly for comparative estimates of any innovations.

[1] At velocities up to 120–150 m/s being very far from the speed of sound, gases behave as non-compressible liquids so that gas and liquid flows are governed by the same laws of motion. Therefore, as it is assumed in aero-hydromechanics when considering low-speed flows the terms "liquid" and "gas" are used as synonyms. Further, in many cases, the common term "fluid" will be used.

Table 3.3 Dependence of kinematic viscosity of water, υ m²/s on temperature

t, °C	10	20	30	40	50	70	90	120	160
$\upsilon \times 10^6$ m²/s	1.31	1.01	0.80	0.66	0.56	0.41	0.33	0.25	0.19

3.3.2 Two Modes of Fluid Motion

According to their physical nature two modes of fluid motion are distinguished: laminar and turbulent. In laminar mode, separate layers of fluid move parallel to the walls and one another and intermix very slowly. Mixing in this case occurs only due to thermal motion of separate molecules in the direction transversal to the direction of the fluid flow. In turbulent mode, micro- and macro-vortices emerge which move chaotically in all directions and intensely mix the entire fluid. The laminar mode occurs at low fluid velocities, while the turbulent mode takes place at high fluid velocities.

As velocity increases transition of laminar mode to turbulent one occurs as soon as the velocity reaches a certain value critical for these conditions. Critical velocity values are different for various fluids and fluxes of different geometry. Virtually uniquely the transition to turbulent mode is determined not by velocity but rather by dimensionless Reynolds' criterion:

$$\mathrm{Re} = v \times d/\upsilon, \tag{3.31}$$

v – flux velocity, m/s;
d – the inner diameter for pipes or specific dimension of a channel
υ – kinematic viscosity of fluid which depends on temperature, m^2/s, Table 3.3.

When fluid flows in pipes and channels the laminar mode is observed at the values up to $Re = 2300$. When $Re > 2300$ the turbulent mode occurs and develops. The fully developed turbulent mode establishes when $Re > 10^4$. Thus, the value Re describes the degree of flux turbulence. As a rule, in tubular panels of EAFs and other water-cooled elements fluxes are always turbulent. From formula (3.31) it follows that in pipes of twice larger diameter the same degree of turbulence is observed at the twice lower velocity.

3.3.3 Boundary Layer

The particles of fluid flow which are in direct contact with the surface of solid body are adsorbed by this surface and quasi-adhere to it. Thus, a very thin immovable layer of the adhered fluid forms on the wall. Since all fluids have viscosity, forces of viscous friction occur between the layers of fluid which move at different velocities; these forces are proportional to viscosity coefficient υ, Eq. (3.31). These forces lead to slowing down of those layers of fluid which are farther away from the wall when they come in contact with the immovable layer. As a result, a layer of slowed-down fluid of thickness δ or so-called hydrodynamic boundary layer forms near the wall. In the turbulent flow, at the major portion of thickness, the boundary layer is turbulent as well. However, in the immediate vicinity of the wall where velocities drop down abruptly approaching zero the turbulent mode transitions to the laminar one. This portion of the boundary layer is called laminar or viscous sub-layer, Fig. 3.8. The thickness of this sub-layer δ_L depends on the degree of turbulence of the main flux which is determined by the criterion *Re*, formula (3.31). As *Re* grows, the thickness δ_L decreases.

The heat flux from the fluid to the wall or in reverse direction should overcome the laminar sub-layer. There is no stirring of the fluid in this sub-layer and heat can spread only by its heat conduction. Thermal conductivity coefficient of fluids and especially of gases is quite small, Fig. 3.3. Therefore, just in the laminar sub-layer, despite its small thickness, almost entire heat resistance to the convective heat exchange between the wall and fluid r_α is concentrated. This resistance is the value reciprocal to α: $r_\alpha = 1/\alpha$. Beyond the laminar sub-layer heat propagation in turbulent flow occurs through stirring of the fluid. This process runs so intensely that the temperature of the fluid virtually does not vary through the cross-section of the flow core, since the thermal resistance of this zone is close to zero. Almost entire temperature change occurs within the laminar sub-layer, Fig. 3.8.

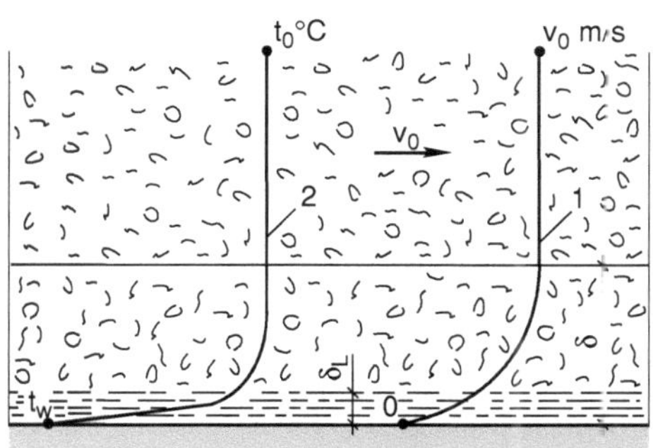

Fig. 3.8 Distributions of velocity (1) and temperature (2) through a cross-section of the flow moving along the wall. δ – thickness of hydrodynamic boundary layer, δ_L – thickness of laminar sub-layer, v_0 – velocity of the flow beyond the boundary layer, t_W and t_0 – temperatures of the wall and flow away from the wall; $t_0 > t_W$

In order to increase the coefficient of heat transfer α from the wall to the fluid flow (formula 3.30) it is necessary to decrease in some way the thickness of the laminar sub-layer and consequently its thermal resistance. To achieve this, the value of *Re* is raised by increasing the velocity of the flow. Increasing of α and intensification of heat transfer can be achieved also by the artificial turbulization of the laminar sub-layer, e.g., by jets directed perpendicularly to the surface of the body. The laminar sub-layer turbulization occurs also when the main flux is swirled and when the fluid boils on the wall and this layer is stirred by vapor bubbles. In all such cases α rises sharply. These and other methods of heat transfer intensification are discussed in detail in Chap. 12 which deals with design and calculation of water-cooled elements for EAFs.

3.3.4 Free (Natural) Convection

Unlike convection under the forced motion of fluid caused by any external agitator such as a pump or a fan, free convection is created by the very heat transfer process. The motion of fluid near the wall in this case does not require any additional energy consumption. The motion is caused by the difference of densities of heated and cooled fluid layers. This so-called free motion develops the more intensive the both greater the difference between the temperatures of the wall and the fluid and fluid's volumetric expansion coefficient is.

Near the hot vertical wall, the fluid (e.g., air) is heated, its density is reduced, and the lifting force appears and makes the heated "lighter" air move along the wall upward. Due to the same reason, near the cold wall the cooled "heavier" air flux appears, which moves downward. In both cases, the convective heat exchange occurs between the wall and the environment. In the former case, the heat flux is directed from the wall to air, which is being heated; in the latter case, the heat flux is directed from air to the wall while air is being cooled.

In comparison with the forced convection, the intensity of the free convection is very low. Its participation in the thermal performance of furnaces is limited mostly to heat losses through refractory lining of walls, roof, and bottom of the furnace to the environment. Hot external surfaces of refractory lining release heat not only by free convection of ambient air but also by radiation; in this case the radiation share increases rapidly as the temperature rises, Sect. 3.4. Usually the calculation formulae do not separate these two types of heat losses but consider jointly using the combined coefficient of heat transfer α_Σ. In the case of the vertical wall, with the temperature t °C of the external surface the following simplified formula can be used for approximate calculations of this value:

$$\alpha_\Sigma = 7 + 0.05\,t, \quad \mathrm{W/m^2 \times {}^\circ C}. \tag{3.32}$$

The obtained value α_Σ is substituted in formula (3.30) where the ambient air temperature is taken as t_{FL}. Heat transfer by free convection to the ambient air depends on the wall orientation. In comparison with the vertical wall, heat transfer from the wall's hot surface of which is directed upward increases by 20–30%. If the surface faces downward heat transfer decreases by the same value.

3.3.5 Convective Heat Transfer at Forced Motion

With reference to electric arc furnaces, the forced turbulent motion of fluid in various channels and pipes is of the greatest interest. In this case, the laminar motion of fluid is encountered seldom. Due to the low heat transfer under these conditions such mode must be avoided. The mode of motion is evaluated by *Re* number.

Under the laminar mode the velocity in the cross-section of the pipe varies parabolically from zero at the wall to the maximum v_0 at the axis. Under the turbulent mode the curve of velocity variation looks like the truncated parabola and away from the walls inside the core of the flux the velocity varies insignificantly, Fig. 3.9. To determine the value Re in formula (3.31) the average velocity v_m should be used, which equals the ratio of fluid flow rate per second V (m^3/s) to the area of the pipe cross-section S (m^2). For the round pipes their internal diameter d is used as the specific size, whereas for the pipes of other shape (e.g., rectangular or elongated narrow slits) so-called hydraulic diameter d_h is used. For non-round pipes $d_h = 4S/P$, where P is the length of the perimeter of the internal cross-section, while for slits $d_h = 2\delta$ where δ is the width of the slit.

A vast number of studies are dedicated to the heat transfer in various fluids with different properties, flowing through pipes. It is established that the results of all these studies for the field of the turbulent motion of any fluids may be summarized by a single formula (3.35) if only three dimensionless criteria are applied as variables: the Reynolds number $Re = v_m \times d_h/\upsilon$, Eq. (3.31); the Nusselt number *Nu* is expressed as follows:

$$Nu = \alpha \times d_h/\lambda \tag{3.33}$$

and the Prandtl number *Pr* is given as follows:

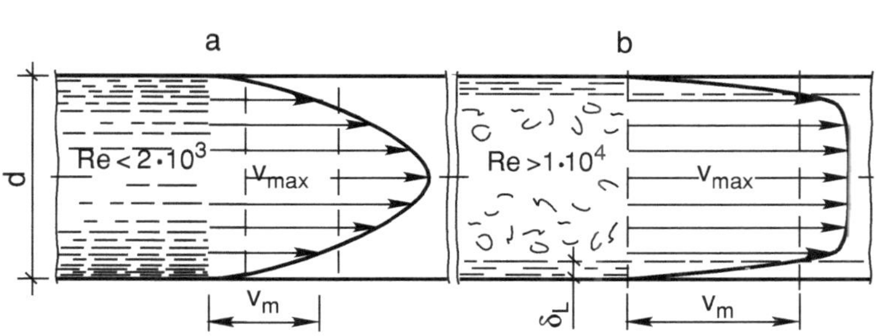

Fig. 3.9 Distribution of velocities through a cross-section at laminar (**a**) and turbulent (**b**) modes of motion of liquid in the pipe. v_m – the mean value

$$Pr = c_p \times \rho \times \upsilon / \lambda, \qquad (3.34)$$

$$Nu = 0.021 \times \mathrm{Re}_{FL}^{0.80} \times \mathrm{Pr}_{FL}^{0.43} \times (\mathrm{Pr}_{FL}/\mathrm{Pr}_W). \qquad (3.35)$$

Formula (3.35) is true in a wide turbulent area when the value Re spans from 1×10^4 to 5×10^6 for channels of any shape, for all dropping fluids and gases. This formula takes into consideration the following physical properties of fluids depending on temperature, kinematic viscosity υ, thermal conductivity λ, heat capacity c_P, and density ρ. The shape of pipe ducts is considered by their hydraulic diameter d_h. The subscript "FL" shows that the corresponding physical parameters are taken at the fluid temperature, while the subscript "W" at the wall temperature. Having determined the value Nu by formula (3.35), the value α is found from expression (3.33).

The great advantage of criterion equations like (3.35) is their versatility. However, for iterative technical calculations of heat transfer coefficients α in similar design elements, it is much more convenient to use simple formulae derived by transformation of the general criterion equation to conditions of the specific given case. In tubular cooled elements used widely in EAFs water, temperature of which varies within rather narrow limits, is always used as a cooling agent. For thermal calculation of such elements (sidewall and roof panels, gas ducts, etc.), Eq. (3.35) is transformed to

$$\alpha = A \times \nu^{0.8} / d_h^{0.2}, \quad \mathrm{kW/m^2 \times {}^\circ C} \qquad (3.36)$$

A – the coefficient which is determined by the physical properties of water which depend on temperature, Table 3.4. The average temperature of water between inlet and outlet in the cooled element should be used.

Formula (3.35) includes the value Pr_W at the wall temperature t_W, which depends not only on the temperature of water but on the heat flux which passes through the wall to water. The value t_W is unknown beforehand. The transformation of formula (3.35) to expression (3.36) is performed at the temperature t_W = 100°C. The real values of t_W may deviate from this value. In practice, these deviations are not significant since the narrow range of water pressure promotes that. Besides, the value Pr_W is present in formula (3.35) in a power of 0.25, which greatly reduces its effect. Therefore, the error of determination of α using expression (3.36) does not exceed 3–4 %. Such accuracy in calculations of water-cooled elements is quite sufficient for practical purposes.

Table 3.4 Dependence of the A-coefficient in formula (3.36) on water temperature

t, °C	20	25	30	35	40	45	50	55	60
A	2.56	2.59	2.63	2.67	2.71	2.74	2.78	2.80	2.83

3.3.6 Heat Transfer Between Two Fluid Flows Through Dividing Wall; Heat Transfer Coefficient k

The need to calculate such more complex processes of heat transfer is encountered rather often. In those processes, heat is transferred by convection, heat conduction, and in many cases by radiation, i.e., by all the three ways simultaneously.

Let us review a homogeneous flat wall with a hot fluid of the temperature t_{FL1} moving along one of its sides, while a cold fluid of the temperature t_{FL2} moves on the other side. The heat transfer coefficient from the hot fluid to the wall is α_1, while from the cold one is α_2. The temperatures of fluids t_{FL1} and t_{FL2} occur at a sufficient distance from the wall. In the boundary layer (Sect. 3.3) they approach the temperatures of wall surfaces t_{W1} and t_{W2} and become equal to them on the very surfaces, Fig. 3.10. The temperature field is stationary and one-dimensional. All the temperatures change in the direction of X-axis and are constant in time. The temperatures t_{W1} and t_{W2} are unknown and should be determined as well as the heat fluxes through the wall from one fluid to the other: the specific q, W/m^2 and the total Q, W.

Under the stationary heat modes the three specific heat fluxes, namely the flux transferred from a hot fluid to the wall $q = \alpha_1(t_{FL1} - t_{W1})$, flux passing through the wall $q = (t_{W1} - t_{W2})\ \lambda/\delta$, and flux released by the wall to a cold fluid $q = \alpha_2(t_{W2} - t_{FL2})$ have to be equal. Only under such conditions the temperatures t_{W1} and t_{W2} can remain constant. Otherwise, the wall will be either heated or cooled.

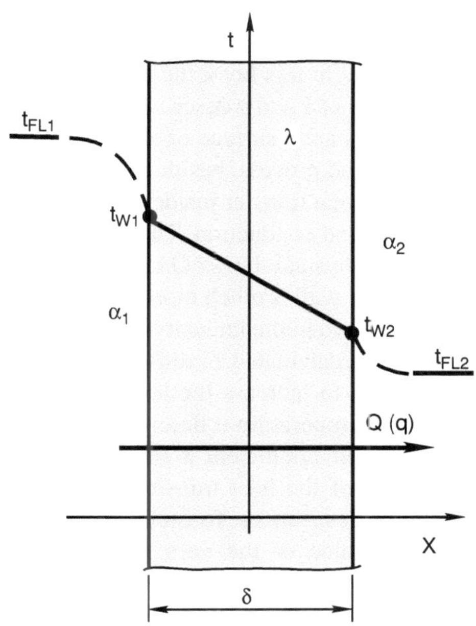

Fig. 3.10 Heat transfer through a flat wall

Having equated these fluxes to each other the unknown temperatures and the heat flux may be easily determined:

$$t_{W1} = t_{FL1} - q/\alpha_1, \tag{3.37}$$

$$t_{W2} = t_{FL2} - q/\alpha_2, \tag{3.38}$$

$$q = \frac{t_{FL1} - t_{FL2}}{1/\alpha_1 + \delta/\lambda + 1/\alpha_2}. \tag{3.39}$$

The value $\delta/\lambda = r_\lambda$ is the wall thermal resistance already known, formula (3.4). Similarly, the value $1/\alpha = r_\alpha$ is called convective heat transfer resistance while the sum of all the three resistances:

$$r_k = r_{\alpha 1} + r_\lambda + r_{\alpha 2}, \tag{3.40}$$

is the overall thermal resistance of the heat transfer. The value inverse to the overall thermal resistance of heat transfer $k = 1/r_k$, W/m^2 × °C, may be called the overall coefficient of heat transfer. This coefficient describes heat transfer process intensity. Formula (3.39) can be written down analogously to Ohm's law as follows:

$$q = (t_{FL1} - t_{FL2})/r_k, \tag{3.41}$$

$$\text{or} \quad q = k(t_{FL1} - t_{FL2}). \tag{3.41$'$}$$

In the case of the multi-layer wall in formulae (3.39) and (3.40) the thermal resistances will be supplemented with additional layers $r_{\lambda n} = \delta_n/\lambda_n$, analogously to formula (3.10).

In a number of countries the overall heat transfer coefficient (3.41′) is indicated as α like the convective coefficient of heat transfer in formula (3.30). In other countries as in this book the overall heat transfer coefficient is indicated as k. The coefficients of k and α describe two quite different processes. In the first process, the participants are a surface of a certain body and a fluid flowing around this surface. In the second process, besides the convective heat exchange of the surface with the fluids, the heat transfer inside the wall from one of its surface to the other takes part due to its heat conduction. The heat conduction and convection are governed by the different physical laws. On the whole, the combined process of heat transfer through the wall is much more complex than the heat transfer from the wall to fluid. Therefore, when the intensity of both processes is expressed by the same coefficient α, which is calculated in different ways, it brings confusion into the concepts.

In order to increase the heat transfer intensity, i.e., to increase the heat flux at the same temperature difference ($t_{FL1} - t_{FL2}$) it is necessary to increase the overall heat transfer coefficient k or, which is the same, to decrease the total thermal resistance of the heat transfer process r_k. The total resistance is the sum of the particular resistances, formula (3.40). The decrease of either lowers r_k. The thermal resistance of the very wall δ/λ may be lowered by either decreasing its thickness δ or using a material with higher thermal conductivity λ. The thermal resistance of the convective heat transfer $1/\alpha$ decreases as the velocity and the degree of fluid flow turbulence increase.

An efficient method of intensification of the convective heat exchange is wall surface ribbing as well. In order to quantitatively evaluate the contribution of ribbing let us write down the equation for the whole flow Q through the entire surface of the wall F. Since $Q = q \times F$, it easy to derive from Eq. (3.39):

$$Q = \frac{t_{\mathrm{FL1}} - t_{\mathrm{FL2}}}{1/\alpha_1 \times F + \delta/\lambda \times F + 1/\alpha_2 \times F}. \tag{3.42}$$

The expression located at the denominator of formula (3.42) is the total thermal resistance of heat transfer through the wall R_k. Correspondingly, the inverse value $1/R_k = k$, W/°C, is the overall coefficient of heat transfer. Hence, the expression analogous to (3.41′) is as follows:

$$Q = k \times (t_{\mathrm{FL1}} - t_{\mathrm{FL2}}).(3.42') \tag{3.42}$$

Ribbing increases the area of flat surfaces of the wall from F to F'. By doing so the thermal resistance R_k decreases, whereas the coefficient k and the flow Q increase, Eqs. (3.42) and (3.42′). However, it should be taken into consideration that heat transfer coefficients α_1 and α_2 on the ribbed surface do not remain the same as on the flat surface. They decrease to the larger degree, the greater is the ratio F'/F called by the coefficient of ribbing. Therefore, in reality, ribbing is less efficient than it could be expected from Eq. (3.42) at the constant α, and it is reasonable to increase the coefficient of ribbing to a certain optimal value only.

Selection of rational methods to intensify heat transfer in every specific case depends on a relative value of particular thermal resistances. If all these resistances $(1/\alpha_1 + 1/\alpha_2 + \delta/\lambda)$ have close values, then increasing of heat flux can be achieved by decreasing any of them or all of them together. However, the cases where one of these resistances is much greater than the rest of them are encountered way more frequently.

As an example the hot gas coolers can be used; they use water as the cooling agent. In these devices, thin metal wall thermal resistance of which can be neglected is flown around by hot gases from one side and by cold water from the other. The heat exchange coefficient from the side of the gases α_1 is approximately 50–100 times less than α_2 from the side of water. Therefore, it is possible to achieve significant increase in the heat transfer coefficient k and the cooling efficiency only by decreasing the thermal resistance from the side of gases. It can be achieved by increasing the velocity of gases and, consequently, α_1 as well as by ribbing the surface of the wall, which is in contact with gases. In this particular case, increasing α_2 by even two or three times as well as ribbing of the second wall practically does not yield any benefit.

The common rule of heat transfer intensification through a wall by convection is that when the heat exchange coefficients α differ significantly a substantial result can be achieved only by increasing the lesser coefficient. The analysis of formulae (3.42) and (3.42′) shows that the heat transfer coefficient k is always less than the

Table 3.5 Typical range of variations of α- and k-coefficients for different processes of heat transfer

Process	α, W/m^2 × °C	Process	k, W/m^2 × °C
Free convection of gases	50–30	Heat transfer:	
Free convection of water	$100–1 \cdot 10^3$	From gas to gas	25–250
Forced convection of gases	50–500	From gas to water	50–500
Forced convection of water	$2 \cdot 10^3–20 \cdot 10^3$	From water to water	$1 \cdot 10^3–10 \cdot 10^3$
Water boiling	$5 \cdot 10^3–50 \cdot 10^3$		

least of the heat exchange coefficients α. The values of α and k can vary within a very wide range. The typical ranges of variation of these values in industrial devices are presented in Table 3.5.

Example 3. Water flows at velocity of 3.0 m/s in a pipe made of low-carbon steel with diameter $d_2 = 60$ mm, $d_1 = 52$ mm, $\delta = 4$ mm. Water temperature t_{WAT} is equal to 30°C. There is an external source of radiation incident on the pipe. Heat flux density to water q is equal to 600 kW/m^2. Temperatures of both inner t_1 and outer t_2 pipe walls should be determined.

The ratio d_2/d_1 is 60/52 = 1.15. Such a cylindrical wall can be considered as a flat wall, Table 3.2.

Let us find criterion Re = $v\times d_1/\upsilon$ at $t_{\mathrm{WAT}} = 30$°C, $\upsilon = 0.8 \times 10^{-6}$ m^2/s, Table 3.3. Hence $Re = 3.0 \times 0.052 \times 10^6/0.8 = 195 \times 10^3$. Such a value of Re indicates to the fact that a high degree of water flow turbulence takes place. This allows using formula (3.36) to determine the α-coefficient.

At $t_{\mathrm{WAT}} = 30$°C, the coefficient A = 2.63, Table 3.4. According to formula (3.36): $\alpha = A \times v^{0.8}/d_h^{0.2} = 2.63 \times 3.0^{0.8}/0.052^{0.2} = 11.5$ kW/m^2 × °C. Rearranging formula (3.30) we obtain $q = \alpha(t_1 - t_{\mathrm{WAT}})$, hence: $t_1 = t_{\mathrm{WAT}} + q/\alpha = 30 + 600/11.5 = 82$°C.

Temperature t_1 could be found using Eq. (3.1). As a first approximation, let us assume that the mean wall temperature $t_{\mathrm{m}} = 100$°C and $\lambda = 55$ W/m × °C, Table 3.1. Then $q = \lambda(t_2 - t_1)/\delta$, hence: $t_2 = t_1 + q\times \delta/\lambda = 82 + 600{,}000 \times 0.004/55 = 126$°C. The calculated mean wall temperature $t_{\mathrm{m}} = (82 + 126)/2 = 104$°C is close to 100°C. Therefore, the value of λ requires no refinement.

3.4 Heat Radiation and Radiant Heat Exchange

3.4.1 General Concepts

As it has been mentioned, Sect. 3.1, at the temperatures of industrial furnaces the heat radiation of heated bodies is concentrated within a range of wavelength λ from 0.4 to 80 μm. In the stated range the solid and liquid bodies radiate the continuous spectrum, i.e., all wavelengths. When approaching the boundaries of this range the heat radiation energy E drops virtually to zero, Fig. 3.11. Let us

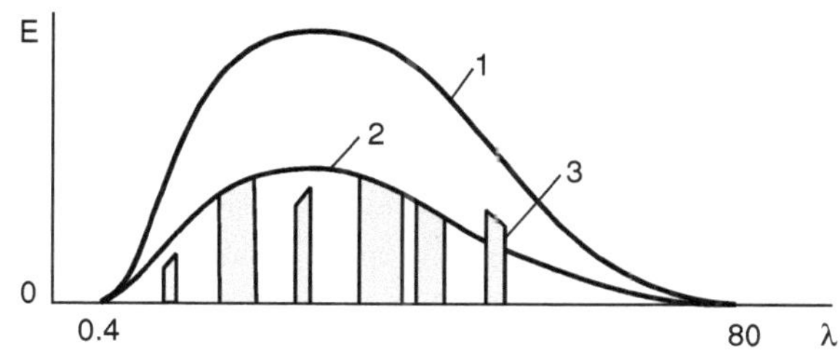

Fig. 3.11 Distribution of energy E in spectrum of heat radiation. 1 – black radiation; 2 – gray radiation; 3 – selective radiation of gas layer

remind that on the left side from the heat radiation region as λ is reduced ranges of ultraviolet radiation, X-rays, and gamma rays are located. On the right side a broad range of radiowaves is located.

Heat radiation cannot be perceived by eye. We can see only narrow part of this radiation spectrum in the region from 0.4 to 0.8 μm. This region has got its special name of light radiation. The heat energy of the light rays comprises only a negligibly minute fraction of the total energy of heat radiation.

The heat radiation of a body falling on other bodies is partially absorbed and partially reflected. The portion of energy, which is reflected, falls again on the bodies considered including the initial one. The reflected radiation, in turn, is absorbed and reflected by all bodies participating in the process. After the series of absorptions and reflections the radiated energy is completely absorbed by the bodies, which increases their internal energy.

As a result of processes associated with the double transformation of energy (the internal energy of the initial body → radiant energy → internal energy of the surrounding bodies) the radiant heat exchange is carried out. Finally, the quantity of the heat released or absorbed by a body is determined by the difference between the quantities of radiant energy emitted and absorbed by the body. Such difference is non-zero if the temperatures of the bodies participating in the mutual heat exchange of the radiant energy are different. At the equal temperatures of these bodies the whole system is in the dynamic thermal equilibrium. In this case, all the bodies of the system are radiating and absorbing, and the input of the radiant energy equals to its output for each body.

3.4.2 Stefan–Boltzmann Law; Radiation Density; Body Emissivity

The density of radiation flux E is the quantity of radiant energy passing in a unit time through a unit area of the body surface. This value is measured in W/m^2. According to Stefan–Boltzmann law the density of the self-radiation of the body is determined by the formula:

$$E = \varepsilon \times \sigma_0 \times T^4, \quad \mathrm{W/m^2} \tag{3.43}$$

T – the absolute temperature of the radiative body surface in Kelvin degrees, K. $\sigma_0 = 567 \times 10^{-8}\,\mathrm{W/m^2 \times K^4}$ – the radiation constant of the black body.
ε – surface emissivity of radiative body.

Formula (3.43) is the basic formula for all calculations of radiant heat exchange. The value σ_0 is quite small while T^4 is very large, on the contrary. Therefore, for engineering calculations formula (3.43) is usually transformed to the more convenient format:

$$E = \varepsilon \times C_0 \times (T/100)^4, \;\; \mathrm{W/m^2} \tag{3.44}$$

C_0 = 5.67 W/m$^2\times$ K^4 – the black body radiation coefficient.

The flux density of radiant energy absorbed by the area unit of the body surface E_{ABS} is determined by the expression:

$$E_{ABS} = \varepsilon \times E_{INC} \tag{3.45}$$

ε – emissivity of body surface absorbing radiation;
E_{INC} – density of radiation flux incident onto the body surface.

In engineering calculations the values ε both for the body self-radiation, formula (3.44), and for absorption of the incident radiation, formula (3.45), are usually taken as equal. It does not introduce any significant errors into calculations.

Let us explain the physical meaning of such concepts as black body and body surface emissivity ε. The value ε determines both the emitting ability of the body surface as well as its absorbing ability. This value can vary within the limits from zero to unity. If $\varepsilon = 1$ then the body absorbs all incident radiant energy according to formula (3.45). At the same time, the body self-radiation reaches the maximum possible value at the given temperature according to formula (3.44). Such bodies are called black bodies. If $\varepsilon = 0$ then the body does not absorb but reflects all the incident radiant energy. Such body can neither absorb nor radiate energy at any temperature according to formula (3.44). Such bodies are called white bodies. The bodies with intermediate values $0 < \varepsilon < 1$ are called gray bodies. Energy distribution curves in heat radiation spectrum of black and gray bodies are similar. At any λ, $E_{GR} = \varepsilon E_{BL}$.

Virtually all engineering materials and substances in solid and fluid state are gray. It is obvious that the higher the emitting ability of a gray body is, the higher its absorbing ability as well. Since the values of ε in both processes are assumed to

[2] The dependence of value ε on body temperature may not be taken into consideration in engineering calculations.

Table 3.6 Typical range of variations of emissivity ε for different materials

Steel non-oxidized	0.55–0.61	Copper oxidized	0.57–0.87
Oxidized	0.86–0.91	Polished	0.02–0.04
Polished	0.07–0.10	Molten	0.15
Molten	0.28		
		Iron oxidized	0.64–0.78
Slag molten	0.55	Molten	0.29
Stainless steel		Refractory lining	
Non-oxidized	0.45–0.49	(in a furnace)	0.80–0.90
Oxidized	0.75–0.80		

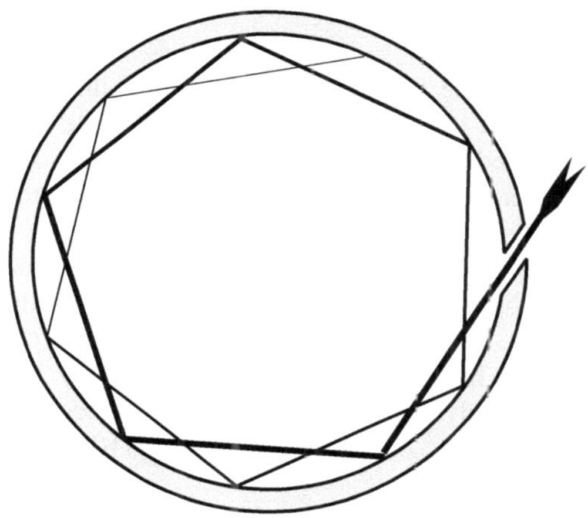

Fig. 3.12 Absorption of a ray inside the black body model

be the same there is a strict mutual correspondence between the discussed physical properties of the bodies.

The emissivity ε of solids and liquids depends on their physical nature as well as on their surface condition.[2] Since at the temperatures of steelmaking furnaces the radiation occurs practically entirely in the invisible (infrared) spectrum range, the real body emissivity does not agree with human visual impressions. For example, the white surface reflects the radiation well in the visible region of spectrum only. The heat (infrared) rays are absorbed by the white fabric and the white dye as intensively as by dark ones. A body surface condition rather than its color is essential for absorbing and reflecting of heat rays. Regardless the color, the emissivity of smooth polished surfaces is many times lower, and reflectivity (1 – ε) is correspondingly higher than that of rough surfaces. When body is oxidized its emissivity greatly increases and when it transforms into liquid state the emissivity sharply drops, Table 3.6.

White or black bodies do not exist in the nature. However, a model of the black body exists and is applied in practice. A small bore in a closed cavity with rough walls possesses all the properties of such body, Fig. 3.12. Radiation incoming

Table 3.7 Values of $(T/100)^4$ and density of black body radiation E_0 at different temperatures

t, °C	$(T/100)^4$	E_0, κ W/m^2	t, °C	$(T/100)^4$	E_0, κ W/m^2
650	7258	41.2	1200	47,080	266.9
700	8963	50.8	1250	53,800	305.0
750	10,953	62.1	1300	61,220	347.1
800	13,256	75.2	1350	69,390	393.4
850	15,903	90.2	1400	78,340	444.2
900	18,933	107.4	1450	88,140	499.8
950	22,373	126.8	1500	98,820	560.3
1000	26,262	148.9	1550	110,450	626.2
1050	30,637	173.7	1600	123,070	697.8
1100	35,537	201.5	1650	136,750	775.4
1150	41,005	232.5	1700	151,540	859.2

through the bore and incident upon the wall is partially absorbed and partially reflected in all directions. After multiple absorptions and reflections radiation is absorbed entirely. If the size of the cavity is rather large compared to the size of the bore, radiation will be absorbed completely even if the emissivity of the cavity wall is not high. According to the law of a mutual correspondence of absorbing ability and emitting ability of a body, which was discussed earlier, if a body absorbs radiation entirely it is the black body and should emit as the black body. Hence, if the walls of a cavity are heated to the temperature T, then the radiation density in all directions inside the cavity is equal to the radiation density of the black body E_0 at this temperature. According to formula (3.44) $E_0 = C_0 \times (T/100)^4$, W/m^2 ($\varepsilon = 1$). The values $(T/100)^4$ and E_0 are given in Table 3.7.

3.4.3 Heat Radiation of Gases

This radiation has a number of substantial specifics. With respect to opaque solid and liquid bodies, it is considered that radiation and absorption of radiant energy occur in a very thin surface layer. Radiation and absorption in gases take place in volume. Only polyatomic gases such as CO_2 and H_2O can emit and absorb. Such gases are present in large amount in fuel furnaces and EAFs. Emitting ability and absorbing ability of diatomic gases such as N_2, O_2, H_2, and other similar ones are negligible. Therefore, the air is practically transparent for heat radiation.

Unlike the continuous radiation spectrum of solid and liquid body the clean gases radiate and absorb the energy selectively, only in certain intervals of wavelengths, in so-called bands located in various sections of spectrum, Fig. 3.11. Outside these bands triatomic gases are as transparent as the air.

When heat rays pass through the mixture of transparent and absorbent gases their energy reduces due to absorption. This reduction and, consequently, the gas emissivity ε_g are determined by the number of CO_2 and H_2O molecules

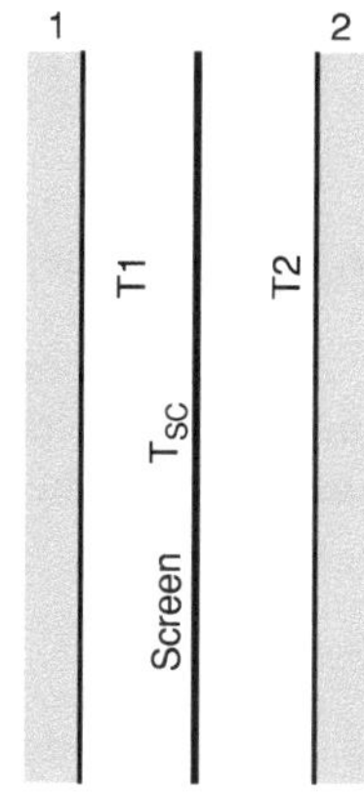

Fig. 3.13 Schematic diagram of plane-parallel bodies

encountered by the ray, i.e., by their concentration in the mixture and the thickness of the gas layer. Usually the ε_g value of pure gases is quite low. However, in EAFs and in other steelmelting units gases contain a lot of highly dispersed dust, which consists mostly of iron oxides. When oxygen is used intensively the dust content in gases reaches 30–50 g/m^3 and higher values.

The dust particles have the gas temperature. The radiation of such severely dust-laden gases is determined mainly not by CO_2 and H_2O concentration but rather solid particles content. Such radiation in its physical nature is much closer to the radiation of solid bodies than to that of gases. Incandescent particles not only emit themselves, but also scatter the radiation. The radiation spectrum of the severely dust-laden gas becomes more filled and approaches continuous one. All mentioned above allows, to a certain degree approximately, to consider a severely dust-laden gas in gas layers of large thickness as a gray body which can have emissivity close to unity. Such approach allows in a number of cases to essentially simplify calculations of gas-aided radiation heat exchange in EAFs. Let us review calculation formulae for the most frequently encountered cases of radiant heat exchange.

3.4.4 Heat Exchange Between Parallel Surfaces in Transparent Medium: Effect of Screens

Let us review a system of bodies with plane-parallel surfaces (1) and (2) of large dimensions compared to the distance between them, Fig. 3.13. In such system, virtually entire radiation of each surface falls onto the opposite surface. If $T_1 > T_2$ then the end result of heat exchange is expressed by the formula which is a consequence of application of Stefan–Boltzmann law, formula (3.44), to this case:

$$q_{1,2} = \frac{1}{1/\varepsilon_1 + 1/\varepsilon_2 - 1} \times C_0 \times [(T_1/100)^4 - (T_2/100)^4], \qquad (3.46)$$

$q_{1,2}$ W/m^2 – the specific net flux of heat exchange by radiation.

The corresponding specific quantity of energy is consumed per unit time for increasing the enthalpy of the body (2). This quantity equals the difference of energies absorbed and emitted by the body (2). The value of

$$\varepsilon_{RED} = \frac{1}{1/\varepsilon_1 + 1/\varepsilon_2 - 1} \qquad (3.47)$$

is called the reduced emissivity of the system of bodies and can vary from zero to unity always remaining less than ε_1 and ε_2. If the areas of surfaces (1) and (2) are identical and equal to F, then the total net flux of heat exchange by radiation Q is determined by the expression:

$$Q_{1,2} = \varepsilon_{RED} \times C_0 \times F\left[(T_1/100)^4 - (T_2/100)^4\right] \qquad (3.48)$$

Dependences (3.48) and (3.49) show that the heat exchange net flux by radiation is directly proportional to the difference of temperatures of body surfaces in the fourth power. In the processes of heat conduction and convection the heat flow is directly proportional to the difference of temperatures in the first power. This can explain the fact that at high temperatures the radiant heat exchange prevails over the other abovementioned processes.

Let us explain the application of screens to protect against radiation. In order to reduce heat flux $q_{1,2}$, formula (3.46), i.e., to partially protect body (2) against the radiation of body (1) the screen can be installed between them, Fig. 3.13. The thin screen under the conditions $\varepsilon_{SC} = \varepsilon_1 = \varepsilon_2$ is heated to the intermediate temperature T_{SC} between T_1 and T_2, which is determined by the expression:

$$(T_{SC}/100)^4 = \frac{(T_1/100)^4 + (T_2/100)^4}{2}. \qquad (3.49)$$

Under these conditions, the value $q_{1,2}$ is reduced by two times. Two screens reduce $q_{1,2}$ by three times, whereas n screens reduce it by $(n + 1)$ times. Even a greater effect can be provided by screens with lower emissivity in comparison with ε_1 and ε_2. Using a system of such screens enables reducing the net heat flux $q_{1,2}$ by many times.

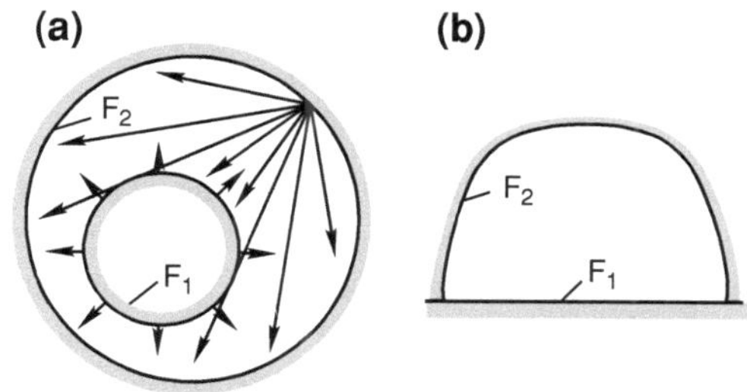

Fig. 3.14 Schematic diagram of bodies with envelope: (**a**) convex body and (**b**) flat body

3.4.5 Heat Exchange Between the Body and Its Envelope: Transparent Medium

In this case, only a portion of energy radiated by the second (larger) surface area F_2 falls upon the first (smaller) surface area F_1. The remaining quantity of this energy by-passing the body (1) falls upon its own surface F_2, Fig. 3.14a, b. The calculation formula for this case looks as follows:

$$Q_{1,2} = \varepsilon_{\mathrm{RED}} \times C_0 \times F_1\left[(T_1/100)^4 - (T_2/100)^4\right], \tag{3.50}$$

$\varepsilon_{\mathrm{RED}}$ – the reduced emissivity of the system is determined by the expression:

$$\varepsilon_{\mathrm{RED}} = \frac{1}{\frac{1}{\varepsilon_1} + \frac{F_1}{F_2} \times \left(\frac{1}{\varepsilon_2} - 1\right)} \tag{3.51}$$

Formulae (3.50) and (3.51) are applicable under condition that the smaller body has a convex shape. It means that radiation of smaller body (1) unlike that of body (2) cannot fall onto its own surface directly, without being reflected from body (2). It should be noted that in the discussed formulae F_1 is always the smallest of the surfaces.

Fig. 3.15 Radiative dust-laden gas in the envelope

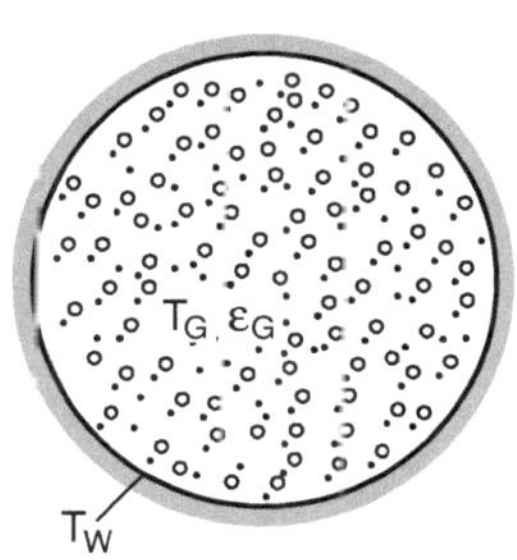

3.4.6 Heat Exchange Between the Emitting Gas and the Envelope

Unlike solid and liquid bodies, the gas emissivity ε_G depends significantly on its temperature. Therefore, strictly speaking, the radiation of gases is not governed by the Stefan–Boltzmann law, i.e., the "law of fourth powers of absolute temperature." The radiation of CO_2 is proportional to $T^{3,5}$, whereas the radiation of H_2O to T^3. However, as to provide uniformity and convenience of engineering calculation the formulae used for determination of gas radiation are based on the "law of the fourth powers."

It is necessary to consider that the gas is usually enclosed by a solid envelope with a temperature T_W and emissivity ε_W, where $\varepsilon_W < 1$. The envelope temperature T_W may differ from the gas temperature T_G, Fig. 3.15. Besides, in the presence of the emitting gas the effective emissivity of the envelope ε'_W increases, i.e., ε'_W is higher than ε_W without gas. In the range ε_W from 0.8 to 1.0 approximately $\varepsilon'_W = 0.5$ $(\varepsilon_W + 1)$. The final formula for the specific net heat flux from gas to the envelope looks as follows:

$$q_{1,2} = \varepsilon'_W \times \varepsilon_G \times C_0 \times [(T_G/100)^4 - (T_W/100)^4]. \tag{3.52}$$

If the terms in the square brackets are re-arranged, this formula is true when $T_W > T_G$ as well. In this case, the gas does not release but rather absorbs energy from the envelope.

Example 4. Heavy dust-laden EAF's offgases travel through a water-cooled gas duct. Parameters of offgas are t_G = 1200°C and ε_G = 0.8 and those of the internal gas duct surface covered by skull are t_W = 650°C and ε_W = 0.9. Density of net heat flux of radiation from offgases to the gas duct wall q must be found.

In this case, an effective emissivity of gas duct surface is determined as $\varepsilon'_W = 0.5(\varepsilon_W + 1)$= 0.5 × (0.9 + 1) = 0.95. The temperature of 1200°C corresponds to the value of $(T_G/100)^4$ = 47,080 and temperature of 650°C corresponds to the value of $(T_W/100)^4$ = 7258, Table 3.7. The C_0-coefficient is equal to 5.67 W/m^2 × K^4. According to formula (3.52) q = 0.95 × 0.8 × 5.67 × (47,080–7258) = 171,601 W/m^2 or q = 171.6 kW/m^2.

Since $C_0 \times (T/100)^4 = E_0$, the calculations may be reduced by using Table 3.7q = 0.95 × 0.8 × (2669 – 41.2) = 171.6 kW/m^2.

With low temperatures of t_W compared to t_G the second term in formula (3.52) might be ignored.

Other examples of application of the formulae and tables presented in this chapter will be discussed in the next chapters of this book where various innovations are analyzed using heat transfer calculations.

Chapter 4
Energy (Heat) Balances of Furnace

4.1 General Concepts

Energy (heat) balances give a general quantitative idea of an electric arc furnace as a thermoenergetical unit. These balances are derived from the law of conservation of energy according to which the total input of energy must be equal to energy output. It is true for the entire unit and for each of its zones. In the EAF the specifics of electrical units are combined with those of thermo-technical ones. Therefore, we can speak about both energy and heat balances. These balances should not be opposed to each other because in the furnaces all forms of energy are utilized or lost ultimately only in the form of heat. Henceforward these terms will be used depending on zone of the EAF under consideration. For instance, with regard to electro-technical zone (b–b) that includes the furnace transformer (1), the secondary electrical circuit (2), and electrodes (3) with electric arcs (4), we will speak about the energy balance, Fig. 4.1. In zone (a–a), which includes the entire EAF excluding the off-gas evacuation system (zone (e–e), different sources of energy are used concurrently. Therefore, it is better to consider the balance of this zone as a heat balance, as well as that of the EAF freeboard, zone (c–c), including the bath, zone (d–d).

Heat (energy) balances are usually written in the form of balance equations, the left side of which is the total sum of the amounts of energy of different kinds coming into the furnace (or zone) while the right side is the total sum of the amounts of all kinds of consumed energy. Heat (energy) balances can be equated for different zones of the EAF, Fig. 4.1, as well as for different time intervals, e.g., for the entire heat, for separate stages of the heat, or for a given moment of time. Equations for the first two cases contain the amounts of electrical and heat energy consumed within a certain time interval. In the last case, the so-called instantaneous balance equations contain the power of the respective energy fluxes. As we did earlier, we will use kWh as a unit of measurement of the amounts of energy. In most cases it is reasonable to examine the specific amounts of energy per ton of liquid steel (kWh/ton) as the weight of liquid steel can be calculated in practice much more precisely than that of the metal charge.

Y. N. Toulouevski and I. Y. Zinurov, *Innovation in Electric Arc Furnaces*, DOI: 10.1007/978-3-642-36273-6_4,

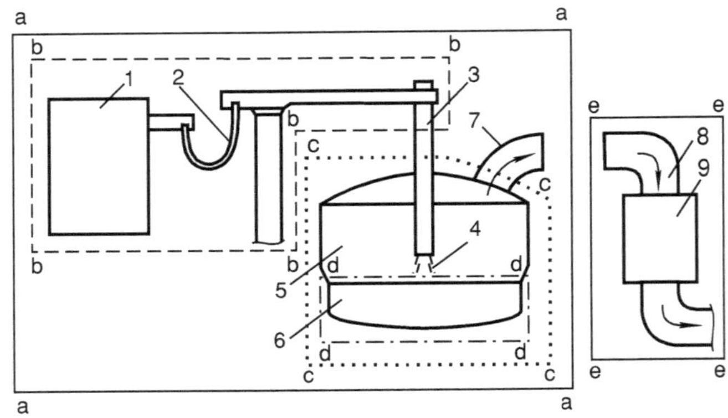

Fig. 4.1 Zones of energy (heat) balances for EAF 1 – transformer; 2 – secondary circuit; 3 – electrodes; 4 – arcs; 5 – freeboard; 6 – bath; 7 – roof elbow; 8 – gas duct; 9 – heat exchanger-utilizer

Equations of the energy (heat) balance for different zones of the furnace are different. They can contain different summands corresponding to different kinds of energy both at the input and at the output of a zone. Some of these summands are enthalpies. There are two kinds of enthalpies in balance equations. The first is enthalpies of heated substances (physical heat) E, for instance, the heat of liquid steel, slag, or off-gases. The second is enthalpies of chemical reactions, which are designated as E_{CH} or H. The enthalpies E are estimated using data on substance heat capacity by formula (2.1), Chap. 2, Sect. 2.1. The enthalpies E_{CH} and H represent the amount of heat emitted or absorbed in a chemical reaction. Methods of calculation of these enthalpies are examined in Sect. 4.5.3 of this chapter. Designation E_{CH} is used in the balance equations where different kinds of energy are summed up. The designation H is used for description of separate chemical reactions.

Heat engineering calculations contain not the absolute values of enthalpies, but their changes corresponding to the changes in temperature $\Delta t = t_1 - t_2$ and to the changes in average heat capacity in this temperature range. For enthalpy E and enthalpy H to be comparable, they would be counted off from the same initial temperature t_1. Traditionally, as an initial temperature, the room temperature of 25°C is used in the tables of H and the temperature of 0°C is used in tables of average heat capacities used for calculation of E. This difficulty can be overcome by a respective conversion of heat capacities. However, in practice it is not necessary, because the temperatures of steelmaking processes are so high that the initial reference point displacement by 25°C does not produce noticeable errors in calculations of heat balances of the furnace.

4.2 Heat Balances of Different Zones of the Furnace

Zone (a–a) includes the entire furnace excluding the off-gas evacuation system. For an EAF operating with oxy-fuel burners, the balance equation for zone (a–a) for the whole cycle of the heat takes the following form:

$$\sum E_{\text{a-a}} = \overbrace{E_{\text{EL}} + E_{\text{SCR}} + E_{\text{NG}} + \sum E_{\text{CH}} + E_{\text{OTH}}}^{\text{ENERGYINPUT}} = \overbrace{E_{\text{MET}} + E_{\text{SL}} + \sum E_{\text{LOS}}}^{\text{ENERGYOUTPUT}}, \tag{4.1}$$

$\sum E_{\text{a-a}}$ – total amount of energy consumed in zone (a–a) per heat, kWh/ton (equal to energy input)

ENERGY INPUT, kWh:

E_{EL} – amount of electrical energy delivered from external high voltage grids
E_{SCR} – enthalpy of metal charge loaded into furnace
E_{NG} – energy of natural gas combusted in burners
$\sum E_{\text{CH}}$ – total chemical energy (enthalpy) of all exothermic reactions

$$\sum E_{\text{CH}} = E_{\text{COK}} + E_{\text{CH.MET}} + E_{\text{CH.SL}},$$

E_{COK} – energy of coke carbon oxidation
$E_{\text{CH.MET}}$ – energy of oxidation of metal (iron and its alloys such as C, Si, Mn, and Cr including those contained in pig iron and hot metal)
$E_{\text{CH.SL}}$ – energy of slag forming from oxides
E_{OTH} – other small sources of energy: oxidizing of electrodes, burning-out of oil in scrap, etc. (with consideration given to endothermic reactions where the heat is absorbed)

ENERGY OUTPUT, kWh:

E_{MET} – enthalpy of liquid steel at tapping temperature
E_{SL} – enthalpy of final slag and of slag removed before
$\sum E_{\text{LOS}}$ – total energy loss

$$\sum E_{\text{LOS}} = E_{\text{EL.LOS}} + E_{\text{OFF.G}} + E_{\text{WAT}} + E_{\text{OTH}},$$

$E_{\text{EL.LOS}}$ – electrical energy loss in transformer and secondary electrical circuit
$E_{\text{OFF.G}}$ – energy loss with off-gases evacuated from freeboard through the roof elbow and electrode gaps
E_{WAT} – energy loss with water cooling the wall and roof panels
E_{OTH} – other unaccountable loss including losses through the bottom.

Electro-technical zone (b–b): in the balance equation of this zone $E_{\text{EL}} = E_{\text{ARC}} + E_{\text{EL.LOS}}$, E_{ARC} is an amount of energy consumed by arcs: $E_{\text{ARC}} = E_{\text{EL}} - E_{\text{EL.LOS}}$

Zone (c–c) is the freeboard including the bath. In the balance equation of this zone the value of E_{ARC} is the input of electrical energy:

$$\sum E_{c-c} = \overbrace{E_{ARC} + E_{SCR} + E_{NG} + \sum E_{CH} + E_{OTH}}^{\text{ENERGY INPUT}} = \overbrace{E_{MET} + E_{SL} + E_{OFF.G} + E_{WAT} + E_{OTH}}^{\text{ENERGY OUTPUT}}. \tag{4.2}$$

Since $E_{ARC} < E_{EL}$ it is obvious that the total amount of energy consumed in zone (c–c) per heat is less than that of zone (a–a).

Zone (d–d) is the bath of the furnace. The stage of the heat after formation of the flat bath is examined. At this stage, natural gas is used only to prevent the burner nozzles from clogging with slag. The power of the burners is sharply reduced; therefore, their effect on the bath heating as well as that of the other small sources of energy can be neglected. For the flat bath, the balance Eq. (4.2) is rearranged in the form:

$$\sum E_{d-d} = \overbrace{E_{ARC} + E_{SCR} + \sum E_{CH}}^{\text{ENERGYINPUT}} = \overbrace{\Delta E_{MET} + \Delta E_{SL} + E_{WAT} + E_{OTH}}^{\text{ENERGYOUTPUT}}. \tag{4.3}$$

The amounts of energy in Eq. (4.3) are calculated not for the whole cycle of the heat as in Eqs. (4.1) and (4.2), but only for the time period from formation of the flat bath to the tapping.

E_{SCR} – in this case, is enthalpy of metal charge loaded directly into the bath like in Consteel process
ΔE_{MET} and ΔE_{SL} – the increases in enthalpy of metal and slag within this time period
E_{WAT} – in this case, is the portion of the total heat losses with water cooling the panels, which is caused by radiation of the flat bath surface to the walls and the roof of the furnace, and not by radiation of electric arcs and dust-laden gases above the surface of the bath
E_{OTH} – heat losses through the bottom and other unaccountable losses

Although the temperature of gases in the freeboard can exceed the temperature of the slag surface, the heat transfer from the gases to the bath can be ignored in comparison with heating of the bath by electric arcs and chemical reactions of oxidation of C, Fe, and other elements.

Zone (e–e) is the off-gas evacuation system of the furnace. If a heat exchanger-utilizer, Fig. 4.1, is installed in the gas evacuation duct, which uses the energy of off-gases, e.g., for scrap preheating or vapor generating, then the heat balance for this zone is expressed by the equation:

$$E_{OFF.G} = E_{UT} + \sum E_{LOS}, \tag{4.4}$$

$E_{OFF.G}$ – total (physical and chemical) energy of gases at the duct entrance
E_{UT} – utilized (useful) energy, e.g., enthalpy of preheated scrap or vapor

$\sum E_{LOS}$ – total heat losses with gases downstream of the heat exchanger and with water cooling the duct walls upstream of the heat exchanger

4.3 Example of Heat Balance in Modern Furnace

As an example, the heat balance for 130-ton EAF is given in Table 4.1 calculated using the data [1]. The furnace operates with the cold charge ($E_{SCR} = 0$) and has all the features of the modern electric arc furnaces, namely, high productivity (tap-to-tap time is 38 min), high electrical capacity (78 MW, 90 MVA), low electrical energy consumption (331 kWh/ton), high oxygen consumption (47 m^3/ton), and fuel consumption (that of coke is 24.7 kg/ton and of natural gas – 10.3 m^3/ton); 16.8 kg of coke is charged together with scrap. The rest of the coke (7.9 kg/ton) in the form of powder is injected into the bath together with oxygen. Jet modules and additional oxy-fuel burners and tuyeres are installed in the furnace walls. All these devices serve for burning of natural gas and injecting of oxygen and coke powder into the bath. These energy carriers are introduced into operation at the points dispersed along the perimeter of the furnace. The total number of oxy-gas burners is 9. The data from Table 4.1 are used in the following sections of the book.

Table 4.1 Heat balance of 130-ton EAF

Energy input		kWh/ton	%
1. Electrical energy,	E_{EL}	331	49.2
2. Natural gas,	E_{NG}	107	15.9
3. Chemical energy of oxidation reactions	ΣE_{CH}	211	31.3
$\Sigma E_{CH} = E_{COK} + E_{CH.MET} + E_{CH.SL}$, including:energy of coke oxidation	E_{COK}	96.1	14.3
energy of metal oxidation (iron and its alloys: C, Si, Mn, Cr etc.),	$E_{CH.MET}$	101.9	15.1
energy of slag formation	$E_{CH.SL}$	13.0	1.9
4. Other small energy sources (oxidation of electrodes, etc.),	E_{OTH}	24.0	3.6
		Σ 673.0	100.0
Energy output		**kWh/ton**	**%**
1. Enthalpy of liquid steel before tapping (useful heat) weight of 115 ton, 1643°C	E_{MET}	391	58.1
2. Enthalpy of slag before tapping (and slag removed earlier),	E_{SL}	58	8.6
3. Enthalpy of off-gases escaping via roof elbow and electrode gapes,	$E_{OFF.G}$	75	11.2
4. Heat losses with water cooling wall and roof panels,	E_{WAT}	77	11.4
5. Electrical energy losses in transformer and secondary electrical circuit,	$E_{EL.LOS}$	23	3.4
6. Other losses including the bottom,	E_{OTH}	49	7.3
		Σ 673	100.0

4.4 Analysis of Separate Items of Balance Equations

Let us examine the separate items of the balance Eq. (4.1) for zone (a–a) as well as the methods of their determination and their affect on electrical energy consumption, Fig. 4.1.

4.4.1 Output Items of Balance

E_{MET} is the enthalpy of metal before tapping or useful heat of the heat, Chap. 2, Sect. 2.2. This item is the greatest on the right (output) side of the balance equation. At the regular tapping temperatures of low and medium carbon steel grades (from 1620 to 1640°C), E_{MET} is equal to approximately 390 kWh/ton of liquid steel. For other temperatures, E_{MET} can be calculated using the formula:

$$E_{MET} = 372 + (t_{TAP} - 1550) \times 0.23,\ \mathrm{kWh/ton}, \tag{4.5}$$

372 kWh/ton – enthalpy of liquid metal after the complete melting of the bath at the temperature of 1550°C
t_{TAP} – tapping temperature, °C
0.23, kWh/ton × °C – average heat capacity of liquid metal within the temperature range from 1500 to 1700°C

For instance, at $t_{TAP} = 1700$°C: $E_{MET} = 372 + (1700 - 1500) \times 0.23 = 406$ kWh/ton.

This example shows that in practice the value of E_{MET} changes within quite narrow limits.

E_{SL} is the enthalpy of final slag and of slag removed before the tapping. The total amount of slag is on average equal to 6–7% of the mass of liquid steel. The ratio between E_{MET} and E_{SL} is the same.

$E_{EL.LOS}$ is the losses of electrical energy in transformer and secondary electrical circuit in zone (b–b). Electrical losses occur due to heat generation during a current passing through transformer, flexible cables, and arms of the electrode holder, as well as due to induction of Foucault currents in metal structures located near the conductors due to electro-magnetic induction. Depending on electric characteristics of the electrode power-supply system, these losses in modern furnaces comprises from 5 to 8% of the electrical energy consumption E_{EL}.

$E_{OFF.G}$ is the heat losses with off-gases. This item of the balance equation is often exaggerated by including into this item the enthalpy (physical heat) of gaseous products of carbon oxidation CO and CO_2 as well as the total chemical energy that could be released with post-combustion of all the carbon to CO_2. In the case when this chemical energy is also included in the input part of the balance, the equality of the input and output parts of Eq. (4.1) is not disturbed. However, such an approach distorts the real picture of the process and leads to the wrong idea about the quantitative ratios between the individual items of the heat balance in the furnace.

In reality, if the thermal effect of carbon oxidation is calculated correctly in full accordance with chemical reactions taking place under given conditions, Sect. 5.3, and if this effect is included in the input side of Eq. (4.1), then neither CO and CO_2 physical enthalpy nor chemical energy of carbon monoxide leaving the freeboard of the furnace should be included in the heat losses with off-gases [2, 3].

The same relates to the heat balance of the bath, Eq. (4.3), zone (d–d), Fig. 4.1. The enthalpy of carbon monoxide CO evolved from the bath should not be included in the output side of Eq. (4.3) as the heat loss with off-gases. It is clear that when analyzing the heat balance of the gas evacuation system with the heat exchanger, zone (e–e), Fig. 4.1, the physical enthalpies of CO and CO_2 as well as the chemical energy of CO should be included in the energy of gases both at the entrance and exit of the heat exchanger, in accordance with the actual progress of the process of carbon monoxide post-combustion in the duct.

In order to correctly determine the evolution of heat produced by oxidation of carbon in zone (a–a), it is necessary to know the gas composition at the exit of the freeboard. Unfortunately, so far only some of the EAFs have been equipped with automatic off-gas analyzing systems. Therefore, the average gas composition varying widely in the course of the heat in most cases is known rather approximately. In the first approximation, it can be assumed that in the modern furnaces operating with intense oxygen bath blowing and with the closed door, carbon oxidizes only to CO in the boundaries of the freeboard [2, 4]. In the furnaces operating with the open door, a great amount of air is infiltrated into the furnace and the ratio CO/CO_2 can vary widely.

E_{WAT} is the heat loss with water cooling the wall and roof panels. These losses increase in direct proportion to the heat duration. They also depend on the average temperature in the freeboard and on the thickness of skull on the panels. The thickness of skull layer can vary from 3–5 mm to several tens of millimeters. The thermal conductivity of this friable layer is much lower than that of the metal walls of the panels, Chap. 3, Table 3.1. As a result, the skull serves as heat insulation reducing the heat losses with water.

The heat is transferred to the surface of the skull mainly by radiation. The sources of radiation are electric arcs, heavily dust-laden gases, which fill the freeboard, and the surface of the liquid bath. If the arcs are submerged in the foamy slag, then for calculation of the heat losses with water in a first approximation, formula (3.52) describing the heat exchange between the radiating gas and its envelope can be used, Chap. 3, Sect. 3.4.6. Based on the data given in Chap. 3, it could be assumed that the mean emissivity of the surface of both the skull on the panels and the bath surface ε_W' as well as that of dust-laden gases ε_G are approximately equal and close to 0.9. The temperatures of gases and the bath surface can be assumed equal to about 1600°C (T_G = 1873 K), and the average temperature at the surface of the skull on the panels equal to 1300°C (T_W = 1573 K). Substituting the assumed values into formula (3.52) and using Table 3.7 of Chap. 3, we will find the heat flux density to the surface of the skull q or, in other words, the average per heat power of the heat losses with water per 1 m^2 of the area of the panel:

$$q = 0.9 \times 0.9 \times 5.67 \left[(1873/100)^4 - (1573/100)^4\right]$$
$$= 284 \times 10^3 \, W/m^2 \, \text{or} \, 284 \, kW/m^2.$$

In accordance with the data on heat balance of the 130-ton EAF presented in Table 4.1, the heat loss with water per heat with duration of 38 min (0.63 h) is 77 kWh/ton. The total area of the wall and roof panels is equal to about 80 m^2. Based on these figures, the heat flux density to water can be determined: $q = 77 \times 130/80 \times 0.63 = 199$ kW/m^2. In accordance with the other data, in the modern furnaces this value can vary from 200 to 300 kW/m^2 [2]. Thus, in spite of substantial simplifications, the calculation using formula (3.52) results in satisfactory results.

The skull surface temperature depends on its thickness and heat flux q. With increase in q, this temperature can exceed the skull melting point. As a result, its thickness starts to decrease. At very high q, the skull can melt completely for a few minutes. The heat losses increase several times in such areas.

E_{BOT} is the heat losses through the furnace bottom. Just like the heat losses with water, these losses are directly proportional to the heat duration. The heat transfer from the external surface of the bottom into the ambient medium occurs by natural convection and radiation. If the temperature t_W and the area of the external surface of the bottom F are known, the power of heat losses Q can be easily calculated using formulae (3.30) and (3.32) of Chap. 3. For the 130-ton furnace, for t_W= 250°C, the ambient air temperature 20°C and F= 70 m^2, we find the following:

$$\alpha_\Sigma = 7 + 0.05 \times 250 = 19.5 \, W/m^2 \times {}^\circ C$$
$$Q = 19.5 \times 70 \, (250 - 20) = 314 \times 10^3 \, W.$$

With the mass of liquid steel of 115 ton and tap-to-tap time of 0.63 h, the specific heat loss through the bottom of 130-ton furnace is $E_{BOT} = 314 \times 0.63/115 = 1.7$ kWh/ton. This value is insignificant. However, it is necessary to consider that the useful heat accumulated by the metal is lost through the bottom, and in order to compensate these losses, it is necessary to consume much greater amount of the outside heat than the amount of heat lost. Therefore, earlier, in the old low-power furnaces with the tap-to-tap time of 2–3 h and low electrical energy efficiency, the heat losses through the bottom were quite significant. The bottoms were much thicker and thoroughly insulated against heat. With intensification of the heat processes and sharp reduction of tap-to-tap time, the thermal insulation is no longer used. The thickness of a bottom is now as minimal as required to just prevent the lining erosion and metal going away through the bottom.

4.4.2 Input Items of Balance

E_{EL} is the amount of electrical energy supplied by the external high voltage network. The amount of electrical energy consumed is measured with high accuracy by a watt-hour meter installed on the high voltage side of the furnace transformer. Depending on the composition of the metal charge and specifics of the process, the share of electrical energy in the total input of heat into the furnace, Eq. (4.1), can vary over a wide range; 25–30 years ago this share was 70–75%, whereas in the modern furnaces it has dropped to 50% and even lower. Increase in oxygen consumption and related increasing use of chemical energy have had the greatest effect on reducing the share of electrical energy.

E_{SCR} is enthalpy of the metal charge loaded into the EAF. It can be increased by preheating of scrap by off-gases before charging it into the freeboard. At present, such preheating is used in the so-called shaft furnaces and in furnaces operating with the Consteel process. The scrap preheating issues are discussed in detail in Chap. 6. The use of hot metal provides for the greatest increase in E_{SCR}. The hot metal is poured into the furnace at the temperatures of 1150–1350°C. It contains on average about 4.2% of C, 1.0% of Si, and 0.5% of Mn. The total amount of energy that the hot metal brings into the furnace is comprised of its own enthalpy and the chemical energy of oxidation of C, Si, and Mn in the liquid bath. This total energy is about 450 kWh/ton of hot metal on average. If the mass of hot metal is $G_{H.MET}$ and the total mass of pig iron and scrap is G_{SCR}, then the input of heat from any other sources of energy in balance Eq. (4.1) could be decreased roughly by 450 $\times\ G_{H.MET}/G_{SCR}$ kWh/ton of steel. Thus, with 40% share of hot metal in the charge, the decrease in electrical energy consumption E_{EL} can exceed $450 \times 0.4 = 180$ kWh/ton of steel, taking into account the electrical energy efficiency. At the same time, the share of electrical energy in the total input of heat would be less than 30%.

E_{NG} is the energy of natural gas combusted in the burners. As already mentioned, 6–9 oxy-gas burners are installed in the modern furnaces. However, as the power of each burner is usually about 3–4 MW only and the period of their active operation is rather short, the amount of natural gas used does not exceed 8–10 m^3/ton and its share in the total input of heat is 14–16%. The issues of using of oxy-gas burners in the EAFs will be discussed in detail in Chaps. 6 and 7.

E_{CH} is the chemical energy of oxidation reactions. This source of energy is one of the most important items in the heat balance. Its significance is enhanced by the fact that, with regard to the bath, it is an internal source of energy and therefore almost all the heat released by this source is useful. Unfortunately, significant errors are often made when determining E_{CH} for the EAF heat balances. Therefore, the methods of determination of E_{CH} should be discussed in more detail.

4.5 Chemical Energy Determination Methods

4.5.1 Utilization of Material Balance Data

To calculate the value E_{CH}, it is necessary, based on material balance of the heat, to determine in what quantities and to what degree the coke carbon, iron, and its alloys of C, Si, Mn, and Cr are oxidized, as well as in what quantities and how the slag forming oxides react with each other. The latter allows calculating $E_{CH.SL}$ As this value is very small, it can be ignored, Table 4.1. In order to obtain all these data, it is necessary to separate out the material balances of the respective chemical elements and their compounds from the total material balance of the heat.

For example, it is well known that silicon is oxidized entirely to SiO_2, and this oxide remains in the slag. If the quantity of silicon in the scrap and in the additions (for example, in pig iron and in coke) as well as the content and quantity of Si in the steel before tapping are known, then by the difference between these quantities it is possible to determine the quantity of silicon oxidized to SiO_2 and the amount of useful heat released in this process. In a similar way, by the difference between the total quantity of carbon charged into the furnace with scrap, pig iron, and coke, and the quantity of carbon in the steel before tapping it can be determined how much carbon has been oxidized. Using the data on average composition of the off-gases exiting from the freeboard, it can be determined how much carbon has been oxidized to CO and how much to CO_2, as well as how much heat has been released in these processes. If it is necessary, some corrections with regard to entrainment of SiO_2, FeO, and other oxides from the freeboard can be made based on composition and dust content in the off-gases. With the modern industrial means and methods of determining the weight and composition of the substances, such as hoisting crane scales, automatic stationary or mobile solids and gas analyzers, radioactive isotope marks and others, a quite precise determination of the value E_{CH} based on the data on material balance of the heat is quite accessible for the companies specializing in this area. Although such tests are rather expensive, their results can be used for a long time since the adjustments due to ongoing changes in production process do not require repeating the tests in full and are easy to carry out.

4.5.2 About the So-Called "Energy Equivalent" of Oxygen

From the beginning of extensive use of oxygen for intensification of the heat, the studies have been carried out in a number of countries. The purpose of these studies was to establish the statistical dependence between the consumption of oxygen V_{O_2}, m^3/ton of steel and E_{CH} as well as E_{EL}. The results of these studies are presented in a way as if E_{EL} and E_{CH} were directly proportional to V_{O_2}, and E_{CH} could substitute for E_{EL} equivalently.

$$E_{EL} = z \times V_{O_2}, \tag{4.6}$$

$$E_{CH} = z \times V_{O_2}. \tag{4.6'}$$

The same proportionality coefficient z in Eqs. (4.6) and (4.6′) has been called the energy equivalent of oxygen. This value is measured in kWh/m^3 of O_2. It is usually determined as the average with regard to consumption of electrical energy and oxygen in a group of furnaces chosen for studies. The number of furnaces in the different studies reaches several tens, and the furnaces may differ in capacity, power, and specifics of the process, including different ways of introduction of oxygen into the furnaces. According to data obtained from such studies, the average values of the energy equivalent z vary from 3.2 to 6.8 [2, 4]. As a result, the use of the values of $z \times V_{O_2}$ as the equivalent substitutes for E_{CH} became widespread, in particular, in composing the heat balance equations. It is recommended to accept the value of z = 5.2 kWh/m^3 of O_2 [2].

This statistical approach to determining E_{CH} and substituting E_{CH} with $z \times V_{O_2}$ in balance equations can be accepted as valid only for quite rough and averaged estimates not quite suitable for the analysis of the specific cases. The summands of the $z \times V_{O_2}$ type, unlike other terms of the balance equations, do not have any clear physical meaning because oxygen itself does not release any energy, and the values of coefficients z are determined by oxidation rates of various components of the charge. In different furnaces with different equipment for introduction of oxygen and different composition of the charge, the quantity of elements oxidized per unit of O_2 introduced can vary within wide limits.

It is observed that an increase in oxygen consumption leads to a substantial decrease in values of coefficients z. Moreover, not only oxygen blown into the furnace but also infiltrated air plays an essential part in the process of oxidation of scrap and coke charged. According to data [2], there may be some considerable differences in efficiency of using oxygen and, consequently, in the value z in the same furnace operated by different operators. As a result, in certain cases the errors in estimating E_{CH} using recommended average coefficients z can reach inadmissible values. Thus, in any case, the main and the only reliable way to determine E_{CH} is to use data of the material balance of the heat.

4.5.3 Calculation of Thermal Effects of Chemical Reactions by Method of Total Enthalpies

Besides using of energy equivalents of oxygen, the methodical errors in calculations of quantity of heat generated by chemical reactions are another source of essential errors in evaluating E_{CH}. This can be explained by insufficient presentation of specifics of such calculations in the literature in the field of steelmaking processes. To the present day, the tables of the so-called standard enthalpies of chemical reactions $-\Delta H^0_{298}$, similar to Table 4.2, are frequently used. Each value of $-\Delta H^0_{298}$ in the table represents the quantity of heat released while carrying out the reaction under standard conditions related to 1 kg of the element or 1 m^3 of oxygen. Hereinafter the quantity of heat released will be called the resultant thermal effect of the reaction and designated as $-\Delta H_{RES}$. Let us recall that by convention in thermodynamics the minus sign is ascribed to heat evolution, whereas the plus sign indicates heat absorption, Chap. 2, Sect. 2.1. However, when adding or subtracting the thermal effects of reactions these conventional signs cannot be used as is customary in algebra. Therefore, as for examples given below the minus sign for the resultant thermal effects (enthalpies) is left out.

It is well known that the enthalpies of reactions depend heavily on temperature conditions under which these reactions take place. In other words, the values $-\Delta H_{RES}$ depend on the temperatures of the original substances and those of the final products. Standard enthalpies $-\Delta H^0_{298}$ given in Table 4.2 have been found by conducting chemical reactions under standard temperature conditions. In such experiments, all the original substances have the standard temperature of 298 K (25°C), and the final products after completing the reaction are cooled to the same temperature. For example, the value $-\Delta H^0_{298} = -9.11$ kWh/kg of C (upper line of Table 4.2) is the quantity of heat obtained from the process of combusting cold carbon in cold oxygen (25°C) and consequent cooling of the resultant CO_2 to 25°C. It is obvious that if CO_2 leaves the reaction zone without being cooled, at a high temperature, then the absolute value of the quantity of the resultant heat actually released in this zone $-\Delta H_{RES}$ will be less than $-\Delta H^0_{298}$.

Under real conditions of the steelmaking processes, the original substances and the products of chemical reactions, as a rule, have quite different temperatures. For instance, carbon dissolved in the liquid bath at the temperatures of 1600–1650°C can be oxidized by both oxygen from the freeboard atmosphere, which has almost the same temperature, and cold oxygen blown into the bath by the injectors. The powdered coke blown together with oxygen is also cold when it enters the reaction zone. Using the same values of $-\Delta H^0_{298}$, Table 4.2, in these different cases results

Table 4.2 Standard enthalpies of some chemical reactions, $-\Delta H^0_{298}$

Chemical reaction	kWh	
	per 1 kg of element	per 1 m^3 of oxygen
$C + O_2 = CO_2$	–9.11	–4.88
$C + 0.5\ O_2 = CO$	–2.55	–2.73
$CO + 0.5\ O_2 = CO_2$	–3.51/$м^3$ CO	–7.01
$Fe + 0.5\ O_2 = FeO$	–1.32	–6.57
$2\ Fe + 1.5\ O_2 = Fe_2O_3$	–2.04	–6.76
$3\ Fe + 2\ O_2 = Fe_3O_4$	–1.85	–6.87
$Si + O_2 = SiO_2$	–8.94	-11.20
$Mn + 0.5\ O_2 = MnO$	–1.93	–9.48
$Cr + 1.5\ O_2 = Cr_2O_3$	–3.05	–9.42

in essential errors in estimating $-\Delta H_{RES}$ and consequently in estimating E_{CH}. Let us recall that both symbols, in essence, mean the same, namely the heat evolution in the case of chemical reactions taking place under certain conditions.

For the cases of deviations of actual temperature conditions from the standard conditions, the methods are known for recalculating the standard enthalpies $-\Delta H^0_{298}$ to the values of $-\Delta H_{RES}$ under actual conditions. However, all these methods are quite unsuitable. They involve composing and solving of additional balance equations requiring the data on the temperature dependence of heat capacity of the reactants. Such data are not always available, especially for the high temperatures of the steelmaking processes. There are also tables of enthalpies of chemical reactions for the temperatures T that are different from 298 K. Such enthalpies are designated as $-\Delta H^0_T$. However, the values $-\Delta H^0_T$ in these tables, as well as the values $-\Delta H^0_{298}$, are valid only for the cases when the temperatures of original substances and final products of reaction are the same. When these temperatures differ greatly (which is the most common case in real practice), the values $-\Delta H^0_T$ and $-\Delta H^0_{298}$ need to be recalculated to the values of $-\Delta H_{RES}$ under actual conditions. All this explains why the standard enthalpies $-\Delta H^0_{298}$ are mostly used in practice despite the fact that it leads to significant errors.

In all cases, the accurate values of $-\Delta H_{RES}$ are easily calculated by using a universal, very convenient method of the total enthalpy. The total enthalpy of the chemical compound $I^0{}_T$ is the enthalpy of its formation at the temperature T from the elements in their standard state at the temperature T_0. The total enthalpy method is used in the different fields of technology, e.g., in complicated calculations of combustion of the propellants. However, this method has not yet become widespread for calculating E_{CH} for the steelmaking processes, perhaps because there are no tables of total enthalpies which would be as suitable and well known to metallurgists as the $-\Delta H^0_{298}$ tables are.

The total enthalpy tables published by the authors earlier [5] were presented to the readers in an inconvenient form. This shortcoming has been corrected in the present edition. Values of the total enthalpies of some chemical compounds, Table 4.3, and individual elements, Table 4.4, needed for determining E_{CH} for the steelmaking processes have been calculated on the basis of the most reliable data [6, 7].

In accordance with the total enthalpy method, the resultant thermal effect of the chemical reaction ΔH_{RES} is determined as the difference between the absolute values of total enthalpies of products of the reaction and total enthalpies of the original substances, Tables 4.3 and 4.4, at the actual temperatures before the beginning and after the completion of the reaction. If these temperatures are the same and equal to 298 K, then $-\Delta H_{RES} = -\Delta H^0_{298}$, Table 4.2. If both temperatures are equal to T, then $-\Delta H_{RES} = -\Delta H^0_T$. The total enthalpies of the elements at the standard temperature $T_0 = 298$ K (25°C), by definition of I^0_T, are equal to zero since it is this temperature which is taken as the initial point of their reading. The application of the total enthalpy method for determining the resultant thermal effects of chemical reactions will be explained by the following examples.

Table 4.3 Total enthalpies of some chemical compounds, $I_T^0 \times 10^2$, kWh/mole

		Chemical compounds								
		CO	CO_2	SiO_2	MnO_2	Cr_2O_3	FeO	Fe_3O_4	Fe_2O_3	CaO
		Mole, kg								
T, K	t,°C	0.028	0.044	0.060	0.071	0.152	0.069	0.232	0.160	0.056
298	25	3.073	10.940	25.318	10.715	31.710	7.392	31.056	22.838	17.656
400	127	2.990	10.829	25.180	10.581	31.368	7.248	30.585	22.517	17.529
500	227	2.908	10.709	25.021	10.446	31.073	7.102	30.086	22.163	17.395
600	327	2.824	10.581	24.846	10.310	30.738	6.950	29.538	21.791	17.256
700	427	2.739	10.446	24.662	10.169	30.396	6.795	28.918	21.390	17.113
800	527	2.651	10.306	24.463	10.027	30.049	6.636	28.231	20.955	16.969
900	627	2.561	10.160	24.248	9.880	29.699	6.480	27.501	20.503	16.822
1000	727	2.470	10.011	24.057	9.731	29.346	6.314	26.927	20.030	16.675
1100	827	2.377	9.859	23.864	9.577	28.994	6.140	26.343	19.621	16.526
1200	927	2.282	9.703	23.668	9.422	28.632	5.978	25.807	19.235	16.375
1300	1027	2.187	9.545	23.407	9.260	28.271	5.809	25.263	18.848	16.225
1400	1127	2.090	9.386	23.207	9.104	27.906	5.638	24.710	18.455	16.073
1500	1227	1.993	9.227	23.005	8.941	27.538	5.464	24.151	18.058	15.920
1600	1327	1.895	9.066	22.802	8.780	27.165	5.289	23.586	17.656	15.766
1700	1427	1.796	8.897	22.598	8.624	26.788	4.228	23.030	17.241	15.612
1800	1527	1.696	8.731	22.392	8.462	26.406	4.027	22.466	16.798	15.457
1900	1627	1.596	8.565	22.185	8.300	26.020	3.814	18.085	16.367	15.301
2000	1727	1.495	8.397	21.708	8.142	25.627	3.622	17.521	15.931	15.144
2100	1827	1.394	8.229	21.476	6.483	25.229	3.422	16.962	15.489	14.987
2200	1927	1.293	8.060	21.244	6.325	24.824	3.227	16.412	15.047	14.829
2300	2027	1.192	7.891	21.012	6.167	24.412	3.032	15.849	14.604	14.670

The thermal effects of dissolution of substances in liquid iron as well as evolving from it are ignored in these examples. This does not produce noticeable errors in calculations.

Example 1. Let us determine the resultant thermal effect ΔH_{RES} of the reaction: $Fe + 0.5\ O_2 = FeO$, at the temperature $T = 298$ K using the total enthalpy method.

The reaction equation shows that one mole of iron (56 g) reacting with half a mole of oxygen (16 g) forms one mole of iron oxide (56 + 16 = 72 g), Tables 4.3 and 4.4. Along with the other units (g, kg, and ton), the mole is a unit that measures the amount of substance as well. The mole is numerically equal to the sum of the atomic masses of all atoms in one molecule of a compound. The atomic masses of elements are given in Mendeleev's periodic table. The mole values of all the considered substances expressed in kg are given in Tables 4.3 and 4.4.

According to the total enthalpy method the value of ΔH_{RES} for the reaction examined is determined by the expression: $\Delta H_{RES} = I^0_{FeO(298)} - I^0_{Fe(298)} - 0.5 I^0_{O_2(298)}$. The total enthalpies of the elements at $T = 298$ K: $I^0_{Fe(298)} = 0$,

Table 4.4 Total enthalpies of some individual elements, $I_T^0 \times 10^2$, kWh/mole

		Individual elements								
		C	O_2	Si	Mn	Cr	Fe	N_2	Ar	H_2
		Mole, kg								
T, K	*t*,°C	0.012	0.032	0.028	0.055	0.052	0.056	0.028	0.040	0.002
298	25	0.000	0.000	0.000	0.000	0.000	0.000	0.000	0.000	0.000
400	127	0.030	0.084	0.060	0.080	0.000	0.074	0.082	0.059	0.082
500	227	0.067	0.169	0.124	0.161	0.069	0.152	0.164	0.117	0.163
600	327	0.111	0.257	0.191	0.247	0.140	0.237	0.247	0.174	0.245
700	427	0.159	0.348	0.260	0.337	0.213	0.329	0.332	0.232	0.327
800	527	0.212	0.440	0.330	0.432	0.289	0.431	0.418	0.290	0.409
900	627	0.268	0.535	0.402	0.532	0.451	0.545	0.506	0.348	0.491
1000	727	0.326	0.631	0.476	0.697	0.537	0.680	0.597	0.406	0.575
1100	827	0.388	0.729	0.555	0.802	0.628	0.840	0.688	0.463	0.659
1200	927	0.451	0.828	0.626	0.907	0.725	0.977	0.781	0.521	0.745
1300	1027	0.515	0.927	0.702	1.014	0.826	1.072	0.876	0.579	0.832
1400	1127	0.581	1.028	0.780	1.190	0.934	1.170	0.971	0.637	0.920
1500	1227	0.648	1.129	0.858	1.371	1.047	1.269	1.068	0.694	1.009
1600	1327	0.716	1.231	0.938	1.907	1.167	1.370	1.165	0.752	1.099
1700	1427	0.784	1.334	2.414	2.035	1.293	1.506	1.263	0.810	1.191
1800	1527	0.853	1.437	2.490	2.158	1.425	1.621	1.362	0.868	1.283
1900	1627	0.922	1.541	2.565	2.291	1.563	2.166	1.461	0.926	1.377
2000	1727	0.991	1.646	2.641	2.419	1.708	2.288	1.561	0.983	1.472
2100	1827	1.061	1.751	2.717	2.547	1.858	2.412	1.661	1.041	1.568
2200	1927	1.130	1.857	2.792	2.675	2.602	2.536	1.761	1.099	1.664
2300	2027	1.200	1.964	2.868	2.803	2.741	2.661	1.862	1.157	1.762

and $I^0_{O_2(298)} = 0$, Table 4.4. Hence $\Delta H_{RES} = I^0_{FeO(298)} = 0.07392$ kWh/mole of FeO, Table 4.3.

Recalculating for 1 kg of iron we will obtain $\Delta H_{RES} = 0.07392/0.056 = 1.3$ kWh/kg of Fe and recalculating for 1 m^3 of O_2 $\Delta H_{RES} = (0.07392/0.5 \times 0.032) \times 1.43 = 6.6$ kWh/m^3 of O_2; $\rho_{O_2} = 1.43$ kg/m^3 is oxygen density.

As the temperature conditions are assumed to be standard (298 K), the values of ΔH_{PE3} obtained are sufficiently close to the standard enthalpies ΔH^0_{298}, Table 4.2. It is easy to verify that the first line in Table 4.3 gives the enthalpies equal to ΔH^0_{298} kWh/mole for all other chemical compounds as well.

Under EAF freeboard conditions the FeO, as a result of oxidizing cold Fe (25°C) by cold O_2 (25°C), is not cooled down to 25°C but has a high temperature, e.g., 1600 K (1327°C). Let us determine the resultant thermal effect of the reaction considered under actual temperature conditions.

$\Delta H_{RES} = I^0_{FeO(1600)} - (I^0_{Fe(298)} + 0.5 I^0_{O_2(298)})$. The expression in the square parentheses is equal to zero as before. $\Delta H_{RES} = I^0_{FeO(1600)} = 0.05289$ kWh/mole, Table 4.3. Hence $\Delta H_{RES} = 0.05289/0.056 = 0.9$ kWh/kg of Fe and $\Delta H_{RES} = (0.05289/0.016) \times 1.43 = 4.7$ kWh/m^3 O_2. Thus, the use in balance equation calculations of standard thermal effects ΔH^0_{298} instead of their actual values ΔH_{RES}, in this case, results in raising too high, approximately 40% (6.6/4.7 = 1.4), an amount of heat obtained from iron oxidation.

Example 2. Let us determine ΔH_{RES} of the reaction: $CO + 0.5\ O_2 = CO_2$ under different temperature conditions. At the standard temperature of 298 K,

$$\Delta H_{\mathrm{RES}} = I^0_{\mathrm{CO_2(298)}} - I^0_{\mathrm{CO(298)}},\ (I^0_{\mathrm{O_2(298)}} = 0.0),$$

ΔH_{RES} = 0.1094 – 0.03073 = 0.07867 kWh/mole or
ΔH_{RES} = (0.07867/0.016) × 1.43 = 7 kWh/m^3 of O_2.

Calculating ΔH_{RES} for 1 m^3 CO: 1 mole CO amounts to 0.012 + 0.016 = 0.028 kg ΔH_{RES} = (0.07867 / 0.028)×1.25 = 3.5 kWh/m^3 of CO (ρ_{CO} = 1.25 kg/m^3). The obtained value of ΔH_{RES}= 3.5 kWh/m^3 of CO is the same as ΔH^0_{298}, Table 4.2.

Post-combustion of CO in an EAF freeboard by cold oxygen, when temperatures of $T_{\mathrm{CO}} \cong$ 1627°C and $T_{\mathrm{CO_2}} \cong$ 1727°C, significantly changes the value of ΔH_{RES}:

$\Delta H_{\mathrm{RES}} = I^0_{\mathrm{CO_2(2000)}} - I^0_{\mathrm{CO(1900)}}$= 0.08397 – 0.01596 = 0.06801 kWh/mole or
ΔH_{RES} = (0.06801/0.016) × 1.43 = 6.1 kWh/m^3 of O_2. This value is less than ΔH^0_{298} by 7.0/6.1 = 1.15 times.

Example 3. Let us determine ΔH_{RES} of oxidation of Si in the EAF bath by oxygen dissolved in the liquid metal according to the reaction $Si + O_2 = SiO_2$ at the temperature of 1800 K.

$$\Delta H_{\mathrm{RES}} = I^0_{\mathrm{CO_2(298)}} - I^0_{\mathrm{CO(298)}},\ (I^0_{\mathrm{O_2(298)}} = 0.0),$$

ΔH_{RES} = 0.2239 – 0.02490 – 0.01437 = 0.1846 kWh/mole. One mole of Si is equal to 0.028 kg. Recalculating for 1 kg of Si ΔH_{RES} = 0.1846/0.028 = 6.6 kWh/kg of Si. The value obtained is less than ΔH^0_{298} by 8.94/6.6 = 1.35 times.

The examples given show that the use in balance equations and other calculations of standard enthalpies of chemical reactions $-\Delta H^0_{298}$ instead of actual values of thermal effects of $-\Delta H_{\mathrm{RES}}$ under conditions of steelmaking processes cannot be justified. The actual values $-\Delta H_{\mathrm{RES}}$ should be determined using the total enthalpy method. This is of special importance for calculations of heat balances for separate zones of EAF.

References

1. Pujadas A, McCauley J, Tada Y et al (2002) Electric arc furnace energy optimization at Nucor Yamato Steel. 7th European Electric Steelmaking Conference, Venice

2. Adams W, Alameddine S, Bowman B et al (2001) Factors influencing the total energy consumption in arc furnaces. 59th Electric Arc Furnace Conference, Phoenix
3. Morozov AN (1983) Modern steelmaking in arc furnaces. Metallurgiya, Moskow
4. Pfeifer H, Kirschen M, Simoes JP (2005) Thermodynamic analysis of EAF electrical energy demand. 8th European Electric Steelmaking Conference, Birmingham
5. Toulouevski YN, Zinurov IY, Popov AN et al (1987) Electrical energy savings in arc steelmelting furnaces. Energoatomizdat, Moskow
6. Elliot DF, Gleizer M, Ramakrishna B (1969) Thermo-chemistry of steelmelting processes. Metallurgiya, Moskow
7. Thermodynamic properties of individual elements (1978) Science Academy of USSR. Science, Moscow

Chapter 5
Energy Efficiency Criteria of EAFs

5.1 Preliminary Considerations

Increasing energy efficiency of EAF is one of the most critical problems of electrical steelmaking. Obviously, to estimate energy efficiency some quantitative indices are needed. But what indices should be used for this purpose? For heating furnaces as well as for boilers and other similar equipment utilizing typically the only source of energy (e.g., fuel combusted by burners), heat energy efficiency coefficient η (or fuel energy efficiency coefficient for fuel furnaces) is a standard criterion which explicitly defines energy efficiency. This criterion can be evaluated as a ratio of useful energy to total energy consumption:

$$\eta = (E - E_{\mathrm{LOS}})/E, \tag{5.1}$$

E – is total energy consumption,
E_{LOS}– is energy loss, and
E–$E_{\mathrm{LOS}} = E^*$ is useful energy, i.e., energy obtained by a heated material.

All the values in Eq. (5.1) are defined by the heat balance of furnace, which gives, in such cases only, complete indication of its energy efficiency. The coefficient η is quite similar to the well-known efficiency coefficients of the combustion engines (e.g., automobile engines), which convert energy of gasoline into the useful work. The only difference is that the useful work of furnace (its thermal work) is not the mechanical work but heating and smelting of materials charged into the furnace, Chap. 2, Sect. 2.1.

The coefficients similar to the coefficient η are not being used in the EAF's practice. The reason for that is the power features of the modern furnaces which utilize concurrently the sources of energy very different by their physical and chemical nature, such as electric arcs, oxy-fuel burners, carbon/oxygen injectors, exothermic reactions of oxidation of the components of solid charge, and liquid bath. As already mentioned, in many modern electric arc furnaces the share of electrical energy in the total heat input has been reduced to 50% and even less.

Y. N. Toulouevski and I. Y. Zinurov, *Innovation in Electric Arc Furnaces*,
DOI: 10.1007/978-3-642-36273-6_5, © Springer-Verlag Berlin Heidelberg 2013

In spite of this, when the energy efficiencies of different EAFs and their modes of operations are being compared, usually the specific consumption of the electrical energy is considered only. It is obvious that such estimation of energy efficiency is insufficient.

Overall energy efficiency is defined by the total cost of all forms of energy being utilized. In order to minimize the cost it is necessary, first, to use each of the energy sources to its maximum efficiency and, second, to replace as many as possible relatively low efficient as well as expensive sources of energy with cheaper and more efficient ones. But what are the energy efficiency criteria of the different sources of energy and of the EAF as a whole? The energy balances of EAF, unlike those of the furnaces with the single source of energy, do not contain any information with regard to these criteria.

Let us get back to the data on the characteristics of the modern 130-ton EAF and on its heat balance, Chap. 4, Table 4.1. The natural gas and coke are utilized in the furnace in the quantities of 10.3 m^3/ton and 24.7 kg/ton. Energy of the gas is 107.0 kWh/ton and that of the coke 96.1 kWh/ton. But what is the efficiency of utilization of these energy carriers? The design and the mode of operation of the oxy-gas burners and those of the injectors for the introduction of the coke into the bath define this efficiency. But how perfect are their design and the mode of operation? Which design and which mode of operation should be given preference to? Is the selected ratio of amount of coke charged together with the scrap to the amount of coke blown into the bath of the furnace together with oxygen, correct? Which furnaces are more energy efficient and why? These are the questions of major interest for the designers of the equipment and for the EAF operating personnel. The balance equations given in Chap. 4 do not answer these questions.

There are different points of view on defining the energy efficiency criteria of the EAF. Since useful energy of the heat – enthalpy of metal at tapping E_{MET} – is a relatively constant value, total energy consumption for heat ΣE is closely related to the total energy loss ΣE_{LOSS}:

$$\sum E = E_{\mathrm{MET}} + \sum E_{\mathrm{LOS}}. \tag{5.2}$$

This equation repeats Eq. (4.1) given in the previous chapter with the only difference: the enthalpy of slag is not shown explicitly isolated from the total heat losses. Since $E_{\mathrm{MET}} \approx$ const, ΣE is lower when the total energy loss ΣE_{LOS} is reduced, Eq. (5.2). This makes it possible, at a first glance, to use total energy consumption ΣE for estimating energy efficiency of the furnace assuming that the lower the value of ΣE, the higher the EAF energy efficiency. In each specific case the value of ΣE can be evaluated accurately enough with the aid of the energy balance of the heat. For the above-mentioned 130-ton EAF ΣE = 673 kWh/ton.

However, regardless of the accuracy of ΣE evaluation, the use of this characteristic for the estimation of EAF energy efficiency has one essential disadvantage. This value summarizes the energies obtained from different sources. From the thermodynamics point of view such a summarizing does not contradict to the law of conservation of energy and is completely acceptable. However, this approach does not consider quite different degree of efficiency of usage and the cost of different kinds of energy. Therefore, the significantly higher consumption of electrical energy and much higher cost might correlate to the lower value of ΣE and vice versa. This correlation can be clearly demonstrated with the help of so-called common energy efficiency coefficient of EAF, η_{EAF}, suggested and used in the work of [1] and a number of others.

5.2 Common Energy Efficiency Coefficient of EAF and Its Deficiencies

The coefficient η_{EAF} is calculated in accordance with the following formula:

$$\eta_{EAF} = E_{MET}/\Sigma E. \tag{5.3}$$

For the 130-ton furnace $\eta_{EAF} = 391/673 = 0.58$.

The analysis of the heat balances of 18 furnaces with the capacity of 80–200 ton carried out by the authors, according to the data published in a number of papers, did not reveal any statistically significant correlations between η_{EAF}, Eq. (5.3), and the main parameters of the EAF, such as capacity, power, tap-to-tap time, usage and the power of burners, and electrical energy consumption. Both the highest and the lowest electrical energy consumption were observed on the furnaces with the lowest η_{EAF} values, Fig. 5.1. The only explicit correlation occurs only between the η_{EAF} and the relative quantity of chemical energy ΣE_{CH}, %. With an increase in the portion of ΣE_{CH} in the total heat input ΣE the η_{EAF} values drop significantly, Fig. 5.2. Thus, η_{EAF} depends mostly not on the design and the mode of operation

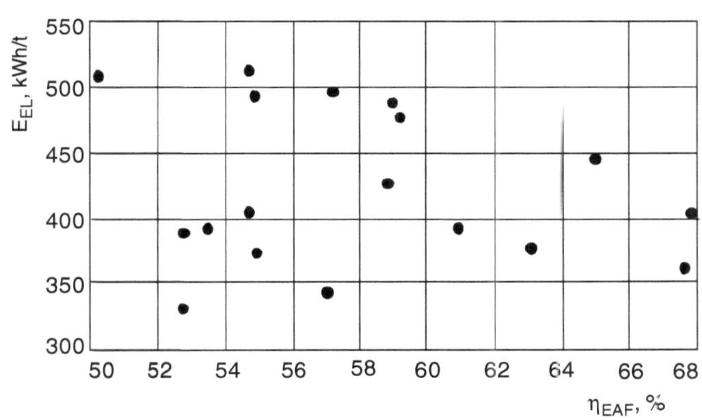

Fig. 5.1 Electrical energy consumption E_{EL} versus η_{EAF}

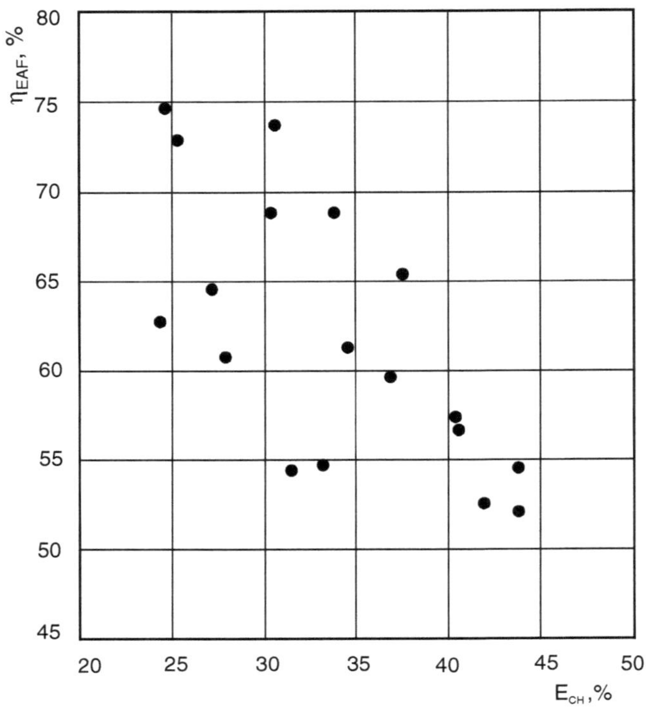

Fig. 5.2 Dependence of η_{EAF} on the share of chemical energy in the total heat input

parameters of the furnace, but on the amount of oxidized elements, i.e., on the iron oxidation, chemical composition of scrap, the amount of coke charged into the furnace, as well as on the grade of steel being produced.

Dependence of η_{EAF} on ΣE_{CH}, Fig. 5.2, can be explained by the fact that a significantly increased share of ΣE_{CH} in the total heat input is achieved in practice mainly as a result of increased use of coke and natural gas. The energy of these sources is used with significantly lower efficiency compared to the energy of electric arcs. Therefore, ΣE_{CH} grows significantly faster than electrical energy consumption reduces due to the increased use of fuel. As a result, the denominator in formula (5.3) increases, and η_{EAF} drops. Use of energy from fuel in EAF is a positive trend. Sharp decrease in η_{EAF} observed with an increase in the ΣE_{CH}, Fig. 5.2, proves only that this efficiency coefficient is not suitable for its purpose. Taking into consideration the complexity of the energy efficiency characteristic of EAF, we have to admit that it is impossible to evaluate it with any single coefficient.

5.3 Specific Coefficients η for Estimation of Energy Efficiency of Separate Energy Sources and EAF as a Whole

As it was stressed earlier, in EAF as well as in the other types of furnaces the processes determining the efficiency of different forms of energy are those of heat transfer other than heat generation. Therefore, coefficients η must characterize specifically the heat transfer processes. They cannot equally relate to different with regard to heat exchange conditions heat sources such as electric arcs, oxy-fuel burners flame, exothermic reactions of oxidation of elements. and post-combustion of CO above the bath. For that reason, instead of the common coefficients η_{EAF}, Eq. (5.3), the specific coefficients η should be considered for each particular form of energy with its specifics of heat transfer from the energy source to the scrap or to the liquid metal. During the heat all energy sources contribute to a preset finite value of useful energy ΣE_{MET}. Each such a contribution is the useful energy consumption of the source. Let us introduce the following indices noting with an asterisk useful energy consumptions:

E^*_{EL}, E^*_{NG} –useful consumption of electrical energy and of energy of natural gas. They are equal to quantities of heat obtained by scrap and liquid metal due to the heat transfer by radiation and convection, respectively, from the electric arcs and from the burners flame.
$E^*_{COK}, E^*_{CH.MET}$ –useful consumption of chemical energy of exothermic reactions. They are equal to quantities of heat obtained by scrap and liquid metal, respectively, from oxidation of coke and of iron and its alloys C, Si, Mn, Cr, etc.
$E^*_{COK} + E^*_{CH.MET} = \sum E^*_{CH}$ –total useful consumption of chemical energy.
E^*_{OTH} –useful consumption of energy from other sources which contribute insignificantly and therefore are not examined separately. Thus, total useful energy consumption for the heat $\sum E^*$ is defined by the following equation:

$$\sum E^* = E^*_{EL} + E^*_{N.GAS} + \sum E^*_{CHEM} = E_{MET}. \tag{5.4}$$

The chemical energy of the reactions of slag formation $E_{CH.SL}$ is not included in Eq. (5.4). Although this energy is being released inside the melt and is consumed by liquid bath completely, it cannot be considered as useful energy similar to $E_{CH.MET}$. Unlike $E_{CH.MET}$, the energy $E_{CH.SL}$ is being released not in the metal, but in the slag. Therefore, it deems appropriate to consider that all this energy is being consumed on heating of slag rather than on heating of metal. This corresponds with the fact that the value of $E_{CH.SL}$ is insignificant in comparison with the enthalpy of slag. In the 130-ton furnace heat balance it makes up only 22% of E_{SL}, Chap. 4, Table 4.1.

In reality, metal and slag get mixed, although not that intense as in the converters. Therefore, $E_{CH.SL}$ as well as electric arcs and other energy sources participate in heating of both slag and metal. However, without adopting an assumption, which will not change quantitative ratios substantially, it is impossible to introduce the clear and consistent system of the efficiency criteria for the separate sources of energy, which would consider as useful the energy E_{MET} rather than $E_{MET} + E_{SL}$. As already mentioned, Chap. 2, Sect. 2.2, the total enthalpy of the bath $E_{MET} + E_{SL}$ does not conform to the concept of useful energy.

Equation (5.4) is correct for the furnaces working without preheating of scrap by off-gases or by burners. If scrap is preheated and charged into the furnace with the enthalpy E_{SCR}, then the required sum of the useful consumptions of all the energy sources decreases by this value, and Eq. (5.4) may be presented as follows:

$$E^*_{EL} + E^*_{NG} + \sum E^*_{CH} = E_{MET} - E_{SCR}. \tag{5.5}$$

When using hot metal, its enthalpy should also be subtracted from E_{MET}, and its useful chemical energy should be added to the value of $\sum E^*_{CH}$.

Let us analyze the values $\sum E^*_{CH}$ and E^*_{COK}. In accordance with the laws of physical chemistry and data obtained from practice, it may be assumed that the carbon dissolved in the liquid metal is oxidized in the bath only to CO and that the heat released by this process is consumed by metal practically completely as well as the heat of oxidation of iron and its alloys. The particles of iron, which are being removed from the furnace along with off-gases, can also be oxidized. But this energy source is insignificant, and it is reasonable not to take it into account.

Not all the carbon of the coke injected into the bath penetrates into the liquid metal and dissolves in it. The part of injected carbon burns in the furnace above the bath. The share of injected carbon dissolving in the metal is characterized by the coefficient $\varphi_{1.COK}$, which can vary from 0.6 to 1.0. The coefficient $\varphi_{1.COK}$ is determined by design of injectors and modes of their operation. The coefficient $\varphi_{1.COK}$ increases if the injectors are being placed closer to the bath surface. If practically all the carbon is injected under the surface of slag near the slag–metal boundary, the coefficient $\varphi_{1.COK}$ becomes equal to unity.

The carbon of the coke charged together with the scrap can be oxidized to CO and partially to CO_2 if there is sufficient amount of oxygen in the atmosphere of the furnace. The same is true in regard to the part of injected carbon, which is oxidized above the bath. Carbon monoxide CO releasing from the bath can also partially burn up in the freeboard. Usually the lumps of coke are placed in the lower part of the charging bucket. When the furnace operates with a "hot heel," part of the coke defined by the coefficient $\varphi_{2.COK}$ approximately equal to 0.5–0.7 is immersed in the melt. This coke is dissolved in the liquid metal and oxidized in it along with injected carbon, thus releasing CO from the bath. The flow of hot dust-laden gases containing CO and CO_2, while rising above the bath, transfers its heat to the unmolten scrap and to the liquid bath by convection and by radiation. Unlike the fully absorbed heat released inside the metal, these heat transfer processes are characterized by very low values of η equal on the average to approximately 25%, just like in case of post-combustion of CO above the bath. Thus, E^*_{COK} is the sum of two useful energies, one of which $E^*_{COK.INJ}$ is obtained from the coke injected into the bath, and another $E^*_{COK.LM}$ from the lump coke charged with the scrap.

Each of the processes of heat transfer from the analyzed above energy sources to the scrap or to the liquid metal can be characterized by their specific energy efficiency coefficients similar to expression (5.1).

$$\eta_{EL} = E^*_{EL}/E_{EL}, \tag{5.6}$$

$$\eta_{NG} = E^*_{NG}/E_{NG}, \tag{5.7}$$

$$\eta_{COK.INJ} = E^*_{COK.INJ}/E_{COK.INJ}, \tag{5.8}$$

$$\eta_{COK.LM} = E^*_{COK.LM}/E_{COK.LM}, \tag{5.9}$$

$$\eta_{CH.MET} = E^*_{CH.MET}/E_{CH.MET}. \tag{5.10}$$

In accordance with what has been said above about $E_{CH.SL}$, the coefficient $\eta_{CH.SL}$ is not included in the system of equations (5.6–5.10), since it has been assumed that $E^*_{CH.SL} = 0$.

Using specific efficiency coefficients η (5.6–5.10), we can write the equation of the energy balance of the furnace as the useful energies' balance equation, similar to Eq. (5.4). For the furnace operating with the cold charge ($E_{SCR} = 0$) we will get

$$\begin{aligned} &\eta_{EL} \times E_{EL} + \eta_{NG} \times E_{NG} + \eta_{COK.INJ} \times E_{COK.INJ} + \eta_{COK.LM} \times E_{COK.LM} \\ &\quad + \eta_{CH.MET} \times E_{CH.MET} = E_{MET}. \end{aligned} \tag{5.11}$$

The useful energies' balance equation (5.11) with consideration of the costs of energy of different sources rather fully characterizes the energy efficiency of EAF. Unlike the balance of energies, Eq. (5.11) shows not only amounts of different forms of energy introduced into the furnace, but also effectiveness of their use.

As it has already been noted, with accuracy adequate for practical purposes, $\eta_{CH.MET} = 1$ and $\eta_{COK.INJ} = 1$ can be accepted for that part of the injected coke, which is dissolved in the metal. Therefore, only three specific energy efficiency coefficients η_{EL}, η_{NG}, and $\eta_{COK.LM}$ and coefficients of coke distribution $\varphi_{1.COK}$ and $\varphi_{2.COK}$ should be defined.

Specific energy efficiency coefficients can be expressed not only through amo unts of useful energy E^* and total energies E consumed in the certain period of time (kWh), but also through the corresponding powers P^* and P, i.e., through the energies per unit time (kW). For example consider the following equations:

$$\eta_{EL} = P^*_{EL}/P_{EL}, \tag{5.12}$$

$$\eta_{NG} = P^*_{NG}/P_{NG}, \tag{5.13}$$

$$\eta_{COK.INJ} = P^*_{COK.INJ}/P_{COK.INJ}. \tag{5.14}$$

Such values characterize the instantaneous energy efficiency coefficients at any given moment in time rather than the average values for the stages of heat or for entire heat. These coefficients are needed for the equations of the instantaneous energy balance of the furnace in the form of the balance of useful energies similar to Eq. (5.11).

5.4 Determining Specific Coefficients η

5.4.1 Electrical Energy Efficiency Coefficient η_{EL}

This coefficient can be presented as a product of two coefficients:

$$\eta_{EL} = \eta_{SEC.C} \times \eta_{ARC}, \tag{5.15}$$

where $\eta_{SEC.C}$ is the coefficient of electrical energy efficiency of the secondary electrical circuit including transformer.

This coefficient defines that part of the electrical energy E_{EL}, which is consumed by the electric arcs after loss of the electrical energy in the secondary electrical circuit $E_{EL.LOS}$:

$$E_{ARC} = \eta_{SEC.C} \times E_{EL}. \tag{5.16}$$

Since, $E_{ARC} = E_{EL} - E_{EL.LOS}$, it is obvious that $E_{EL.LOSS} = (1 - \eta_{SEC.C}) \times E_{EL}$.

η_{ARC} is the thermal efficiency coefficient of the electric arcs. This coefficient characterizes the process of heat transfer from the arcs to the scrap or to the liquid metal.

In the modern furnaces, $\eta_{SEC.C}$ is the relatively constant value, which does not change significantly during the heat, and amounts approximately from 0.90 to 0.95 [1]. On the contrary, coefficient η_{ARC} can vary within a wide range during the heat or for the different furnaces, depending on the specifics of technological process. In the furnaces operating on the scrap the values of η_{ARC} are quite low at the very beginning of the melting of each basket when the electrodes are not yet deepened into the scrap pile, since the substantial part of the energy of arcs is radiated onto the furnace roof and walls. Then the arcs bore-in scrap, the electrodes are lowered, and the arcs are partially shielded by the scrap, hence their thermal efficiency grows.

When the electric arcs are burning onto the liquid bath, the values of η_{ARC} are mainly determined by the depth of immersion of the arc into the slag. In order to ensure the most complete shielding of the arcs, the slag is being foamed. However, even when the arcs are completely immersed into the slag they become the internal source of energy only for the slag layer, but not for the metal. Thus the heat transfer from the arcs to the metal depends on the metal–slag stirring intensity in the arc zone. And, as a result of heat transfer conditions changing during the liquid bath heating, the value of η_{ARC} may vary widely from 0.6 to 0.9 [2].

For the short heat in the modern EAF only the average values of η_{EL}, which are the most important to the practice, can be evaluated with sufficient accuracy. Apparently, the most reliable are the average per heat values of η_{EL} shown in the work [2]. They are related to E_{MET} and are from 0.6 to 0.8, on average 0.7. Using this value, data on $\eta_{SEC.C}$ given above, and formula (5.15) the average value of η_{ARC} can be evaluated: η_{ARC} = 0.7/0.93 = 0.75.

The average values of η_{EL} = 0.70 and η_{ARC} = 0.75 will be used in the following chapters, in particular for the analysis of the electric arcs and the oxy-gas burners operating in tandem. These values are substantially lower than those which recently are frequently given as evidence of extremely high efficiency of use of electric energy in the modern arc furnaces. This is relevant to the EAF with the continuous scrap charge into the liquid bath as well as to the furnaces using hot metal. In many cases, the utilization of the published values of η_{ARC} and η_{EL} presents the problems, since it is not specified whether these values reflect the useful heat E_{MET}, or the enthalpy of entire liquid bath $E_{MET} + E_{SL}$. Corresponding values of these coefficients differ significantly from each other.

Determining the coefficients $\eta_{SEC.C}$ with the aid of the known electro-technical methods is not difficult. The value of η_{ARC} for the liquid metal heating stage can be found from the bath heat balance equation (5.17) after discontinuing carbon injection for the short period of time $\Delta\tau$:

$$\eta_{ARC} \times \eta_{SEC.C} \times \Delta E_{EL} + \Delta E^{*}_{CH.MET} = \Delta E_{MET}. \tag{5.17}$$

During this stage, all the sources of metal heating other than the electric arcs and the chemical reactions of oxidation of iron and its alloys $E_{CH.MET}$ can be disregarded. In Eq. (5.17): ΔE_{EL} is electrical energy consumption during the time period $\Delta\tau$, and $\Delta E^{*}_{CH.MET}$ and ΔE_{MET} are, respectively, the heat release in the metal due to chemical reactions and the increase in the enthalpy of metal during the time period $\Delta\tau$.

The value $\Delta E^{*}_{CH.MET}$ can be determined based on the composition of metal samples taken in succession ($\eta_{CH.MET}$ = 1), and the value of ΔE_{MET} can be determined based on the results of the consecutive metal temperature measurements. By repeating such measurements during several heats, the average value of η_{ARC} can be calculated with sufficient accuracy. The value of $\eta_{EL} = \eta_{SEC.C} \times \eta_{ARC}$ can also be determined by the method of inverse heat balance if the values of η_{NG}, $\eta_{COK.LM}$, and $\eta_{COK.INJ}$ are known, Sect. 5.4.4.

5.4.2 *Fuel Energy Efficiency Coefficient of Oxy-Gas Burners η_{NG}*

This coefficient can be determined by comparison of the results of the heats carried out with the burners turned on with those carried out with the burners turned off. If as a result of such experiments it is found that with the consumption of the natural gas equal to E_{NG}, kWh/ton, the burners decrease the electrical energy consumption by ΔE_{EL} kWh/ton, then with the known η_{EL} the value of η_{NG} can be calculated according to the formula:

$$\eta_{NG} = \eta_{EL} \times \Delta E_{EL}/E_{NG}. \tag{5.18}$$

Heat transfer from the burners' flame to the scrap occurs mainly through convection. The products of complete combustion of natural gas H_2O and CO_2 at high temperatures can oxidize iron and, in so doing, they are reduced to H_2 and CO. As a result, besides the design of the burner, there are two factors that affect the value of η_{NG} strongly: the surface area of the scrap covered by the flame and the temperature of this surface.

The highest values of η_{NG} are observed during the first minutes after charging of each basket of scrap, while the surface area of the lumps of scrap contacting with the flame is the greatest, and the temperature of this surface is still low. As the heated scrap settles down, the area covered by the flame decreases as well as the temperature difference between the flame and the scrap. At the temperatures close to or above 1400°C oxidation of iron by-products of gas combustion intensifies sharply. The content of H_2 и CO in the off-gas increases quickly which is equivalent to the significant fuel underburning. As a result, η_{NG} decreases sharply, and it becomes necessary to turn the burners off.

The power of the sidewall burners used in EAF usually does not exceed 3.5 MW. When charging is carried out with three baskets, the duration of burner operation is approximately 70% of the time interval between the first and the second baskets and 50% of the time interval between the second and the third baskets. After charging a third basket the burners are only used for a short term or not used at all. A time-average value of η_{NG} does not exceed 50–60%. Practice showed that an increase in the power of such burners is inexpedient. With an increase in the power the reasonable duration of burner operation is reduced almost proportionally, whereas a quantity of useful heat transferred to the scrap E^*_{NG} grows insignificantly. Therefore, in order to increase E^*_{NG} the number of burners has to be increased up to six or even nine. A significant increase in the power of the burners and in E^*_{NG} can be achieved through the use of the rotary burners which can change the direction of flame during their operation within the wide range. The utilization of different types of burners in EAF is described in detail in Chaps. 6 and 7.

5.4.3 Energy Efficiency Coefficient of Coke Charged Along with Scrap

As already mentioned, the value of the coefficient $\eta_{COK.LM}$ is close to 0.25. Although it is affected by the large number of factors, it is necessary to consider that the errors in its determination do not have any significant influence on the results of calculations, since obtained in practice energy $E_{COK.LM}$ is relatively low.

To a considerable extent, the same is true for determination of the coefficient $\eta_{COK.INJ}$ for that part of the injected coke, which does not dissolve in the liquid metal.

The large amount of carbon monoxide formed by the intensive oxygen blowing of the bath only partially burns out within the freeboard. Even in the furnaces equipped with special oxygen tuyeres for post-combustion CO large amount of carbon monoxide escapes into the gas duct. For that reason, in order to simplify the calculations a number of authors consider as the first approximation that all coke is oxidized in the furnace only to CO. In reality, this is not true, and in each specific case it needs to be determined more accurately.

5.4.4 Determining the Specific Coefficients η by the Method of Inverse Heat Balances

This simple method is used for determining of an unknown term of the total sum of energy input or output in the equation of the heat balance when the total sum is known from the independent source. The unknown term is calculated as a difference between the total sum and the sum of the remaining terms. Any of the four coefficients η_{EL}, η_{NG}, $\eta_{COK.INJ}$, and $\eta_{COK.LM}$ can be determined by this method if the other three coefficients are known.

5.5 Tasks of Practical Uses of Specific Coefficients η

Increasing energy efficiency of EAFs is achieved at present basically by increasing use of chemical energy. The designs of burners, injectors, tuyeres, combined jet modules intended for the introduction into the furnace of chemical energy carriers and oxygen come in a growing assortment. Reduction of electrical energy and total energy consumption of EAF steelmaking is, to a great extent, assured by these new designs developed and supplied by numerous companies. Equipment selection for the furnaces, considering their specific operating conditions, requires the use of the objective, scientifically substantiated quantitative criterion of their effectiveness. Today the decisions are being made based on the advertising materials issued by the suppliers, which usually contain the performance indices of the furnaces before and after the installation of new equipment. It is typical that these indices are usually improved insignificantly and are approximately the same in the advertising brochures of different companies. If one considers that these indices such as productivity, and electrical energy consumption depend on very many factors changing over time, then "afterward" does not necessarily mean "as a result of."

In many cases when the estimation of innovations is carried out by comparing the results of an improvement in the furnace performances, all of the above said makes the comparative estimation not objective enough, and sometimes even impossible. The proposed method of estimation of the efficiencies of different sources of energy and of furnace as a whole utilizing objective criteria, such as η coefficients, offers new and better possibilities. This method gives strictly scientific basis for selection of energy innovations for EAF.

A customer judges energy economy performance of automobile to gasoline consumption per 100 km. The equipment designed for introduction and utilization of chemical energy in the EAF must also be characterized by similar objective performance indices. Coefficient φ_{1COK} is such an index for a device injecting the coke powder into the bath. It is obvious that the preference must be given to the injectors with higher coefficient φ_{1COK}, since the efficiency of chemical energy of coke not dissolved in the metal is very low. The index of energy efficiency of oxy-gas burner is the coefficient of η_{NG}. If the value of $\eta_{EL}$for the furnace is known, then the coefficient η_{NG} can be used to reliably determine the expected electrical energy savings due to utilization of a certain burner, formula (5.18). A device for blowing the bath with oxygen, along with the other indices, must be characterized by oxygen consumption needed to ensure the required rate of decarburization of the metal. Unfortunately, the companies developing injectors, burners, oxygen tuyeres, jet modules, and other similar devices do not provide such data for their products. It happens, apparently, due to the fact that these data are simply absent.

The values of η_{EL}, η_{NG}, and other coefficients η given above as well as the factors of coke distribution φ_{1COK} and φ_{2COK} are approximate and need to be determined more accurately. These coefficients may vary quite widely in the different furnaces, depending on design and technological factors. Conducting thermo-technical studies under specific operating conditions of different furnaces is necessary for more accurate determination of these coefficients. The method of these studies must be totally different from the studies conducted for the purpose of the energy balance calculations.

Balance studies are just the passive recording of the thermo-technical and technological parameters of a furnace under the normal day-to-day operating conditions. In the studies conducted to determine coefficients η and the factors of coke distribution needed for estimation of efficiency of different energy equipment, the active experiments must be conducted. In such experiments the usual mode of operation of a furnace must be purposely changed. Several examples of such changes are turning the burners off, stopping injection of coke, etc.

Conducting active experiments with the purpose of determining the efficiency coefficients of different forms of energy should become the main direction of further studies of EAF's thermal work. Unfortunately, at present such studies are not conducted systematically, although accumulation of the results of such studies will contribute not only to the objective evaluation of different types of the existing energy equipment, but will also help determine the new and promising trends in development and fabrication of the new devices ensuring the most complete and efficient utilization of electrical and chemical energy in EAF.

References

1. Pfeifer H, Kirschen M (2002) Thermodynamic analysis of EAF energy efficiency and comparison with a statistical model of electric energy demand. 7th European Electric Steelmaking Conference, Venice
2. Pfeifer H, Kirschen M, Simoes JP (2005) Thermodynamic analysis of EAF electrical energy demand. 8th European Electric Steelmaking Conference, Birmingham

Chapter 6
Preheating of Scrap by Offgases in Combination with Burners

6.1 Potentials and Limiting Factors

6.1.1 Expediency of Heating

Despite of a scrap shortage in some parts of the world and partial replacement of scrap with hot metal or reduced iron, it is still the scrap that both now and in the distant technical perspective should be looked at as the prime metal charge for the majority of arc furnaces. Preheating of scrap is the important technical task and numerous research projects are dedicated to finding a solution to this problem. Preheating the scrap prior to the heat allows increasing the enthalpy of the metal-charge E_{SCR} by tens times and, consequently, sharp reducing the consumption of useful energy required for the heat from E_{MET} to $E_{MET} - E_{SCR}$, Chap. 2, Sect. 2.2, and Eq. (5.5), Chap. 5. The total consumption of useful heat includes the useful consumptions of electrical energy and of other energy-carriers. The sum of these useful consumptions decreases by the value of E_{SCR}. The total consumptions of energy-carriers decrease respectively. Reduction of electrical energy consumption is usually accompanied by a number of advantages, in particular, by a decrease in electrode consumption. At the same time, power-on-time shortens almost directly proportional to the decrease of the required consumption of the useful heat, and productivity increases accordingly. This assures the same or close productivity level of EAF while its electrical power can be significantly reduced.

Scrap preheating can be carried out by total thermal energy of off-gas which includes chemical energy of its combustible ingredients CO and H_2 as well as fuel energy. Compared to using heat of off-gases for getting hot water or steam, heating of scrap is much more attractive option since return of lost heat straight back to the technological process assures not only reduced energy consumption, but also a significant increase in productivity of EAF.

If natural gas or other fuels are used for heating of scrap, then the economy of this process depends both on price lesvel of different energy-carriers and on the ratio of the energy efficiency coefficients η for different forms of energy. However, since an increase in productivity decreases operational personnel costs as well as

DOI: 10.1007/978-3-642-36273-6_6, © Springer-Verlag Berlin Heidelberg 2013

overall-factory costs per 1 ton of output, heating of scrap is usually economically profitable, if a satisfactory technical solution has been found for the problems related to scrap heating. In order to get the right idea of possible effectiveness of scrap preheating, it is necessary to quantitatively estimate all three components of the useful energy of the heat E_{MET}.

6.1.2 Comparison of Consumptions of Useful Heat for Scrap Heating, Scrap Meltdown, and for Heating of Metal up to Tapping Temperature

At tapping temperatures 1620–1640°C the value of E_{MET} is approximately equal to 390 kWh/t. This useful heat is accumulated during the heat at the stages of scrap heating, scrap melting, and heating of liquid metal to tapping temperature. The heat capacities of solid and liquid iron and its melting heat are such that the energy needed for heating of scrap to the melting point of 1530–1540°C is approximately 294 kWh/t, that of for melting is 75 kWh/t, and only about 25 kWh/t is needed for heating of melt from the melting point to the tapping temperature, Table 6.1.

Thus, three-fourths of all the useful heat are used for heating of scrap to the melting point, and only one-fourth for the melting of scrap and heating of liquid metal. As scrap heating requires the highest energy consumptions, the productivity

Table 6.1 Heat capacity c and enthalpy E of solid and liquid iron

t, °C	c, Wh/kg × °C	E, kWh/t	t, °C	c, Wh/kg × °C	E, kWh/t
25	0.124	3.1	900	0.190	170.7
50		6.5	950		181.0
100	0.130	13.0	1000	0.190	189.7
150		19.8	1050		198.4
200	0.134	26.8	1100	0.188	206.9
250		34.1	1150		215.9
300	0.139	41.6	1200	0.187	224.8
350		49.5	1250		233.9
400	0.144	57.7	1300	0.187	242.9
450		66.5	1350		254.4
500	0.151	77.5	1400	0.190	265.9
550		85.0	1450		275.1
600	0.159	95.1	1500	–	286.7
650		106.1	1536 solid		294.2
700	0.169	118.7	1536 liq.	–	369.5
750		131.7	1600		384.0
800	0.181	145.0	1650	0.23[a]	394.7
850		158.0	1700		405.6

Notes The data can be used for low-carbon steels having enthalpies E higher by 3.6% at 100°C; by 2.9% at 500°C; by 1.5% at 1000°C; by 1.8% at 1300°C

[a] The average value of heat capacity in the range from 1550 to 1700°C, Chap. 4, Sect. 4.4

and effectiveness of EAFs depend to the greatest extent on methodology and energy efficiency of this process. If the above-mentioned processes went on sequentially, for instance, in three different baths, then in the last bath only about 6% of total useful heat would be used for heating of the liquid metal obtained after the melting of all the scrap.

However, in reality, in conventional EAF all these processes take place simultaneously. When lumps of scrap are heated by electric arcs, their partial or complete melting occurs. The melt formed flows down to the bottom of the furnace, the level of the liquid bath rises, and the substantial part of the scrap is immersed in the melt with the temperature lower than the melting point. This is very typical, especially for the furnaces operating with a so-called hot heel which means that the substantial part of the liquid metal and slag from the previous heat is being left in the bottom after tapping.

Further heating and melting of the lumps of scrap occurs in the liquid bath, which temperature grows due to intensive heat transfer from the external and internal sources of energy. The need for melting the significant masses of scrap after their immersion into the liquid bath makes this stage of the heat relatively time-consuming. On average it takes about 40% of power-on time. In order to shorten this stage, it is necessary to intensify the scrap heating to highest possible temperatures prior to immersion of scrap into the melt. This can be achieved by high-temperature preheating of scrap.

6.1.3 Reduction in Electrical Energy Consumption with High-Temperature Heating of Scrap: Calculation of Potentials

In the rolling mills steel billets are heated by the burners in the furnaces to medium mass temperatures of 1150–1250°C. At 1250°C the enthalpy of iron is 234 kWh/t, Table 6.1. If utilization of the burners made possible to heat scrap to the same temperature level, it would lead to a radical change and an improvement in power engineering of modern EAF.

The aforesaid can be explained by the approximate calculation utilizing data presented in the Sects. 5.3, 5.4.1 and 5.4.2 in Chap. 5. Let us take a look at the Eq. (5.4) in Chap. 5: $E^*_{EL} + E^*_{NG} + \sum E^*_{CH} = E_{MET}$. In this case, when the scrap is heated by burners to 1250°C, useful heat of natural gas is equal to the enthalpy of scrap at this temperature, $E^*_{NG} = 234$ kWh/t. It is known that $E^*_{MET} =$ 390 kWh/t. Substituting these values into the equation gives: $E^*_{EL} + \sum E^*_{CH}$ $= 390-234 = 156$ kWh/t. $\sum E^*_{CH} = E_{CH.MET} + E^*_{CH.COK}$. In the modern EAFs heat released from oxidation of iron and its alloys $E_{CH.MET}$ is approximately equal to 100 kWh/t, Chap. 4, Table 4.1. This heat is completely absorbed by the bath. On the contrary, absorbing of chemical heat of coke is characterized by quite low values of the η coefficients. But even if we completely disregard the value of

$E^*_{CH.COK}$, from the expression $E^*_{EL} + E_{CH.MET} = 156$ kWh/t we will obtain: $E^*_{EL} = 156 - 100 = 56$ kWh/t. Considering $\eta_{EL} = 0.7$, electrical energy consumption in the case is $E^*_{EL} = 56/0.7 = 80$ kWh/t.

In the modern EAF operating without heating of scrap, electrical energy consumption can be assumed to be equal to 345 kWh/t. Therefore, heating scrap to 1250°C could ensure reduction in electrical energy consumption E_{EL}, as a minimum, by $345 - 80 = 265$ kWh/t, or by $345/80 = 4.3$ times. It might be assumed that natural gas with the calorific value of 9.5 kWh/m^3 is utilized for heating of scrap, and $\eta_{NG} = 0.6$. Then the consumption of natural gas V_{NG} required for heating of scrap to 1250°C (enthalpy is equal to 234 kWh/t) is: $V_{NG} = 234/9.5 \cdot 0.6 = 41$ m^3/t.

6.1.4 Sample of Realization of High-Temperature Heating: Process BBC-Brusa

The calculation given above is not an abstract one. In principle, the technical capability and energy efficiency of the high-temperature heating of scrap were confirmed in the seventies on an industrial scale using the installations which combined an EAF and a rotary tube-type heating furnace. One of the first installations of this kind was the steel melting unit of BBC–Brusa (Italy) with a capacity of 36 t equipped with a low-power transformer of 7.2 MW [1]. In this unit, a 13 m long rotary heating furnace (1) is installed above an EAF, Fig. 6.1. The gases escaping through an opening in the EAF roof are drawn into the tube-type furnace. The scrap is continuously charged into the bath of the furnace through the same opening. When passing through the tube furnace, the gases are heating the scrap coming from a batcher (2) equipped with vibrator (3). At the upper end of the furnace the cooled gases are drawn into a fume hood (4) and are removed for purification.

Gas burners (5) are located at the bottom of the rotary furnace. The scrap passes through the furnace for 6–10 min. During this time the scrap is heated up to medium mass temperature of $t_{SCR} = 1000$°C. This temperature was reached not quite due to offgas heat but mostly due to the burners which account approximately 73% of all heat coming into the rotary furnace. Rotation of the furnace assures that heat from the refractory lining is being used for scrap heating. This enhances the advantages of countercurrent system of gas and scrap motion which makes gases exit the furnace at low temperature. The thermal efficiency of the rotary furnace calculated for the total heat input reached approximately 45%.

Performance of the BBC-Brusa unit for three years of service shows great potentialities and principle advantages of high-temperature scrap heating. Its performance indices correspond well to the calculation results given above. With natural gas consumption of 30 m^3/t, electric energy consumption was cut by 220 kWh/t. Furnace productivity increased up to 100000 t per year which at that

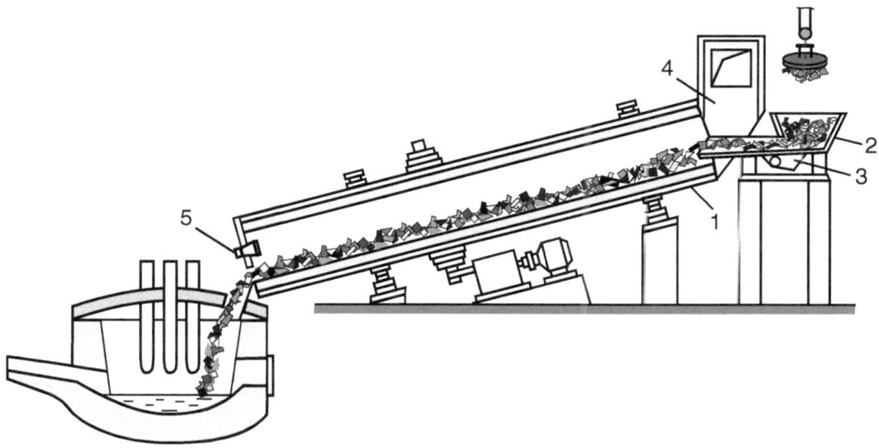

Fig. 6.1 BBC-Brusa unit (designations are given in the text)

time was equal to the productivity of a furnace with the same capacity, but equipped with high-power transformer. Continuous charge of scrap assured very quiet burning of arcs and low noise level (less that 80 db). Durability of the refractory lining of the rotary furnace was 1500 heats.

Despite the advantages attributed to the high-temperature scrap heating, such units had quite limited use and only for a short period of time. This can be explained mostly by the fact that for the modern high-productivity EAF the dimensions of a rotary furnace required call for really too big size and height of the buildings for EAF shops. Besides, rotary furnaces can operate only using properly prepared fragmentized scrap. This narrows raw material supply base and increases cost. The units with rotary furnaces also have other significant drawbacks which prevent them from being used in the modern steelmaking shops. Nevertheless, the impressive results obtained on BBC-Brusa units promoted a search for new options of high-temperature scrap heating. At a later date similar results have been achieved on the 6-t plasmarc furnaces when scrap was preheated by a burner up to the average mass temperature of more than 1000°C, Chap. 7, Sect. 7.4.2.

6.1.5 Specifics of Furnace Scrap Hampering Its Heating

In EAF, as a rule, the cheapest light scrap is used. It has low bulk density of approximately 0.79–0.8 t/m^3. Such a scrap consists mostly of lumps with relatively small mass and thickness. The length and shape of these lumps vary widely. The denser, cleaner and more expensive scrap is used in converters which are not suitable for melting light scrap. Intent of metallurgists to use cheap scrap in EAF is determined by the fact that cost of scrap accounts for approximately 70% of total cost per heat of materials, energy and personnel.

Depending on the source of scrap supply and the method of its preparation for melting the thickness of scrap lumps varies from a few millimeters (sheet bushelling) to 100–150 mm. Internal thermal resistance of such lumps is so low that each single lump can be preheated at any practically achievable rate. The temperature difference between the surface of a lump and its centre remains negligible and can be ignored. This is not true for the relatively large bales. The bales are heated through quite slowly, and therefore their use should be avoided.

Though the scrap for EAF is preselected, it always contains some amounts of rubber, plastics and other flammable organic materials including oil. The chips from metal cutting machines are especially contaminated with oil. The chips are produced in large amounts and require utilization. Oil and other flammable contaminants present in the scrap emit a lot of heat while burning out. This causes quite undesirable consequences. Even when moderate-temperature (1000–1200°C) gas is used for pre-heating of scrap, pockets of burning and melting of small fractions can be formed in the heated layer. When this occurs, the separate scrap lumps are welded together forming so called "bridges" which obstruct normal charging of preheated scrap into the furnace. Because of this preheating of metal chips is usually avoided.

At temperatures higher than 800–900°C the fine scrap is oxidized intensely due to its very large surface area. This decreases the yield and can create dangerous situations during charging scrap into the furnace. Charging of large quantities of fine strongly oxidized scrap into the liquid bath can cause an explosion-like release of CO.

When preheating of scrap, the varying in composition highly toxic compounds of halogens with hydrocarbons, known under the general name of dioxins, are formed when organic materials are burning out. In the majority of the countries, the permissible content of these compounds in the emissions into the atmosphere is legislatively restricted to the quite low level of the order of 10^{-10} g/m^3. This makes the problem of cleaning of the off-gases from the electric furnaces difficult to solve which is considered in Chap. 14.

Let us note thereby only following specifics of the behavior of dioxins during heating and cooling the off-gases. If gases containing dioxins are heated up to the temperatures exceeding 1100°C, then the latter decompose rapidly and practically completely. However, during further relatively slow cooling of these gases, decomposed dioxins can synthesize once again. The secondary synthesis occurs in the range of temperatures from 600 to 300°C.

It is known that in case of smelting of alloy and special steels, when clean scrap is used, there are practically no dioxins in the furnace off-gases. However, for mass production of electric steel, avoiding the use of cheap although contaminated scrap, as well as ensuring its thorough cleaning, does not seem possible due to economic reasons. Therefore, cleaning off-gases from the dioxins is, at the present time, an obligatory element of production process. Thus, the specifics of the steel scrap utilized in EAF create certain difficulties for its preheating, especially for the high-temperature preheating.

6.2 Heating on Conveyor

6.2.1 Consteel Furnaces with Continuous Scrap Charging into the Bath

An idea of scrap heating on a conveyor is not new. However, commercially this idea was first successfully implemented by Tenova Company. The first Tenova's conveyor furnace was started up in the USA in the 1980s. At present, more than forty conveyor furnaces named Consteel® operate in a number of countries of America, Europe, and Asia.

To operate as per Consteel scheme, an electric arc furnace is equipped with a special vibrating conveyor (1) using which scrap is transported to the furnace at a preset rate and discharged into the liquid bath through the door in the sidewall of furnace freeboard, Fig. 6.2. On its way, scrap passes through the tunnel (2) approximately 30 m long. The furnace off-gas flows along the tunnel at a low velocity towards the scrap. The gas heats the scrap and then passes via the gas duct (3) for purification. Principal energy features as well as main advantages and shortcomings of continuous scrap melting in the liquid bath were considered in detail in Chap. 1, Sect. 1.3.8. Let us remind that in such a process the power of electric arcs is limited by the scrap melting rate. The rate of scrap melting is a bottleneck of the process and does not depend on the power of electric arcs.

Such limitations of the power do not exist in the EAFs. Both the rate of scrap melting and the hourly productivity of furnaces increase with an increase in the power of the arcs. Therefore, the transformer power of the EAFs is higher than that of the Consteel furnaces. However, the lower power should not be considered as an advantage of the Consteel process. The power reduction is a necessity which ultimately leads to the decrease in productivity.

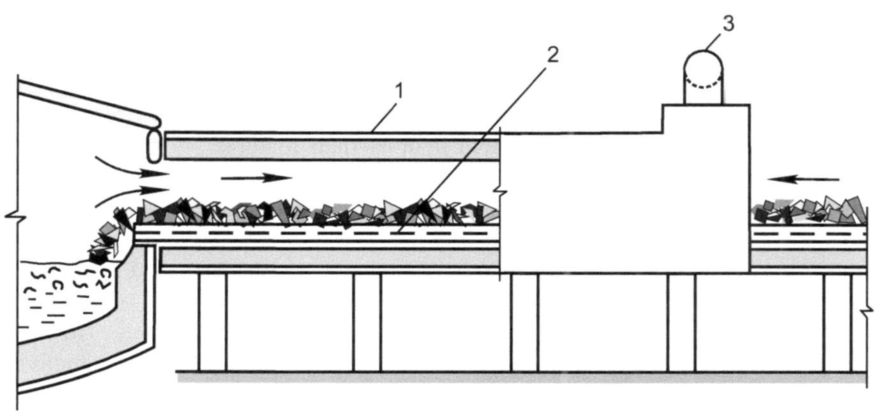

Fig. 6.2 Consteel system (designations are given in the text)

In the EAFs, the scrap melting process may be divided into two periods. During the first period, the main body of scrap pile is melted down with the electric arcs in the furnace freeboard above the hot heel surface. Plasma of the arcs has a temperature of more than 5000°C and possesses high kinetic energy. The heat energy of the arcs is transferred to the scrap by both radiation and convection. The intensity of these heat transfer processes is much higher than that in the liquid metal. The power of the arcs is also higher than that of the Consteel furnaces. Furthermore, as opposite to the Consteel furnaces, the oxy-gas burners and the process gases formed during post-combustion of CO in the freeboard contribute to the heating and melting of scrap.

During the second period, upon completion of both forming of the flat bath and submerging of the arcs into a foaming slag, the rest of the scrap melts down in the liquid metal as in the Consteel process. However, just like in the first period, the rate of melting is considerably higher. It could be explained by the fact that the scrap lumps before submerging into the melt are heated up to a temperature close to the melting point in the EAF freeboard. Only a small portion of fine scrap placed on the bottom of the basket submerges in the hot heel not being preheated. In addition, scrap is melted not in a limited area of the bath as it takes place in the Consteel furnaces but in the entire volume of the bath, which accelerates the process. All the factors listed above along with those considered in Chap. 1, Sect. 1.3.8, have to increase the scrap melting rate and productivity in EAFs in comparison with those in Consteel furnaces. Let us pay attention to the actual data.

6.2.2 Comparison of Melting Rates, Productivities, and Electrical Energy Consumptions Between the Consteel Furnaces and EAFs

The actual performances of Consteel furnaces put into operation in recent years are compared with those of the state-of-the-art modern EAFs. This allows a comparative evaluation of the two types of furnaces to be most objective especially when evaluating their prospects. Furnaces operating with hot metal have not being reviewed since performances of such furnaces are mainly defined by the percentage of hot metal in a charge rather than by their features.

Numerous Tenova publications in the periodicals "MPT International", "Iron and Steel Technology", the proceedings of the European and American conferences, etc. have been dedicated to the Consteel furnaces. Unfortunately, these publications contain data needed for evaluating the melting rate and hourly productivity for only a few of Consteel furnaces operating without using of hot metal. Data on electrical energy consumption are provided even more rarely. All these data have been used for a comparison. Advertised data on the furnace productivity not confirmed by the actual results were not taken into consideration.

The rate of melting was determined by dividing the total mass of scrap and pig iron by melting time. In the cases when the data on composition of the charge were absent, the mass of solid metal-charge was determined using the tapping weight with consideration of the liquid steel yield. The melting time was determined using the power-on time with due correction for the time needed for heating of the melt to the tapping temperature.

The scrap melting rate depends on the furnace capacity M, which was assumed to be equal to the sum of weights of both the hot heel and tapping. With the increase in the M from 100 up to 350 tons, the scrap melting rate in all the furnaces increases by more than 2 times. However, for the entire range of M, the rate of melting in the Consteel furnaces is lower than that of the EAFs by approximately 1.4 times. The comparison of furnaces operating under similar conditions at the same region in Russia after reducing them to the equal M gives the same result, Fig. 6.3.

The melting rate S increases in direct proportion with the quantity of heat Q transferred by convection to the scrap from liquid metal per unit of time. The value of Q, kWh/t, is determined by the following expression:

$$Q = \alpha \times F \times \Delta t \tag{6.1}$$

Δt, °C	The average temperature difference between the liquid metal and scrap during the whole melting
F, m^2/t	Specific surface area of scrap lumps submerged into the bath per their weight

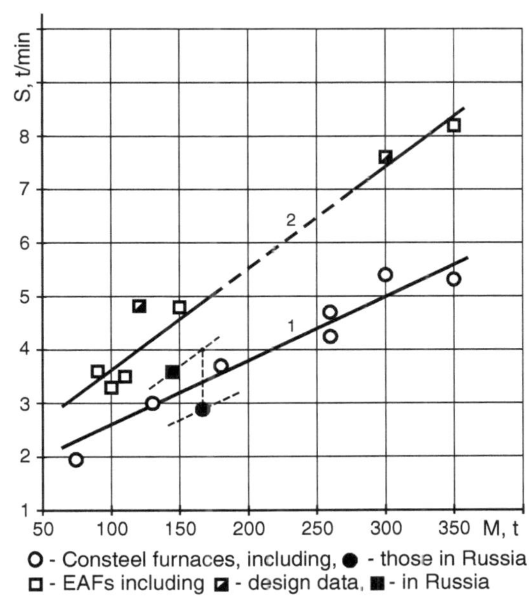

Fig. 6.3 Scrap melting rate S, t/min versus furnace capacity M, t. *1* Consteel furnaces [2–7], *2* EAFs [8–14]

α, kW/(m^2 · °C) The coefficient of convective heat transfer from metal to the scrap

Diameter, depth and volume of the bath increase with the increase in the bath capacity M. The conveyor width, and consequently, the specific surface area F increase as well. Criterion Re defining the extent of turbulence of the bath also grows. With the increase in both the M and Re the intensity of circulating and pulsating bath stirring grows, Chap. 9, which increases the α-coefficient. The temperature difference Δt increases as well since with increasing stirring intensity the temperature of metal at the scrap charging area rises and approaches the average values of temperatures outside of this area. As a result, both the heat transfer intensity Q, expression (6.1) and the scrap melting rate S increase which explains a tendency to increase the Consteel furnace capacity as much as possible. According to data from Tenova studies, the optimal hot heel weight in Consteel furnaces comprises 50–60% of the total capacity [6] which exceeds that in the EAFs by two to three times. The necessity to keep an enormous hot heel in Consteel furnaces increases the heat losses.

Except for the first heat in series the Consteel process eliminates the expenditure of time required for charging scrap by baskets from the top, which shortens the power-off time τ_{off}. Due to this advantage, a decrease in tap-to-tap time and an increase in hourly productivity were achieved in a number of cases when the existing EAFs were replaced with the Consteel furnaces, despite the decreased melting rate. At present, however, the situation has changed. Due to increased capacity of the new generation of furnaces scrap charging is carried-out once per heat on small and medium EAFs and twice per heat on the biggest EAFs. This, along with an increased speed of operation of cranes and other mechanisms, personnel training and an improvement in the production organization, made it possible to reduce the power-off time of the best EAFs to close to the typical values of the power-off time of the Consteel furnaces.

For example, a 110-t EAF at Nucor Steel Texas operates with a single charge even at the minimal scrap density of 0.65 t/m^3. Time of charging is 50–70 s including opening and closing of the furnace roof [8]. A combination of these achievements with the higher scrap melting rate results in an increase in productivity of the EAFs as compared to that of the Consteel furnaces by 20–30% and lower electrical energy consumption as well, Table 6.2.

It should be noted, that the hourly productivity per 1 MW (t/h/MW) of Consteel furnaces is higher than that for EAFs. And this creates the false impression of higher productivity in the Consteel furnaces. In fact, the productivity defined by t/h/MW is higher for the Consteel furnaces but only because their power is lower. This factor does not mean an increase in actual ton per hour.

Table 6.2 Performances comparison between EAFs and Consteel furnaces

Furnace type	Country	Capacity/ tapping weight (t)	Power (MVA)	Tap-to-tap time (min)	Productivity (t/h)	Electrical energy (kWh/ t)
EAF	Europe:USA	120–130	150–160	<35	>200	345–355
	Russia	146/128	95	55	140	374
	Turkey [14]	380/320	240	60	320	359
	USA [8]	110/82	110	32	155	317
Consteel	Norway [3]	123/83	75	41	122	384
	Greece [4]	180/130	120	48	162	395
	Russia	168/118	90	61	116	416
	Thailand [6]	300/187	125[b]	65	173	363[c]
	Italy [7]	350/250[a]	209	63	238	–

[a] tapping weight, expected
[b] the values calculated per active powers of 95 MW, 34 MW
[c] for 40% of pig iron in a charge which reduces electrical energy consumption

6.2.3 Scrap Preheating Temperature

When developing the Consteel process it was supposed that the scrap on the conveyor could be heated by off-gases up to a temperature of 700–900°C. Such a heating could provide both an increase in productivity and sharp reduction in electrical energy consumption in Consteel furnaces in comparison with EAFs. Curve 1, Fig. 6.4, shows that when raising the medium mass temperature of scrap heating from 0°C up to 800°C the productivity of furnaces with a continuous scrap charging into the liquid bath increases by 1.4 times, and the electrical energy consumption reduces by about of 170 kWh/t, for example, from 400 kWh/t down to 230 kWh/t, curve 2, or from 350 kWh/t down to 180 kWh/t, curve 3, respectively. When using these dependences for furnaces of various capacities the inaccuracy does not exceed 3–4%.

However, the heating of scrap with offgases on the Consteel conveyor proved to be inefficient. This is mainly explained by the low effectiveness of heat transfer from gases to scrap. In the tunnel the gases move at quite low velocity and heat only the surface layer of the scrap. Radiation from the gases and the tunnel refractory lining acts likewise. As per the energy balance of the 350-t Consteel furnace with the thickness of the layer of scrap on the conveyor of 900 mm the enthalpy of scrap is 18 kWh/t [15]. The average mass temperature of about 137°C corresponds to this enthalpy. In furnaces with a medium capacity, the thickness of the scrap layer decreases and the temperature of scrap rises up to approximately 250°C. At such low temperatures the content of generated dioxins is insignificant, and no special measure for environmental protection requires.

In the Consteel process, possibilities for increasing of temperature of the scrap due to improving of utilization of the heat of the offgases have practically been exhausted. The heating of scrap up to higher temperatures requires using the burners. Let us examine the design using the burners along with the heat of

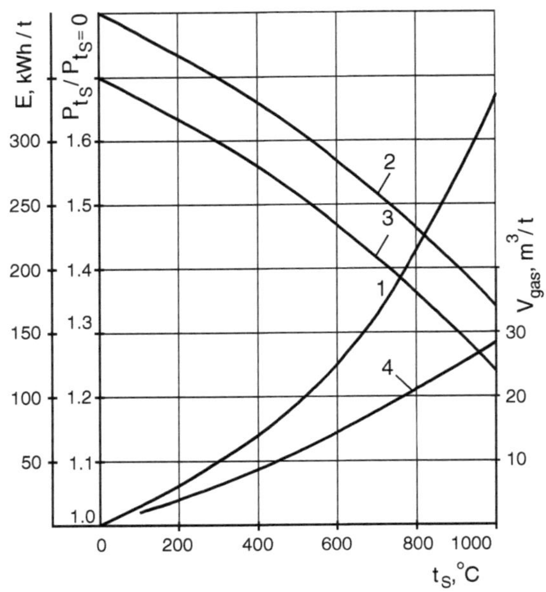

Fig. 6.4 Performances of the Consteel process versus scrap heating temperature t_S *1* relative productivity, Pt_S productivity at a temperature of t_S, *2*, *3* electrical energy consumption E, *4* gas flow rate V_{gas}

off-gases developed by the Tenova [6]. As per this design, another chamber through which goes the conveyor is constructed in front of the existing tunnel of the Consteel furnace. In the roof of this chamber the burners are installed. Evacuation of gases from both the tunnel and the additional chamber is carried out via a common gas duct located between them. In this gas duct, off-gases after the heating of scrap in the tunnel mix with fuel combustion products from the burners. It is supposed that this will provide decomposing dioxins generated when heating scrap with the burners. However, because of air infiltration into the tunnel and the additional chamber as well as expenditure of the heat for the scrap heating, it could be expected that the temperature of the gases after their mixing in the common gas duct will be insufficiently high to completely decompose dioxins. In this case, additional fuel consumption will be required to heat off-gases up to a temperature of about 1000°C.

Furthermore, when increasing the temperature of scrap at the entrance into the tunnel the difference between temperatures of scrap and gases moving through the tunnel decreases. Correspondingly, very low effectiveness of the heating of scrap in the tunnel decreases additionally. Thus, instead of using the burners as per design [6] it is reasonable to do away with the heating of scrap by offgases altogether. This will allow to develop more efficient systems for the heating of scrap with the burners, Chap. 7.

In the Tenova's design, the burners can be used for removing snow and ice from the scrap before it enters the tunnel. In countries where the winter is cold, operation of the Consteel systems meets with additional difficulties. Snow and ice on the conveyor thaw in the tunnel. Because of a low temperature of the scrap water has no time to evaporate and enters the furnace. To avert this danger an additional preparation of the scrap for the heat is required.

6.3 Heating Scrap in a Large-Thickness Layer

6.3.1 Heat Transfer Processes

Off-gases can pass through a scrap layer either from top to bottom, as it takes place in scrap charging baskets, or in opposite direction as in shaft furnaces. A device for the first variant is shown schematically in Fig. 6.5. The container (1) with scrap is installed in heating chamber (2) and equipped with a cover (3). The high-temperature off-gases come under the cover (3). The cover has holes (4) through which air from the atmosphere can infiltrate. Both a volume of the infiltrated air and a temperature of gases mix entering the scrap can be changed by opening and closing these holes. A cover design (3) ensures uniform distribution of gases over a cross-section of a layer of scrap. Passing through the scrap, the gases heat the scrap, cool down and leave the container (1) through openings (5) in its lower part. The off-gases are removed through EAF evacuation system.

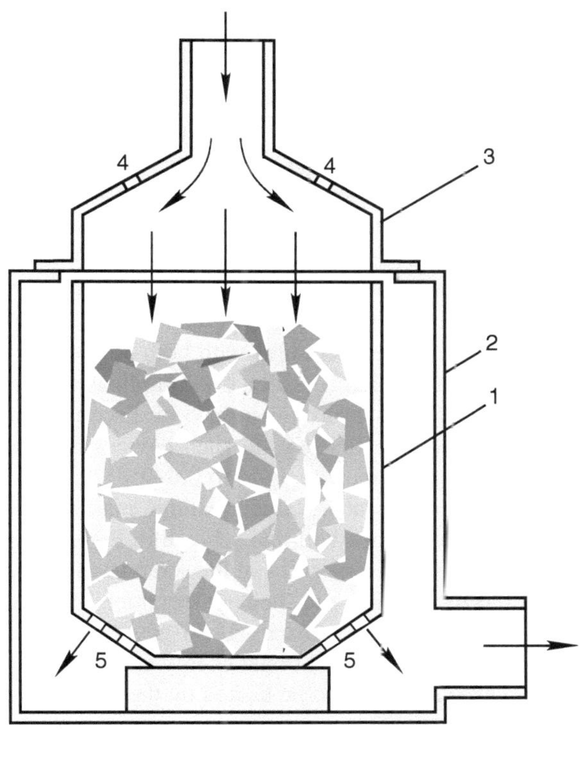

Fig. 6.5 Schematic diagram of scrap heating in a large-thickness layer (designations are given in the text)

The basic thermo-technical principles of heating the scrap in accordance with the Fig. 6.5 can be discussed without carrying out complex calculations for this non-stationary process. A thickness of separate scrap lumps is small while their thermal conductivity is quite high. Therefore, at actual scrap heating rates the difference of temperatures across the whole thickness of lumps or, what is the same, their thermal resistance can be neglected. Unlike that of separate lumps of scrap, thermal conductivity of a layer of lumps caused by the contact between the lumps both in vertical and horizontal directions is negligible and can be ignored in calculations as well. Therefore, the heat losses through the walls of the container are also so small, that the walls can be cooled by air or even by water without reducing the efficiency of scrap heating. The heat transfer by radiation from the heating gases can also be disregarded, since the gaps between the lumps are small and the emissivity of the gases in them is low.

The basic forms of heat transfer in the layer of scrap are convective heat transfer between the high-temperature gases and the surface of the lumps and heat radiation between the heated lumps. Radiation becomes essential when the lumps temperature is higher than 300–350°C. Because of the complete uncertainty in estimation of the free surface of scrap lumps contributing to the heat exchange, it is impossible, in this case, to separate these two forms of heat transfer. Therefore, in the approximate calculations of heat transfer intensity in the layer of scrap, as in the layer of other bulk materials, usually the so-called volumetric overall heat transfer coefficient, α_V, $kW/m^3 \cdot °C$, is used. This coefficient considers both the convection and radiation, and is related not to the surface area unit, like α-coefficient, but to the volume unit of the layer, although this does not correspond to the process physics.

During the heating, medium mass temperature of the scrap increases from the initial near-ambient value to the finite value, t_{SCR}. In doing so, at any given moment of time the temperature of the upper layer t'_{SCR} is considerably higher than the temperature of the lower layer t''_{SCR}. In order to avoid overheating of the upper layer the temperature of gases entering the scrap preheater t'_G is limited by the dilution of gases with cold air. Usually t'_G is maintained in range of 800–1100°C. The temperature of gases exiting the preheater t''_G grows as the scrap heats up, but does not usually exceed 300–400°C if the layer thickness is sufficiently large. All the variations of temperatures with both height of a layer and heating time have a nonlinear way, which considerably complicates the calculations. Because of a very large area of heat-absorbing surface of a scrap layer, this type of heating devices has high thermal efficiency. In such devices the heat efficiency coefficient of the gases η is approximately equal to 0.6–0.7.

The maximum achievable medium mass temperature of the scrap $t_{SCR.max}$ heated in accordance with the scheme in Fig. 6.5 can be limited by three factors: the discussed above specifics of the scrap itself; the amount of heat in the off-gases available for heating and their temperature, as well as the heating time which is very short in modern furnaces. If off-gases contain insufficient amount of heat for the scrap heating the additional energy of the fuel combusted in the burners can be used. The scrap can also be heated by the burners only. This type of scrap heating devices has the advantage that makes possible the utilization local fuels which were not used before. The electrical steelmelting shops which are the part of the integrated plants can utilize coke gas or converter gas as such a fuel.

The scrap heating process is described by the Eq. (6.2):

$$q^* = \alpha_V (t_G - t_{SCR}) \cdot \tau / \rho, \text{ kWh/t} \tag{6.2}$$

q^*	Amount of useful heat transferred from gases to scrap during heating, kWh/t
α_V	Volumetric heat transfer coefficient, $kW/m^3 \cdot {}^\circ C$
t_G	Averaged over height of the layer as well as over heating time temperature of gases, °C
t_{SCR}	Averaged over height of the layer as well as over heating time the average mass temperature of scrap, °C
$t_G - t_{SCR}$	Average temperature difference in the heating process Δt_{AVE}, °C
τ	Heating time, h
ρ	Bulk density of the scrap, t/m^3

Time τ needed for heating the scrap to the average mass temperature of t_{SCR} can be calculated from equation (6.2):

$$\tau = \rho \cdot q^* / \alpha_V \, (t_G - t_{SCR}) \tag{6.3}$$

In magnitude of q^* the final average mass temperature of scrap heating is readily determined by utilizing data from Table 6.1. Unfortunately, studies needed for creating a calculation technique of heat transfer processes in a layer of scrap heated with off-gases are not so far carried out and values of α_V coefficients were not determined. The necessity of such studies is obvious.

If during the heat all the scrap for the next heat is heated, then the achievable duration of its heating τ is approximately equal to or is less than the power-on time. The duration of heating τ can be no more than 20–25 min on the modern furnaces of small and medium capacity and no more than 35–40 min on the furnaces of large capacity. It should be considered that amount of off-gases is sharply reduced by the end of the heat. Under such conditions only relatively low temperatures of scrap heating with the off-gases are achievable.

The possibilities of speeding up the scrap heating process according to Fig. 6.5 are quite limited. In order to shorten τ, the temperature difference Δt_{AVE} has to be increased through an increase of t_G, formula (6.3). It is impossible to increase the temperature of gases at entry t'_G, because the overheating the upper layer of the scrap is unacceptable. Therefore, t_G can be raised substantially only due to an increase in the exit temperature of gases t''_G. It can be achieved by using the burners and increasing the fuel consumption in them, at the same time, increasing an amount of air which dilutes the products of combustion. An increase in the amount of gases passing through the scrap with the same temperature t'_G will increase not only t''_G, but α_V as well. This, however, will sharply reduce the fuel efficiency coefficient η, which will considerably worsen the economy of heating.

At different times various devices such as baskets and special buckets for scrap charging as well as twin-shell and shaft furnaces are utilized for scrap preheating with off-gases in a large-thickness layer. The obtained results are discussed below.

6.3.2 Heating Scrap in Baskets and Special Buckets

Due to its seeming simplicity, heating of scrap in baskets according to Fig. 6.5 was one of the first quite widely used scrap heating methods. It has all the thermo-technical characteristics of heating scrap in the layer of a large thickness. Mainly the off-gases were used for heating. Autonomous units equipped with gas burners were used in the rare cases. Such a unit operated at the Electrostal plant, Russia, in the 1970s. This unit was utilized for preheating of scrap before the heat in 20-t EAFs producing mainly high-alloy steels.

The unit consisted of two identical chambers. The scrap in a charging basket with a capacity of 10–11 t was heated in each chamber. The basket made of stainless steel was installed in the chamber with a flap cover with air-gas burners built in it. The burners were designed for natural gas combustion, with the air consumption coefficient close to one, followed by diluting of combustion products with cold air. This allowed maintaining the temperature of gases entering the basket within a range from 800 to 1150°C.

If the specific consumption of natural gas was 10–11 m^3/t and heating time was 1.5 h, the medium mass temperature of heated scrap was in a range of 500–550°C, and the efficiency coefficient of gas η_{NG} was approximately 0.7. Scrap preheating reduced tap-to-tap time by 9–10% and the electrical energy consumption of the furnace by 72–78 kWh/t. Taking into account the electrical energy consumption of the scrap heating unit itself (approximately 12 kWh/t), overall reduction of the electrical energy consumption was 60–66 kWh/t. The plant of Electrostal used clean scrap only. Therefore, there was no need of environment protection from dioxins.

In the eighties, in a number of countries of Europe, Asia and America, the units for heating scrap in baskets by off-gases became widespread on the furnaces with a capacity of from 25 to 150 t. At that time more than 30 of such units operated in Japan alone. Usually, the regular charging baskets were utilized. The temperature of gases at the unit entry was about 1000°C, and at the exit it was about 200–300°C. The medium mass temperature of heated scrap did not exceed 300–350°C. Because of the low productivity of these units, usually only a part of the scrap was heated, for instance, two baskets out of three. Utilization of these units ensured reduction of the electrical energy consumption by 30–40 kWh/t. Tap-to-tap time was shortened by 7–9%.

Two methods of heating scrap by the off-gases were used: with or without recirculation of gases. As per without recirculation method, the off-gases flow is directed into a basket with scrap placed in heating chamber as this is shown in Fig. 6.5. After passing through a layer of scrap in a basket the gases are drawn off from the chamber by an exhauster and directed for cleaning.

However, very soon it became clear that such simple schemes were suitable for heating of sufficiently clean scrap only. As has already been mentioned in Sect. 6.1.5, the scrap usually contains big amounts of organic materials, and when being heated, off-gases are saturated with dioxins. In order to avoid the emission of toxic gases into the atmosphere of a shop and into the environment, two technical solutions were introduced. One of them was to sort out the dirty scrap and place it into a separate basket. Only the clean scrap was heated. Although preparation of scrap for the heat was complicated considerably, such units existed at some plants until recently.

The second solution was to use the units with recirculation of gases. Despite their lower effectiveness and higher energy consumption, these units became more widespread because of increasing environmental protection requirements. In the units with recirculation, Fig. 6.6, the gases after heating scrap in a heating chamber (1) return to a combustion chamber (2) installed in a duct of the direct off-gases evacuation from the furnace freeboard. In this chamber the gases are mixed up with the high-temperature gases extracted from the furnace and are additionally heated by burners, that leads to quite complete decomposition of dioxins. Approximately 60% of gases from the combustion chamber are returned to the chamber (1) for scrap heating in the basket. The rest of gas is directed to a gas duct (3) for cleaning. Thus, with the help of an additional exhauster (4), the most of the gases exiting the furnace continuously circulates between the combustion chamber and the scrap heating chamber.

It is necessary to have in mind that not only heating of scrap in the baskets, but transporting the baskets with the heated scrap to the furnace as well as charging

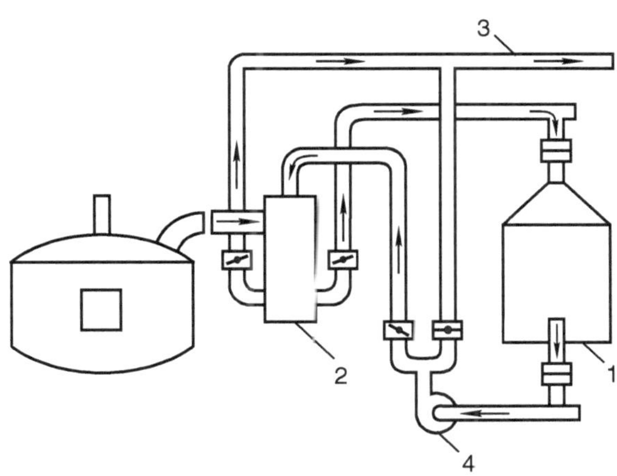

Fig. 6.6 Schematic diagram of scrap heating in charging basket with recirculation of gases (designations are given in the text)

hot baskets with fresh scrap cause air pollution both inside and outside of the shop. All these operations result in substantial emissions of smoke and toxic gases, which are practically impossible to capture. Therefore, considering an increased productivity of furnaces and, at the same time, more demanding requirements for environmental protection, the method of heating scrap in the baskets, being relatively low productive and insufficiently effective with regard to reduction of energy consumption, ceased to meet the current production requirements. Gradually, the application of this method has stopped.

Subsequently, the attempts to eliminate or at least diminish its shortcomings were undertaken. One of such attempts is a development of Danarc Plus arc furnace. The basic requirements for heating of scrap by off-gases under the operation conditions of the modern highly productive furnaces with the short heat are to the greatest extent considered in the design of this 90-t steelmelting unit developed by the Danieli Company (Italy) [16]. One of the most important features of the Danarc Plus furnace is tall furnace shell with a capacity sufficient for charging the entire amount of scrap needed for one heat (100 t) in a single charge. There are 12 oxy-gas burners in the furnace shell installed in two rows one above another.

The scrap is heated in a special water-cooled bucket (1) which, just like the furnace shell, can contain all 100 t of scrap. The bucket is moved on a carriage (2) on rails located at each way from the furnace (3), Fig. 6.7. The bucket can be placed in two positions: either above the furnace for discharging the heated scrap or aside the furnace. In the position aside the furnace the bucket is charged with fresh scrap with the help of standard fast opening baskets. During this operation, the offgases are directed, using a special valve, through the bypass into the combustion chambers without passing through the bucket and then on cleaning. Thereupon the same valve changes the direction of flow, and the gases enter the bucket again to heat the scrap. Since the carriage with the bucket and another carriage with the furnace roof (4) move simultaneously on the same rails, the duration of the entire operation of charging the furnace with scrap is less than 45 s, which considerably decreases unorganized emissions of gases and heat losses. All the unorganized emissions are captured by the hood located above the furnace under the roof of the shop.

The gases pass through the layer of scrap in a bucket from the top to the bottom. To increase the temperature of scrap heating, there are three air-gas burners with the power of 4.5 MW each installed in the upper part of the bucket. The temperature in the uppermost layer of scrap is maintained by these burners at the level of 1100–1200°C, while the temperature of the lowest layer is close to 200°C. According to the published data [16], the medium mass temperature of heated scrap t_{SCR} is about 600°C.[1] The burners also ensure partial post-combustion of CO in the offgases. The heat efficiency coefficient of all the gases passing through the scrap η is 0.55–0.65.

[1] This figure seems somewhat overstated.

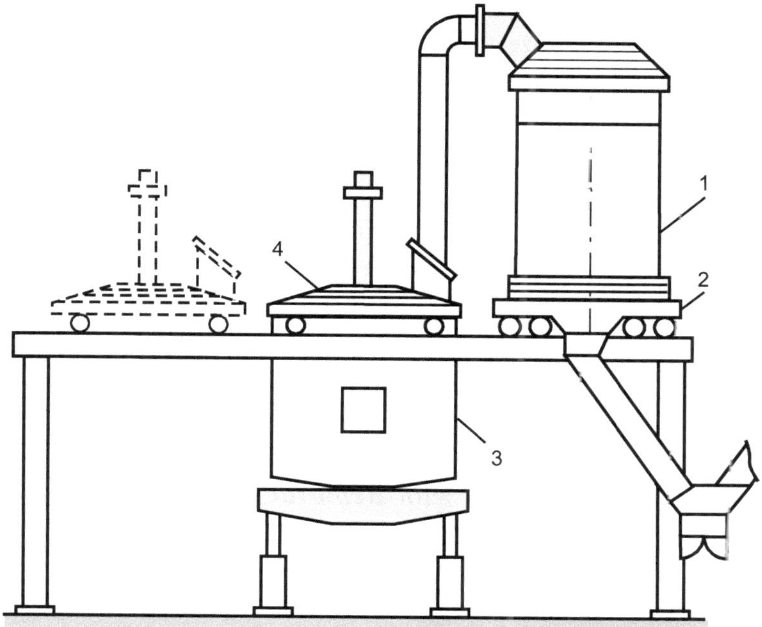

Fig. 6.7 Danarc Plus unit with scrap heating by offgases (gas duct system providing for a bucket movement is not shown, designations are given in the text)

After heating the scrap in a bucket, the gases with a temperature of 300–400°C are directed into two combustion chambers placed in tandem. These chambers are equipped with the burners with the total power of 21 MW, which heat the gases up to 950°C. In combination with intensive stirring, high turbulence level of gas flows and their sufficiently long stay in the chambers, this ensures practically complete decomposition of dioxins and post-combustion of CO.

In the Danarc Plus furnace the following results were obtained, Table 6.3. With scrap heating, tap-to-tap time has shortened by 3 min, or by 7.1%, the productivity increased by 8.7%, and electrical energy consumption decreased by 21%. This confirms the significant effect of burners on scrap heating as well as the high heat energy efficiency of gases passing through the scrap.

Table 6.3 Some performances of Danarc Plus furnace operation [16]

Performances	With scrap heating	Without scrap heating
Electrical energy consumption (kWh/t)	260	330
Oxygen flow rate (m^3/t)	35	41.6
Tap-to-tap time (min)	39	42
Liquid steel weight (t)	90	90
Productivity (t/h)	138	127

In order to get quite a complete idea about energy efficiency of this steelmelting unit, it is necessary to analyze its heat balance for the zone which includes both the furnace with the bucket and the combustion chambers of offgases. Although the published data are insufficient for the heat balance compilation, nevertheless they allow drawing important conclusions.

The temperature of the gases at outlet of chambers is about 950°C, which is close to the temperature of the gases at the entry into the bucket. This means that the amount of heat equal to almost three-fourth of the total heat of offgases used for scrap heating had to be brought back to the process. Thus, a complex and expensive heating system becomes inefficient in terms of energy. Furthermore, during the operation the reliability of a number of the basic units of this system was found to be not quite sufficient. As a result, the Danarc Plus furnace was switched to the mode of operation without scrap preheating.

6.3.3 Twin-Shell Furnaces with Removal of Off-Gas Through the Second Bath

Twin-shell furnaces are the unit consisting of two furnaces placed next to each other, Fig. 6.8, with common furnace transformer (1) connected by a secondary electrical circuit with a rotary system current conducting arms (2) and with electrodes. This system allows to place electrodes alternately in each of the baths. Several similar furnaces were built in a number of countries, sometimes for the purpose of implementation of the special technological processes such as the units combining the functions of EAF and oxygen converter. This type of units is not discussed here. Both baths of the twin-shell furnaces were equipped with oxy-gas burners (3) for heating of scrap and with devices for oxygen and carbon injection.

Technological operations in the twin-shell furnaces are carried out in a certain order. For simplicity sake, one might discuss, as an optimal version, the operation of furnaces with single scrap charging. Charging by two baskets does not change the principle of their operation. While in the first bath the electric arcs melt down scrap and heat liquid metal to the tapping temperature (operations with total duration of τ_1), in the second bath the other operations not requiring the use of electric arcs are carried out, namely the operations of tapping, closing a tap hole, preparing the furnace for the heat, scrap charging, and scrap preheating by burners. The total duration of these power-off operations is τ_2.

As soon as in the first bath the temperature of metal and the carbon content reach the required final values, the electrodes are moved to the second bath. The tapping is begun in the first bath, and the melting by the arcs of the preheated scrap in the second. Then the same operations are repeated in that same order. Since the tapping from the both baths has to follow in the identical time intervals, the operation of a twin-shell furnace must satisfy the condition: $\tau_1 = \tau_2$. Tap-to-tap time for both baths τ_1 is defined by such basic parameters as electric arc power,

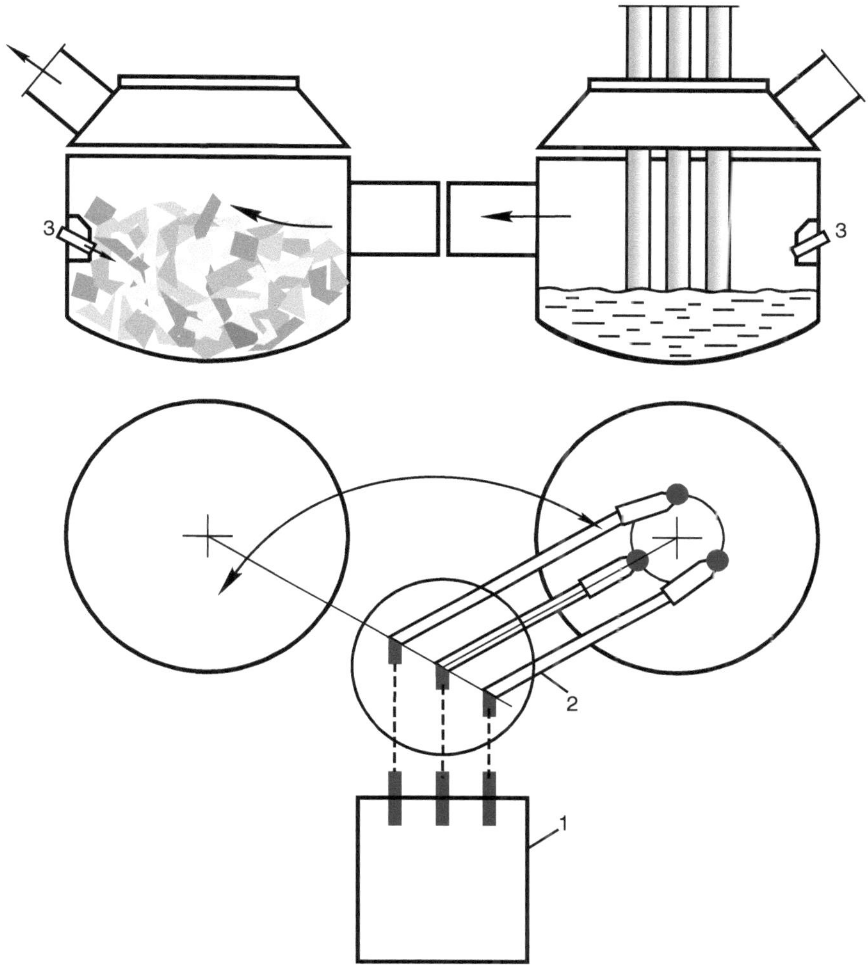

Fig. 6.8 Twin-shell furnace. Scrap heating with off-gases (designations are given in the text)

oxygen consumption, etc. Maintaining the total duration of all power-off operations at the level of τ_1 is easily achieved by controlling the scrap heating time.

Along with heating of scrap by the burners[2] in the twin-shell furnaces, the attempts to utilize the off-gases for the purpose of heating were repeatedly undertaken. In order to achieve that, the freeboards of both shells were connected by gas duct, Fig. 6.8. The off-gases evolving in the first bath during the oxygen blowing were directed into the second bath where they heated the scrap charged

[2] The heating of scrap with burners is reviewed in Chap. 7.

into it. Then off-gases were evacuated for cleaning through the fourth opening in the roof. When the oxygen blowing of the second bath started after the tapping from the first bath and its charging with scrap, the direction of off-gases evacuation was reversed by the valves in the gas duct.

The try-out of this method of scrap heating demonstrated its rather low effectiveness. This can be explained by the fact that the gases passing through freeboard at a low speed do not penetrate inside the scrap pile charged into the furnace and heat only its surface mainly by radiation. Later this method of using the heat of the off-gases has been abandoned.

6.4 Heating Scrap in Shaft Furnaces

6.4.1 Shaft Furnaces with Fingers Retaining Scrap

The development of shaft furnaces is associated with the G. Fuchs Technology Company's activities, Germany (former G. Fuchs Systemtechnic) and with the name of G. Fuchs. In the first shaft furnaces the water-cooled shafts were installed above the furnace roof and did not have fingers holding the scrap in the shaft. The scrap was charged into the furnace through the shaft. While the lower part of the scrap pile was located on the bottom of the furnace, its upper part was in the shaft. Gases from the furnace were evacuated through the shaft and heated the scrap located in it. As the scrap melted in the furnace, the entire scrap pile settled down. This created the free space in the shaft, which allowed charging of additional portions of the scrap.

Later G. Fuchs has developed and put into operation at several plants the furnaces with one row of fingers in the lower part of a shaft, Fig. 6.9. The scrap is charged into the furnace by two baskets. At the tapping the scrap of the first basket heated by the offgases during the previous heat lies on the fingers in the shaft. After the tapping the fingers split apart, and the heated scrap is charged into the furnace. After that, the cold scrap from the second basket is charged into the empty shaft. The share of the scrap from the second basket remaining in the shaft is heated by the offgases passing through the shaft. As the scrap melts in the furnace, the scrap in the shaft rapidly caves in and the shaft clears. The fingers are then shut, and the first basket of scrap for the following heat is charged on the fingers. By the tapping time this portion of scrap is already preheated by the off-gases up to relatively high temperature.

With this heating method when gases pass through the scrap from the bottom to the top, the overheating and even the partial melting of the lower scrap layer do not create any problems. The melt and the liquid slag formed flow down into the furnace and do not obstruct scrap discharging. This design concept does not require reducing in a temperature of gases entering the layer of scrap by means of dilution them with cold air. On the contrary, at those periods of the heat when the

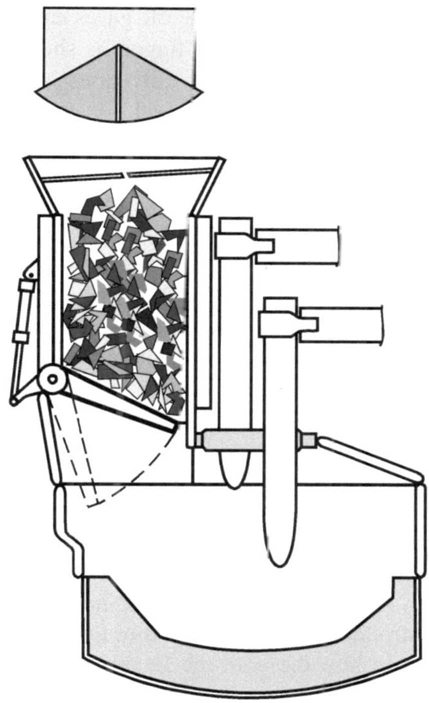

Fig. 6.9 Shaft furnace with fingers retaining scrap (before the beginning of the heat)

offgas temperature reduces, a lack of heat can be compensated by using of burners. All this allows increasing the medium mass temperature of heated scrap and reducing electrical energy consumption. Another advantage of shaft furnaces is that a substantial part of the dust carried out from the freeboard settles down in the layer of scrap filling the shaft. This increases the yield by approximately 1%.

Although there is no reliable data on average-mass temperatures of scrap heating for the finger shaft furnaces, this temperature can be estimated with the help of the heat balance data for one of such furnaces. In accordance with these data for the 135-t shaft EAF with the 120 MVA transformer and tap-to-tap time of approximately 38 min, the first basket of heated scrap introduces 55 kWh/t of steel [17]. The total amount of scrap per heat is 148 tons. Assuming that the first basket contains 89 t (60%) of scrap, we can find its enthalpy $E = 55 \times 135/89 = 83.4$ kWh/t of scrap. Corresponding to this enthalpy is the average-mass temperature of scrap of 550°C, Table 6.1. It can be assumed that only half of the scrap from the second basket, i.e. 30.0 t, is in the shaft and that this additional amount of scrap is heated by the off-gases to the temperature of 550°C as well. In this case, the average enthalpy of the entire amount of scrap heated by the off-gases will be $(89.0 + 30.0) \times 83.4/148 = 67.0$ kWh/t of scrap, and respective heating temperature of 450°C. Such relatively low temperature can be explained by the fact that only a part of the scrap equal to approximately 80% of its total amount is heated by the off-gases and that the duration of heating of the scrap from the first basket on the fingers is shorter than the power-on time τ_{on}.

Heating by the off-gases returns to the process 83.4(89 + 30)/135 = 73.5 kWh/t of steel. Assuming the electric energy efficiency coefficient $\eta_{EL} = 0.75$, Chap. 5, Sect. 5.4.1, we will find that reduction of the required useful heat consumption by 73.5 kWh/t can reduce electrical energy consumption by 73.5/0.75 = 98 kWh/t. The minimum electrical energy consumption on the shaft furnace under consideration [17] was 285 kWh/t. In comparison with conventional EAFs operating without scrap heating, the reduction of electrical energy consumption is about 90 kWh/t, which corresponds to the results of calculation.

Later design of a finger shaft furnace was persistently improved. The latest variation of the furnace, named Siemetal EAF Quantum, developed by Siemens VAI Metals Technologies, Germany, along with other innovations comprises a new system of shaft charging and improved design of fingers retaining scrap [18].

Scrap is charged into the shaft from above with a special chute moving up and down on inclined elevator rather than baskets with a crane. The chute is loaded on scrap yard and contains one-third of the scrap required for the heat. Such a system allows covering the shaft with a hood and considerable decreasing uncontrolled gas-dust emissions from the shaft during the charging of scrap.

As opposite to the fingers shown in Fig. 6.9 the fingers of the EAF Quantum are introduced into the shaft through its sidewalls. To charge a batch of the preheated scrap into the bath the fingers are pulled out of the sidewalls of the shaft. After charging they are immediately introduced into the shaft to receipt the next batch of scrap. Thus, all the scrap is heated on the fingers what allows increasing its average mass temperature. The prior system, Fig. 6.9, did not enable to close the fingers at once after falling of the first batch of scrap into the furnace freeboard. It was necessary to wait for caving-in of scrap. Therefore, only the first basket of scrap was heated on the fingers [17].

A new 100-t finger shaft furnace with a transformer of 80 MVA will start into operation in 2013. Designed furnace productivity is 182 t/h when charging 3 chutes, and when charging 4 chutes is 162 t/h. Expected electrical energy consumption is 280 kWh/t [18]. Since approximately even electrical energy consumption has been achieved at the G. Fuchs' shaft furnace in the late 1990s [17] it can be supposed that in the new furnace an average mass temperature of scrap preheating with off-gases will amount of about 500–540°C.

6.4.2 *Shaft Furnaces with Continuous Scrap Charging into the Liquid Bath by Pushers*

The heat efficiency coefficient of off-gases when scrap heating in a shaft is higher by approximately three times than that when scrap heating on a Consteel conveyor. On the other hand, continuous furnace operation with the flat bath is an advantage of the Consteel process. To combine these both advantages G. Fuchs has developed and implemented real-life shaft furnaces with the continuous charging of scrap into the liquid bath. The design of such a furnace is schematically shown in Fig. 6.10.

A rectangular shaft (1) is installed on the cart (2) next to the furnace. The shaft is connected to the furnace with a short tunnel (3). A sliding gate (4) opens for charging of scrap into the shaft. The charging is carried out with the power-on and does not interrupt the operation of the furnace. A gas duct (5) is placed under the sliding gate (4) in order to reduce gas-dust emissions into the shop during charging of the shaft. The shaft is lined with the massive steel segments and has no water-cooled elements which could be damaged during the scrap charging. The mass of scrap in the shaft is gauged by the measuring elements on which the shaft rests. This makes it possible to control the rate of charging of scrap into the furnace.

The scrap is charged continuously into the liquid bath with the help of the pusher (6) which is moved forth and back by the well protected hydraulic cylinder (7). During the entire period of charging and melting of scrap in the liquid metal its temperature is kept by the electric arcs at the constant level of 1560–1580°C, just like in the Consteel process. All the fundamental features of this method of scrap melting, its advantages and shortcomings are the same as those of the Consteel

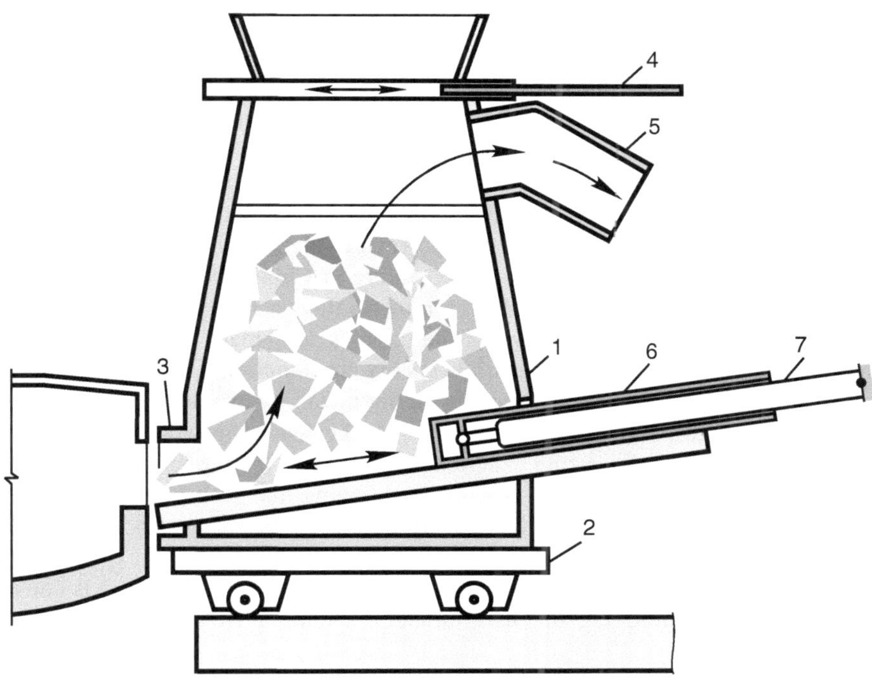

Fig. 6.10 Shaft furnace with a pusher. Continuous scrap charging into the bath (designations are given in the text)

furnaces, Chap. 1, Sect. 1.3.8. Unlike in the first finger shaft furnaces, in the furnaces with the pushers the off-gases heat all the scrap. This made it possible to expect the higher temperatures of heating. Unfortunately, the reliable data on the temperatures for the furnaces operating without hot metal are absent.

As reported by G. Fuchs,[3] the hourly productivity of the three 140–150 t furnaces put into operation recently was 125–149 t/h with the minimum electrical energy consumption of 300–34 kWh/t on the better days. With regard to productivity, these furnaces are considerably inferior to the modern EAFs which can be explained not only by the relatively low melting rate of scrap in liquid metal, Sect. 6.2.2, but by the low power of transformers (80 MVA) as well. The electrical energy consumption does not differ significantly from that of the best results achieved on the modern EAFs operating without scrap preheating during some of the short periods of their operation. Obviously, that the high-temperature scrap heating has not been achieved in these shaft furnaces either. This can be associated with both the relatively low heat capacity of the off-gases flow in furnaces operating without use of hot metal and insufficient duration of the heating.

In all of the shaft furnaces, the off-gases leave the shaft with a temperature of 300–400°C. These gases are saturated with dioxins. Subsequent heating of gases up to approximately 1000°C by the burners in the special chambers is required for decomposing of the dioxins. The data on additional energy consumption due to the necessity of such heating are lacking in the publications on the shaft furnaces. This consumption, however, cannot be lower than that of the Danarc Plus furnace, Sect. 6.3.2, which considerably reduces energy advantages of the shaft furnaces. It is worth mentioning that the shaft furnaces gained some spread mainly in such countries as China, Indonesia, and others where the strict government environmental regulations do not exist as opposed to North America and Western Europe.

When operating with hot metal, the heat capacity of the off-gas flow increases sharply what makes it possible to achieve the high-temperature heating of scrap in the shaft furnaces. As an example, let us review the performances of the shaft furnace of 150-t capacity with pusher operating in China [19]. For this furnace, share of hot metal in the charge is 40%, tap-to-tap time is 35 min, oxygen flow rate is 40 m^3/t, and electrical energy consumption is less than 100 kWh/t.[4] Scrap melting is carried out with low electrical energy consumption, mainly by means of physical and chemical heat introduced with hot metal, by exothermic reactions of oxidation of C, Si, Mn, and P, and by highly heated scrap which temperature reaches 800–1000°C. Operating experience proves that the pusher works reliably at such temperatures of scrap. It is worth mentioning that such performances can also be achieved on the EAF without scrap heating if the share of hot metal in a charge is significantly increased.

Elimination of gas-dust emissions when charging scrap into the shaft is one of topical tasks of improving of shaft furnaces operation with continuous scrap

[3] Presentation: Ekaterinburg, Russia, 2012.

[4] This figure seems somewhat understated.

melting in the liquid bath. In order to solve this problem an original scrap charging system named EPC [20] has been developed by the Company KR Tec GmbH, Germany. Operating principle of this system is explained by Fig. 6.11.

The main component of the system is a movable hopper (1) with an opening bottom placed in a bunker (2) adjacent to the shaft (3). By means of hydraulic cylinders of the bunker the hopper can be moved into the shaft through an aperture in its sidewall. When the hopper is placed in the bunker its front wall is closing the aperture. When the hopper is positioned in the shaft the aperture is closed with the hopper back wall. At this position the bottom of the hoper is opened and the scrap falls smoothly into the shaft.

In the course of feeding of the scrap into the bath by means of a pusher (4) the scrap caves in inside the shaft and the hopper can be moved backward to the

Fig. 6.11 EPC system of scrap charging into the shaft (designations are given in the text)

bunker. Then the hopper is charged again with the help of a scrap basket (5). A slide gate (6) is opened for a while to charge the hopper only. Thus, during the heat the scrap can be charged into the shaft by separate batches without loss of airtight of the system. Therefore, gas-dust emissions from the shaft into the shop atmosphere are almost completely eliminated.

Melting of the scrap starts when the scrap batch preheated in the shaft during the prior heat is feeding into the hot heel. The system allows realizing different variations of furnaces operation. During the heat one or several of scrap batches can be charged into the shaft depend on a furnace capacity, shaft and hopper dimensions. This requires the certain number of hopper movements and charging of the hopper with baskets. Figure 6.11 shows a stage of the heat preceding the next charging of the scrap from the hopper into the shaft. Off-gases passing through the scrap layer heat it and are evacuated via off-gas duct (7). The new systems of scrap charging into the shaft of both EPC and Quantum (Sect. 6.4.1) make shaft furnaces close to furnaces with continuous scrap charging.

6.5 From Utilizing Off-Gases to Scrap Preheating by Burners Only

The idea as such of utilizing the heat of the off-gases for the purpose of scrap heating in the arc furnaces seems quite attractive. However, it cannot be considered in isolation from the actual possibilities of its realization. Persistent attempts to implement this idea in practice are being carried on for about 30 years. Nevertheless, they did not yield the results which, in comparison with EAF, could justify the usage of sophisticated equipment requiring additional maintenance. In comparison with state-of-the-art EAFs conveyor and shaft furnaces with scrap preheating by off-gases have a lower productivity. Therefore, despite a number of substantial advantages related to continuous charging and melting of scrap in the liquid bath these furnaces gained only limited spread.

In order to achieve the level of the modern EAFs productivity or even exceed it is necessary to increase the scrap preheating temperature up to at least 800°C. However, such a high-temperature scrap preheating without use of a large amount of hot metal could be implemented neither in the conveyor furnaces, nor in the shaft furnaces. There is a main hindering factor: the insufficient heat capacity of the off-gases flow.

Utilizing of off-gases in combination with powerful burners for scrap preheating could eliminate this shortcoming. However, in this case, it is necessary to heat off-gases up to approximately 1000°C for decomposing of dioxins which are contained in off-gases after their passing through the scrap. Such a process requires additional high energy consumptions; as a consequence the scrap heating with off-gases in combination with the burners becomes inexpedient in terms of energy. The environment regulations will be stricter all over. Therefore, the utilization of off-gases polluted by dioxins without their decomposing in order to heat the scrap does not have the future even in those countries where it is possible today.

The promising trend resolving the problem in complex is scrap preheating by the burners only. The mentioned trend allows avoiding additional energy consumptions for decomposing of the dioxins, eliminating of limitations related to insufficient heating capacity of off-gas flow and opens new possibilities for developing highly productive and energy efficient steelmaking furnaces operating with scrap. These possibilities are considered in the next Chap. 7.

The heating of scrap with the off-gases should thoroughly be abandoned. The most reasonable utilization of these gases is for producing of steam with energy parameters for the purpose of electrical power production. Lately, such utilization of the off-gases has gained increasing attention.

References

1. Neumann F, Leu H, Brusa U (1975) BBC-Brusa is a steelmaking process. Stahl und Eisen 95(1):16–23
2. Miranda U, Lombardi E, Bosi P (1999) Saving energy and protecting the environment the first consteel plant in Europe. 6th European Electric Steelmaking Conference, Düsseldorf
3. Giavani C, Guzzon M, Picciolo F (2010) Start-up and results of the EAF consteel plant of Celsa Nordic-Mo i Rana (Norway). AISTech Conference, Pittsburgh
4. Bouganosopoulos G, Papantoniou V, Sismanis P (2008) Start-up experience and results of consteel at the SOVEL meltshop. AISTech Conference, Pittsburgh
5. Andrews S, Powers J, Fox M et al (2003) Start-up and operating results of the cojet® sidewall injectors on a consteel furnace. AISE Annual Convection, Pittsburgh, pp 1–14
6. Memoli F, Guzzon M, Giavani C (2011) The Evolution of preheating and the importance of hot heel in supersized consteel in Europe. AISTech Conference, Indianapolis
7. Ferri MB, Giavani C (2009) World's largest consteel® plant feeds the new thin slab mill at Arvedi. MPT Int, Italy 6:26–28
8. Cottardi R, Miani S, Partyka A et al (2008) Faster, more efficient EAF. AISTech Conference, Pittsburgh
9. Patrizio D, Pesamosca A (2012) Successful start-up and benchmark operating results of 100-t FastArc EAF at Kosco, Korea. AISTech Conference, Atlanta
10. Salomone Ph, Legrand M, Fabbro M (2004) Increasing productivity at the Duferco La Louviere meltshop. MPT Int 2:52–56
11. Alzetta F, Poloni A, Ruscio E (2006) Revolutionary new high-tech electric arc furnace. MPT Int 5:48–55
12. Fabbro M, Chiogna A, Cappellari G (2009) Steelmaking and casting at the new minimill of Siderurgica Balboa. MPT Int 1:30–36
13. Sellan R, Fabbro M, Burin P (2008) The 300 t EAF meltshop at the new Iskenderun minimill complex. MPT Int 2:52–58
14. Abel M, Hein M (2008) The breakthrough for 320 t tapping weight. MPT Int 4:44–48
15. Arvedi G, Manini L, Bianchi A et al (2008) Acciaieria Arvedi: A new giant consteel in europe. AISTech Conference, Pittsburgh
16. Sanz A, Lavaroni G (2000) Advanced approaches to electric arc furnace offgas management. MPT Int 4:72–82
17. Manfred X, Fuchs G, Auer W (1999) Electric arc furnace technology beyond the year 2000. MPT Int 1:56–63

18. Abel M, Dorndorf M, Hein M et al (2011) Highly productive electric steelmaking at extra low conversion costs. MPT Int 3:92–96
19. Fuchs G, Rummler K, Hassig M (2008) New energy saving electric arc furnace designs. AISTech Conference, Pittsburgh
20. Rummler K, Tunaboylu A, Ertas D (2011) Scrap preheating and continuous charging system for EAF meltshop. MPT Int 5:32–36

Chapter 7
Replacement of Electric Arcs with High Power Oxy-Gas Burners

7.1 Attempts for Complete Replacement

EAF is a high-power energy unit which includes complex electrical and mechanical equipment as well as expensive electrodes. In comparison with EAF, the oxy-fuel burners with the same power are simple and inexpensive devices. In addition, for a long time, the cost per kWh of natural gas was significantly lower than that per kWh of electrical energy. It is not surprising that, under these conditions and due to insufficient understanding of differences in principles of heat transfer from the electric arcs and that of the flames of burners to the solid metal-charge and to the liquid bath, the idea of not only high-temperature preheating of scrap by burners, but also of avoiding completely the use of electrical energy in the process of steelmaking from scrap seemed quite attractive and reasonable. It has been presumed that the flames of oxy-fuel burners using natural gas or oil with temperatures reaching 2800°C could replace for electric arcs in these processes.

One of the first attempts of this kind was undertaken in nineteen sixties when BISRA Company in England developed and tried out so-called FOS (Fuel Oxygen Scrap) process [1]. The process was implemented in the EAFs of small capacity. In these furnaces, the vertical oxy-gas or oxy-oil burners were inserted in the roof ports instead of electrodes. As the scrap melted and a flat bath formed, the burners, just like the electrodes, were lowered into the freeboard at the optimum depth.

Prolonged industrial testing of FOS process have shown that during the scrap melting, the burner flames have very high oxidizing capacity, being the highest when natural gas was used and slightly lower when oil was used. Despite the fact that in the course of the heat oxygen consumption was gradually reduced to 60–75% of amount required for complete fuel combustion, scrap was oxidized so intensely that the *FeO* content in the slag exceeded 50% and the yield dropped below 90%. In order to assure the required carbon content in the bath and reduce the yield drop, the charge contained over 15% of pig iron in addition to 2% of coke. Due to a high degree of underburning of fuel in the freeboard of the furnace, the natural gas consumption was approximately 150 m^3/t and the oxygen consumption was 200 m^3/t.

DOI: 10.1007/978-3-642-36273-6_7,

At the final stage of the heat, the FOS process was very difficult to control because of the discrepancy between the rate of carbon oxidation and the rate of liquid bath heating. The decarburization of metal was taking place quite unevenly. Often, for short periods of time decarburization rate jumped up by a number of times with a relatively stable and limited rate of heating metal with the burners. In order to make the process more controllable, it was necessary to raise input of chemical heat into the bath through the increase of pig iron share in the metal-charge. But even when the share of iron was significant, the process was suitable only for obtaining of semi-finished product having a composition quite different from that of the finished steel. The problem was insufficient effectiveness of heating the liquid metal with oxy-fuel burners. Obtained energy and technological characteristics of FOS-process made it absolutely unacceptable for practical use.

Until quite recently, further attempts to develop an effective process of steel-making from scrap without use of electrical energy were made again and again. One of these attempts is so-called NSR-process of oxygen melting of scrap developed in Japan [2]. In this process only oxy-fuel burners and coke are used as the sources of energy. Steelmaking unit designed for NSR-process includes a shaft installed over the bath divided into heating and refining zones.

Scrap, coke and fluxes are continuously charged into the shaft from above. Through the same shaft the off-gases are removed from the unit. The tuyeres for post-combustion of CO are installed in the upper section of the shaft, and oxy-oil burners are installed in its lower section. The scrap is heated in the upper section of the shaft and is melted in the lower section. The melted and carburized metal flows down into the bath together with the slag. In the bath the melt is heated with additional oxy-fuel burners. In order to increase yield, carbon powder is injected into the bath. The semi-finished product flows down to the refining zone where it is blown with oxygen and reaches the required final temperature and carbon content. In the refining zone, as well as in the heating zone, the metal is stirred by inert gas blown into from below.

The pilot NSR unit with productivity of 6 t/h operated with the following energy-carriers consumption: oil—50 L/t, coke—40 kg/t, and oxygen—100 m^3/t. Due to utilization of heat of gases in the shaft, these consumptions are significantly lower than those in FOS-process. There are no data on further development of NSR-process and its industrial try-out.

As other methods of steelmaking from scrap without the use of electrical energy, the processes discussed above have a common fundamental feature. At the final stage of the heat, the heating of the liquid metal to the tapping temperature is achieved, as in converters, by means of chemical heat generated in the bath by the oxygen blowing. The heat is generated by the oxidation of carbon and by the partial oxidation of iron. There is no alternative to this method other than the use of electric arcs since the intensity of heat transfer from the flames of oxy-fuel burners to the surface of the liquid bath is too low. The immersion burning of natural gas or liquid fuel, as mentioned above, can be effective only in non-ferriferrous slag but not in the steel bath.

In the converter-type processes, the rates of decarburization and heating of the bath are strictly interrelated. In order to simultaneously achieve the final values of the bath temperature and carbon content, stabilization of all the process conditions, first of all chemical and granulometric composition of charge-metal, is required.

Electric arc furnaces operate with the cheapest light scrap of widely-varying composition. This is quite justified. However, as this kind of metal-charge is used, the level of stabilization of heat conditions typical for converters is unachievable. Therefore, complete avoidance of use of electrical energy including the final stage of the heat makes process control extremely difficult and leads to reduced productivity. In the electric arc furnaces, such avoidance can not be justified by any means since only no more than 5–6% of the total electrical energy consumption is utilized for the heating of liquid metal. Nevertheless, substantial though partial replacement of electrical energy with fuel supplied by oxy-gas burners seems to be scientifically-based and promising innovation direction.

7.2 Potentialities of Existing Burners: Heat Transfer, Limiting Factors

Low-power oxy-gas burners are widespread in EAFs. Further they are called as conventional burners. Unit power of such burners does not exceed 3.5–4.5 MW. They are installed in the wall panels, usually about 500 mm above the bath level, as well as in the oriel covers and in the slag doors. In the past, three sidewall burners used to be installed in the furnace in the so-called cold zones between the electrodes where the scrap melting required longer time. The sidewall burners equalized the temperature field along the whole perimeter of the furnace. The oriel burners eliminate the cold zone at the oriel, and the door burners do the same at the slag door sill area. The latter makes possible an earlier metal sampling and temperature taking, which allows shortening a heat. As burners had low unit power their use did not significantly affect electrical energy consumption.

Further practice has lead to understanding the necessity of increasing the fuel consumption in the burners not so much for the purpose of saving electrical energy as for intensification of the process. With tap-to-tap time being continuously reduced, this required a significant increase in the power of the burners. However, all attempts made in this direction have not given positive results. At present, unit power of burners, due to the reasons discussed in detail below, remains at the same level as 20–30 years ago. Therefore, in order to increase overall power of the burners, the number of burners has been increased. The number of burners in the furnaces reached six to nine, and even 12, as in the Danarc Plus furnace, Chap. 6, Sect. 6.2.2.

Since the early nineties installing the burners inside water-cooled copper boxes, which protected the burners against both high temperatures influence and damages to those when charging scrap, has been started. Thereat the burners have been combined with nozzles for the bath blowing with oxygen. Operation of such combined burners is in detail reviewed in Chap. 11, Sect. 11.2.

Despite the increase in the number of burners, specific consumption of natural gas in the furnaces did not grow significantly. Usually, it does not exceed 6–7 m^3/t. This is a result of the further reduction of the tap-to-tap time and, correspondingly, burners' operation time. The effectiveness of the burners did not change as well. As before, they ensure reduction of tap-to-tap time and electrical energy consumption by not more than 6–8%.

In the vast majority of cases, natural gas is used in the burners of EAF. All conventional burners are similar in general principles, regardless of a furnace size and their location. The design of these burners provides for intense mixing of gas and oxygen firstly in the combustion chamber of the burner and then does that close to its orifice. When used for scrap heating, the burners operate with oxygen flow rate coefficient of approximately 1.05. Usually, they form a narrow high-temperature flame. Initial flame speeds are close to the sonic speed or exceed it; maximum flame temperatures reach 2700–2800°C. When the burners are used for scrap cutting or for post-combustion of CO, the oxygen excess is increased to 2–3.

Heating of liquid bath with burners is ineffective. However, small amounts of both gas and oxygen have to be supplied to the burners to maintain the so-called pilot flame. This allows to avoid clogging of the burners with splashed metal and slag. These forced non-productive consumptions of energy-carriers noticeably worsen burners' performance indices including energy efficiency coefficient of gas η_{NG}. A pilot flame only diminishes but does not completely eliminate clogging of the burners. Periodically, they have to be purged, which poses certain difficulties considering the number of burners.

Let us review the causes hindering the increase in power of burners and their application for high-temperature scrap heating. During the operation of these burners, the direction of flame remains constant. Burner flames attack the scrap pile from the side, in the direction close to radial. The kinetic energy of the flames of conventional burners is low due to their low power. Penetrating into a layer of scrap these flames quickly lose their speed and are damped out. Therefore, their action zones are quite limited.

Since emissivity of oxy-fuel flame in the gaps between the scrap lumps is low, heat from flames to scrap is transferred almost completely by convection. With convection heat transfer, the amount of heat transferred to scrap per unit time is determined by the following aerodynamic and thermal factors: the surface area of the scrap lumps surrounded by gas flow; the speed of gas flow which determines the heat-transfer coefficient α; and the average temperature difference between gases and heat-absorbing surface of the scrap, Chap. 3, Sect. 3.3. In the action zone of the burners, at high temperatures of oxy-gas flame the light scrap is heated very quickly to the temperatures close to its melting point. Then the scrap settles down and leaves the action zone of the flame which loses the convective contact with the scrap. In the course of the burners operation, the area of the heat-absorbing surface of the scrap lumps and the temperature difference between the scrap and the flame diminish progressively. The heat transfer remains high only during a short period after the start of the burners operation. Then the heat transfer reduces gradually and finally, drops so low that the burners must be turned off, as their operation becomes ineffective.

Besides, potential duration of conventional burners operation is also limited by the physical-chemical factors. At the scrap temperatures approaching 1400°C and especially during the surface melting of scrap, the rate of oxidation of iron by the products of complete combustion of fuel sharply rises. In doing so, the products of fuel combustion are reduced to CO and H_2 according to the following reactions:

$$CO_2 + Fe = FeO + CO \qquad \text{and} \qquad H_2O + Fe = FeO + H_2$$

The fuel underburning increases, and CO and H_2 burn down in the gas evacuation system. The temperature of the off-gases rises sharply which, along with the other signs of reduced effectiveness of the burners operation, requires turning the burners off.

The described above processes in the scrap pile attacked by a narrow high-temperature flame explain comprehensively the futility of attempts to increase the power of conventional burners. Indeed, in accordance with well-known aerodynamic principles, the length and the volume of the flame and, therefore, its action zone increases insignificantly as the burner power increases. As a result, the critical temperatures causing fuel underburning and settlement of the scrap in this zone are reached in a shorter time. Respectively, approximately proportionally to the increase in power of the burner, the potential effective burner operation time is shortened, whereas the amount of heat transferred to the scrap increases insignificantly. Only a relatively small portion of scrap pile is heated, which has little effect on energy characteristics of the furnace.

Attempts to expand an action zone of the burners due to changing the shape of the flame during the heat are known. Both the narrow and flat or fan-shaped flames can be used in such burners depending on changeable conditions of the heating of scrap. The action zone of the fan-shaped flames as well as that of flat flames is expanded but their kinetic energy drops. An attempt to increase the power of such burners has been failed and they did not become spread.

How to increase power of the burners? How to spread their action over the entire scrap pile in the furnace and assure relatively uniform high-temperature heating of scrap without significant oxidation of iron? It seems that there is only one way to actually solve this complex technical problem. It is necessary to replace the burners with fixed direction of flame with the rotary burners capable of changing direction of the flame during the operation and over a wide range.

7.3 High-Power Rotary Burners (HPR-Burners)

7.3.1 Fundamental Features

The main fundamental feature of the HPR-burners is that in the course of the heat the flames of HPR- burners can be moved from those already heated to the relatively cold zones of the scrap. This new feature enlarges considerably the action zone of the flames and allows to increase by several times the effective power of the burners. The flames of HPR-burners, in comparison with the conventional burners, are also characterized by lower temperatures and by several times higher kinetic energy. The latter allows the flames to penetrate through the scrap pile down to the bottom. In this case, the heating gases pass the maximum distance in the layer of scrap, which considerably increases the heat transfer and the natural gas efficiency coefficient η_{NG}.

The number, the location and the power of HPR-burners must ensure the sufficiently rapid and relatively uniform heating of the entire scrap pile. Reduction in the temperature of the flames is necessary for preventing the intensive oxidation of iron. For this purpose, the final temperature of the scrap heating must be limited as well.

7.3.2 Slag Door Burners: Effectiveness of Flame-Direction Changes

The past long and extensive experience of the use of conventional slag door burners is of great interest for the application of the HPR-burners. Specifically, it was the door burners to clearly show for the first time the need for changing the direction of the flame in the course of scrap heating.

In the eighties, at many plants in the former Soviet Union, a large number of 100-ton EAFs and the furnaces with smaller capacity were equipped with non-water-cooled slag door burners. These burners were mounted on the two-arm brackets which allowed changing the flame-direction in horizontal and vertical planes within sufficiently wide limits. In the course of heating of each basket of scrap, a steelmaker-assistant repeatedly turned a burner by hand directing the flame towards those sections where the settling down and melting of scrap were taking too long. Only apparent and significant shortening of the heat resulting from such burner manipulations could force the steelmakers to stay for a long time at the half-open door enduring the highest levels of noise and radiation. Operation with the changing flame direction allowed increasing the power of the burners up to 7–8 MW, i.e. by 1.5–2 times in comparison with the conventional burners. This range of power was not justified if the burners were not being rotated. Later, the rotary slag door burners have been mechanized in some furnaces.

The experience of developing and application of a mechanized slag door burner in the 60-ton EAF at the mini-mill in Akko (Israel) [3] is of special interest for the developments in the field of HPR-burners. Maximum actual power of the furnace is 32 MW. The charging of 65 t of scrap is carried out by the three baskets. The furnace is equipped with a BSE manipulator with consumable pipes for blowing oxygen and injecting carbon into the bath through the slag door, Fig. 7.1. Oxygen is blown through the two pipes located in the lateral boxes (1); carbon is injected through the pipe in the middle. The oxygen burner operating on liquid fuel is installed under the middle box (2) of the manipulator. For heating the scrap, the burner is pushed forward by the pneumatic cylinder. When the operation is finished, the burner is brought back to the "off" position. In this position, the burner head (3) is closed with the guard (4) which protects the nozzles of the burner from splashes of metal and slag during the liquid bath.

The mechanisms of the manipulator allow an operator to change the position of the lateral boxes by inclining them and turning to the left and to the right. This way, the optimum directions of the oxygen jets to the scrap or into the melt are being chosen, Chap. 11, Sect. 11.1.2. The burner can be inclined together with the middle box and moved in the lateral directions together with the manipulator. In order to increase the possible rotation angles of the flame in the horizontal plane limited by the water-cooled arch of the slag door of the furnace, the burner head (3) was made rotary and supplied with special rotating mechanism. This considerably enlarged the zone of flame action in the lateral directions and increased the effectiveness of the burner.

The possibility to change the flame direction within wide limits allowed increasing the power of the burner up to 12–15 MW, which is approximately 50%

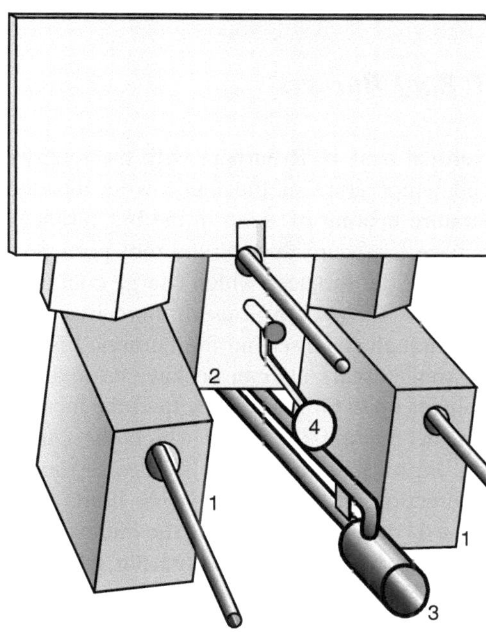

Fig. 7.1 Slag door HPR-burner on BSE manipulator (designations are given in the text)

of the actual power of the electric arcs during the stage of scrap heating and melting. Thus, taking into account its power, this burner can be considered as an HPR-burner. Another important feature of this burner is a relatively low temperature of the flame resulting from the dilution of oxygen with air. This burner is supplied with approximately half of the amount of oxygen required for complete combustion of the fuel. The rest of the oxygen comes with the compressed air used for the atomizing and with the air infiltrated from the atmosphere into the flame on its way from the burner to the freeboard of the furnace. Reduction in the flame temperature together with periodical change of its direction contributed to a significant increase in the reasonable duration of the burner operation and in the amount of heat transferred to the scrap.

The burner operation gets started immediately after the charging of another basket of scrap. At the same time, blowing of the furnace with the oxygen pipes begins. By directing the flame and the oxygen jets into the zones of insufficiently heated scrap, an operator accelerates the settling down and melting of the scrap. On the other hand, the oxygen jets clear the passages for deep penetration of the flame into the scrap pile. As a result, the efficiency of both fuel and oxygen use increases.

Depending on the fractional composition of the scrap charged, the duration of the burner operation was 5–7 min after charging the first and the second baskets, and 2–3 min after charging the third one. A continuous use of a burner with consumption of liquid fuel of 4.1 L/t reduced the tap-to-tap time by 6.4% and the electrical energy consumption by 5.2% [3]. Approximately the same effect is produced by installation of three conventional sidewall burners.

7.3.3 Roof Burners

The vertical roof HPR-burners with the retarded mixing of gas and oxygen and reduced temperature of the flame were developed by the authors for the high-temperature heating of scrap in the big furnaces. These burners are inserted into the furnace freeboard through the roof ports near the roof ring. The burners were designed for the furnaces which charge contained a large amount of heavy rolling scrap. When such scrap is used, there is a free space left near the walls of the furnace enough for inserting the burners, Fig. 7.2.

The roof burners (1) can be lowered and raised as well as turned around the vertical axis up to 60° with the help of the mechanisms mounted on the carriage (2) and on the column (3) along which the carriage moves. The gas and oxygen nozzles are located on the side surface of the burner at an angle to the bath. The flame direction changes within wide limits when the burner is moved along the vertical axis or turned. Changing the flame direction of the burners in the course of the melting period can be carried out either with the help of automatic device according to the specified program, or manually. In the latter case, an operator directs flames into those zones of the freeboard where the scrap settlement is going

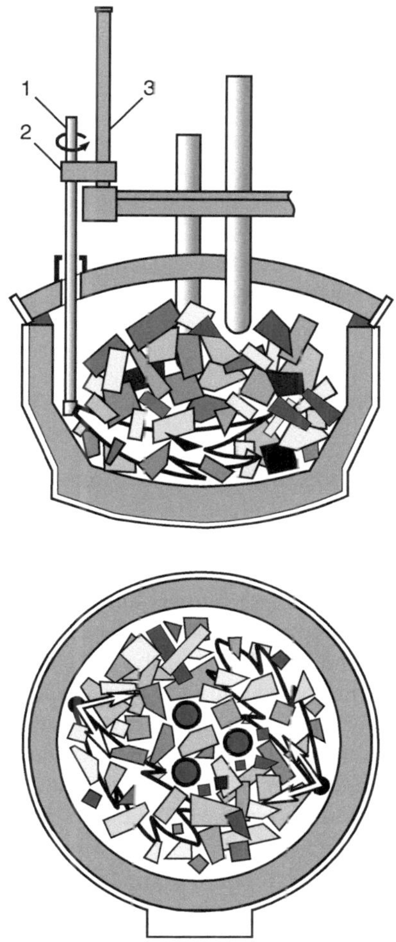

Fig. 7.2 Roof HPR-burners (designations are given in the text)

slowly. This allows to consider the specifics of the heats and to accelerate the process essentially. Prolonged industrial trials showed that two roof HPR-burners, when being properly controlled, ensure quite uniform heating of the entire scrap pile charged into the furnace. This is promoted by high kinetic energy of the flames which allows them to penetrate deep into the layer of scrap practically reaching the bottom, as well as by the design of the burners providing the retarded mixing of gas and oxygen and reduced combustion temperatures. The burners are lowered into the freeboard only for the duration of their operation. This excludes a necessity to consume gas and oxygen for keeping up a pilot flame.

The tests of the HPR-burners were carried out in the old 100-ton furnaces at Chelyabinsk's Metallurgical Complex (CMC), Russia, with the power of 32 MVA and in the 200-ton furnaces at the plant "Red October" (city of Volgograd) with the power of 60 MVA [4, 5]. The furnaces possessed by lined walls and roofs.

The combined maximum power of two burners in the 100-ton furnaces was 25 MW. Two roof burners, 15 MW each, and a slag door burner with the power of 5 MW were installed in the 200-ton furnaces. At the Orsko-Khalilovo Metallurgical Combine (OKMC, city of Novotroitsk), the 100-ton EAFs with the power of 80 MVA with the water-cooled wall and roof panels were equipped with two roof HPR-burners with the power of 20 MW each. However, because of the limited resources of oxygen at the Combine and insufficient throughput of the furnace gas evacuation system, only the restricted in scope tests of these most powerful burners could be conducted. The results of the tests of the roof HPR burners are reviewed below in Sect. 7.4.3.

7.3.4 Oriel Burners

The roof HPR-burners cannot be used under the normal conditions of EAF operation with utilization of the cheap light scrap. This scrap fills the entire freeboard of the furnace with no space left near the walls for lowering the burners. For the heating of light scrap, the oriels HPR-burners have been developed by the authors. These burners are designed for the installation in the covers of the main and supplementary oriels outside of the furnace.

The installation of the HPR-burner in the main oriel is shown in Fig. 7.3a. The water-cooled burner (1) is fixed to the bracket (2). It can be turned in both directions from the mid position around its axis with the help of the hydraulic cylinder (3). The gas and oxygen nozzles of the burner are located on its lateral surface near the lower end, at an angle to the bath. The vertical displacements of the burner are carried out with the help of the hydro-plunger (4) serving as a stand for the bracket (2) fixed to it in a manner allowing the rotation in a horizontal plane. The rotation of the bracket (2) with the burner is carried out with the help of the hydraulic cylinder (5). The cooling and shielding from the radiation are provided for all the mechanisms including the gas and oxygen lines of the burner.

The burner operates as follows. After charging of the first basket of scrap, the burner with the help of the mechanisms of horizontal and vertical displacement is inserted into the technological opening in the cover of the oriel intended for maintenance of the tap hole of the furnace. Then the burner is brought down into the chamber of the oriel to the lowest position, and the flame is ignited. Due to the fact that only insignificant amount of the scrap gets into the oriel chamber, it becomes possible to lower the burner almost to the bottom of the furnace or to the level of the hot heel left in the furnace bottom from the previous heat. Thereat, the gases pass the maximum distance through the scrap from the bottom to the top.

In the course of scrap heating, the burner is periodically turned around its axis from one end position to another and back with the help of the cylinder (3). By combining the heating of the layer of scrap from below and the periodic change of the direction of flame in the horizontal plane, the zone of flame action is being enlarged considerably, and the local overheating in the scrap are being eliminated.

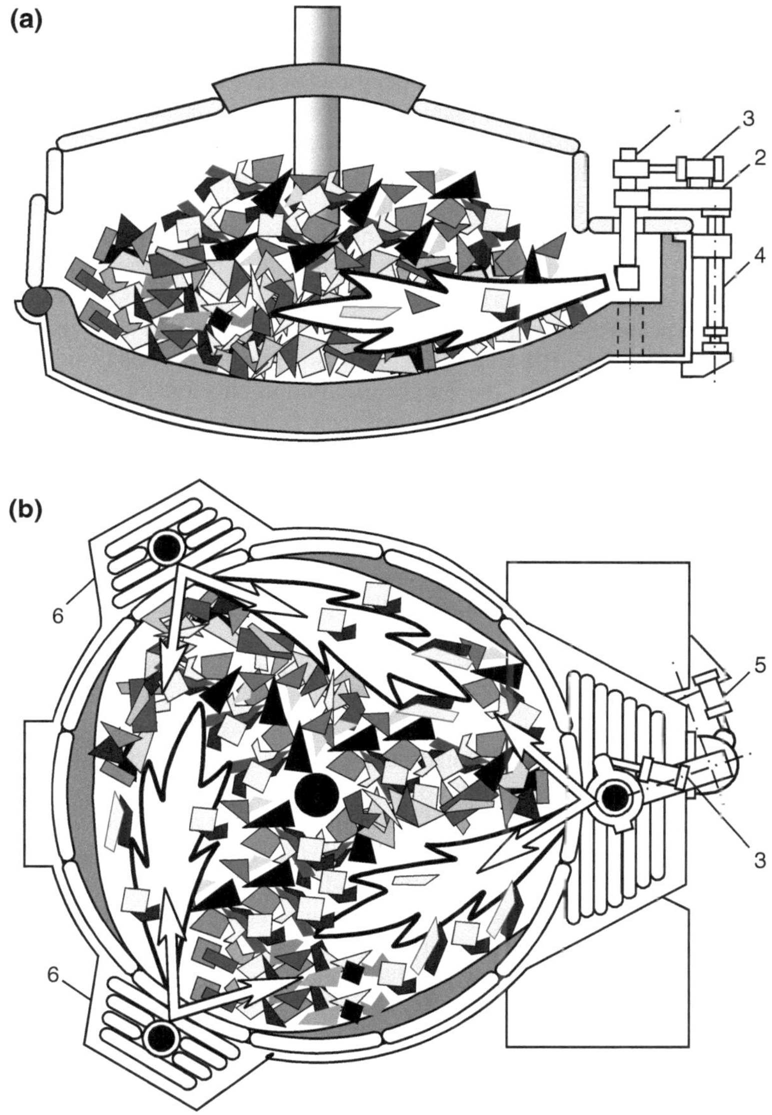

Fig. 7.3 HPR-burners in main and extra oriels of EAF (designations are given in the text). Patent of Russian Federation, No. 1838736 A3

As a result, the optimum heat-transfer conditions are being ensured as well as, consequently, the possibility to increase sharply the power of the burner, the gas efficiency coefficient η_{NG} and the medium mass temperature of scrap preheating.

As the liquid metal accumulates on the bottom, the burner is gradually pulled up so that it does not immerse into the slag. At the end of the heating of the first portion of scrap, the burner is switched off and raised above the oriel cover. This way the nozzles of the burner are protected from clogging with splashed metal and slag with no use of a pilot flame. The operations of heating of each new portion of scrap charged are repeated in the same order. At the end of the heating of the last portion, the burner is raised to the upper position and is swung together with the bracket (2) in the horizontal plane to the off-position on the side of the oriel with the aid of hydro cylinder (5). The technological opening in the cover of the oriel is freed for conducting the maintenance of the tap hole of the furnace.

The usage of just one HPR-burner located in the main oriel is enough for heating of the entire scrap charge in small furnaces only. The additional oriels in the form of special niches (6) in the side walls of the EAF are required for installation of several HPR-burners, Fig. 7.3b. Such niches opened from the side of the bath make free space for the burner installation and for the changing of the flame direction. The number of niches and their positioning in the furnace can be different. The version with one of side niches intended for the installation of a HPR-burner is shown in Fig. 7.4. The lining of walls and bottom of the niche is the extension of the lining of the banks and the bottom of the furnace. From the top, the niche is closed with the water-cooled cover (1) with the opening for inserting the burner. The burner is turned by using cylinder (3) installed on the bracket (4). In order to preclude the gases emission, the opening is equipped with the jet-type gas-dynamic seal (5).

The application and the function of the mechanisms of each burner (2) are completely analogous to those of the oriel burner shown in Fig. 7.3. The main difference is that the shaft-type mechanisms with supporting rollers (6) installed on the operating platform of the furnace are used for raising and lowering of the burners. The burner displacement to off-position is carried out with the help of the mechanism (7) turning the bracket (4) with the burner in a horizontal plane. This brings the burner outside the boundaries of the furnace. This layout of the mechanisms of the burner facilitates the conditions of their operation and maintenance considerably in comparison with the installation shown in Fig. 7.3. Setting-up of the additional oriels requires a change in the furnace design, which is not always acceptable. The sidewall HPR-burners do not have this shortcoming.

7.3.5 Sidewall Burners

The general arrangement of the sidewall HPR-burner is schematically shown in Fig. 7.5. The burner (1) is located in the water-cooled box (2) installed in the aperture of the sidewall panel (3). The same type of the boxes is used for the installation of the combined sidewall burners, Sect. 7.2. The burner in the box can be moved vertically and turned around its axis. The burner is moved into the lower operating position through the opening in the base of the box. In the upper "off" position, the end of the burner shuts this opening. Being placed in the box, the burner does not require a pilot flame for the protection of its nozzles. The box

Fig. 7.4 Installation of HPR-burner in an extra oriel (designations are given in the text)

works as a reflector beating back the scrap lumps during the charge, leaving free space under the box big enough for lowering the burner, for the formation of the flame, and for the changing the flame direction within the required limits. The proximity of the flame to the box is not dangerous. As already mentioned, the characteristic feature of the flames of the HPR-burners is their reduced temperature. As a result, they do not damage the water-cooled elements.

The hydraulic cylinder (4) carries out the vertical displacement of the burner by generating the force sufficient to exclude the burner jamming the opening of the box (2) due to the skull adhered to the burner. When the burner is raised, the skull

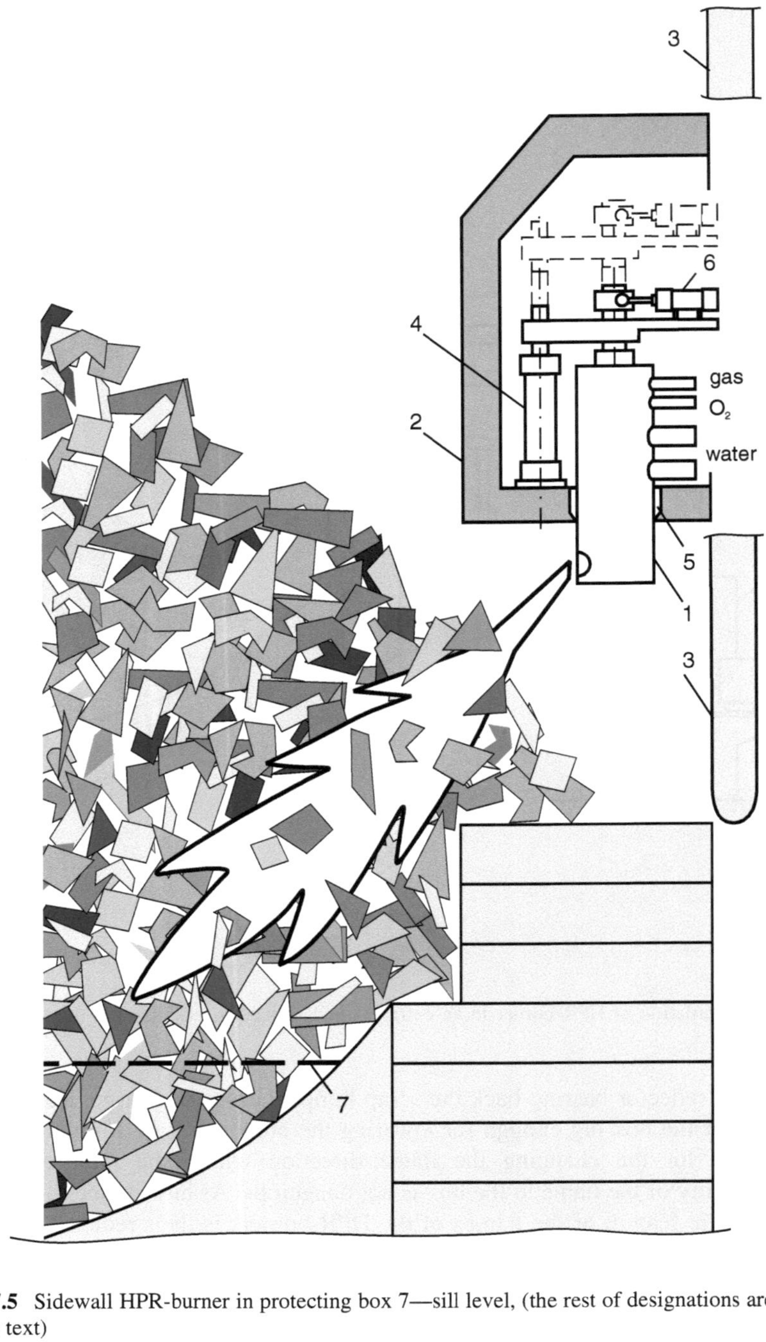

Fig. 7.5 Sidewall HPR-burner in protecting box 7—sill level, (the rest of designations are given in the text)

is purged off from its side surface by the special insert (5). The burner is turned around its vertical axis with the help of the hydraulic cylinder (6).

The sidewall HPR-burners have the essential advantage. They give options in selecting the number of the burners and their placement location along the perimeter of the furnace freeboard depending on dimensions of the furnace, as well as the power of the transformer and the technological parameters of the heat. In addition, there is no need for substantial changes in the furnace design. The power of the sidewall burners, as well as that of the oriel burners, can vary over wide limits up to 30 MW for each burner. The mechanisms other than those already described can also be used in the installations of the oriel and sidewall HPR-burners.

7.4 Two-Stage Process of the Heat with Use of HPR Burners: Industrial Trials

7.4.1 General Energy Ratios

The high power of HPR- burners opens up fundamentally new possibilities for replacing electrical energy with natural gas. The amount of useful heat transferred to the scrap by the burners per unit time, i.e. the useful power of the burners P^*_{NG}, is determined by the expression: $P^*_{NG} = \sum P_{NG} \cdot \eta_{NG}$., where $\sum P_{NG}$ is the total power of the burners. The analogous expression for the useful electrical power is: $P^*_{EL} = P_{EL} \cdot \eta_{EL}$, Chap. 5, Sect. 5.3, formulas (5.12) and (5.13). It is obvious that for the duration of time when the equality $P^*_{NG} = P^*_{EL}$ is true, a complete replacement of electric arcs with burners can be implemented.

Let us assume that at some period of the heat the efficiency coefficients of benefit use of electrical energy η_{EL} and energy of natural gas in the burners η_{NG} are equal. Then, from the expressions shown above, it follows that for the duration of this period the furnace can operate, without the loss in productivity, with the burners only without the electric arcs, if the power of the burners $\sum P_{NG}$ and the power of the arcs P_{ARC} are equal. In order to ensure the same productivity with $\eta_{NG} < \eta_{EL}$, the total power of burners $\sum P_{NG}$ must be proportionally higher than the power of the arcs.

The values of η_{NG} close to η_{EL} can occur only during the period of scrap heating. Thus, with the sufficiently high power of HPR-burners, it becomes possible to implement the so-called two-stage heat. During the first stage the scrap is heated only by the burners without the arcs, and at the subsequent second stage both the burners and the arcs or the arcs only are used. Such a process ensures maximum possible replacement of electrical energy by fuel in the furnaces. The HPR-burners make possible not only to radically increase the effectiveness of two-stage process in the twin-shell steelmelting units, where heating of scrap occurs in the second bath, Chap. 6, Sect. 6.2.3, but also to implement this process on the regular single bath furnaces.

Let us examine the influence of energy parameters on the difference $\Delta\tau$ between the tap-to-tap time of the regular heat τ_1 and of the two-stage heat τ_2:$\Delta\tau = \tau_1 - \tau_2$. The duration of operations carried out without arcs and the burners for both types of the heat may be considered equal and ignored in further calculations. Let us introduce the designations:

τ'_{EL} the duration of arcs operation in one-stage heats

τ''_{EL} and τ_{NG} the duration of arcs operation and of burners operation in two-stage heats

In these designations,

$$\tau_1 = \tau'_{EL};\ \tau_2 = \tau_{NG} + \tau''_{EL};\ \Delta\tau = \tau'_{EL} - (\tau''_{EL} + \tau_{NG})\ \text{or:}\quad \Delta\tau = (\tau'_{EL} - \tau''_{EL}) - \tau_{NG} \tag{7.1}$$

By dividing both left and right sides of the equation by τ_{NG}, we get:

$$\Delta\tau/\tau_{NG} = (\tau'_{EL} - \tau''_{EL})/\tau_{NG} - 1 \tag{7.1$'$}$$

The amount of useful heat absorbed during the scrap heating is equal for the both types of the heat:

$$P_{EL} \cdot \eta_{EL} \cdot \tau'_{EL} = P_{NG} \cdot \eta_{NG} \cdot \tau_{NG} + P_{EL} \cdot \eta_{EL} \cdot \tau''_{EL} \tag{7.2}$$

P_{EL} and P_{NG}—the power of the arcs and of the burners. Dividing both left and right sides of this equation by $P_{EL} \cdot \eta_{EL}$ gives us: $\tau'_{EL} = (P_{NG} \cdot \eta_{NG}/P_{EL} \cdot \eta_{NG}) \cdot\tau_{NG} + \tau''_{EL}$, or:

$$(\tau'_{EL} - \tau''_{EL}) = (P_{NG} \cdot \eta_{NG}/P_{EL} \cdot \eta_{EL}) \cdot \tau_{NG} \tag{7.2$'$}$$

By substituting the expression for $(\tau'_{EL} - \tau''_{EL})$ from (7.2′) into the formula (7.1′), we find:

$$\Delta\tau/\tau_{NG} = P_{NG} \cdot \eta_{NG}/P_{EL} \cdot \eta_{EL} - 1 \tag{7.3}$$

In comparison with the regular heats, the tap-to-tap time of the two-stage heats can increase ($\tau_2 > \tau_1$ and $\Delta\tau < 0$), if the duration of scrap heating by the burners τ_{NG} exceeds shortening power-on furnace operation time $(\tau'_{EL} - \tau''_{EL})$, formula (7.1), or if the ratio $P_{NG} \cdot \eta_{NG}/P_{EL} \cdot \eta_{EL}$ is less than one, formula (7.3). The tap-to-tap times will decrease ($\Delta\tau > 0$), if $\tau_{NG} < (\tau'_{EL} - \tau''_{EL})$ or $P_{NG} \cdot \eta_{NG}/P_{EL} \cdot \eta_{EL} > 1$. As already mentioned, the tap-to-tap time of the two-stage process will not be different from that of the regular heat ($\Delta\tau = 0$) under the condition of $\tau_{NG} = (\tau'_{EL} - \tau''_{EL})$ or $P_{NG} \cdot \eta_{NG} = P_{EL} \cdot \eta_{EL}$.

The difference between electrical energy consumptions ΔE_{EL}, upon switchover to the two-stage process is determined by the formula analogous to the formula (5.18) in Sect. 5.4.2 of Chap. 5:

$$\Delta E_{EL} = E_{NG} \cdot \eta_{NG} / \eta_{EL} \tag{7.4}$$

In accordance with (7.4), electrical energy consumption for the two-stage process is reduced directly proportional to the fuel consumption E_{NG}, and the higher the ratio η_{NG}/η_{EL} the greater the reduction. The parameters of the burners and of the two-stage operating mode of the furnace which reduce both electrical energy consumption and tap-to-tap time are of the greatest practical value. The potentialities of both traditional and two-stage processes have been compared based on the results of the industrial trials with the use of several types of the HPR-burners developed by the authors.

7.4.2 Process with a Door Burner in 6-ton Furnaces

In the studies of the two-stage process, the 6-ton plasmarc furnaces of the Chelyabinsk Metallurgical Combine (CMC), Russia [4] were used as the pilot units. These furnaces were utilized in the production of special high-alloy steels and alloys by the method of remelting of clean materials in the neutral atmosphere. The special features of plasmarc furnaces allowed to most fully and with high accuracy determining the potentialities and the basic laws controlling the two-stage process. For this purpose, the doors of the furnaces were equipped with two shutters: one regular shutter for insulating the freeboard and the second additional shutter with a powerful rotary oxy-gas burner installed in it. The second shutter had an exit aperture for the combustion products. In order to reduce the flame temperature the burners with a retarded mixing of gas and oxygen are used.

The studies were conducted during the charge melting period which was divided into two stages: the stage of the burner operation with the duration of τ_{NG} and the stage of the plasmatron operation with the duration of τ_{EL}. The end of the melting period with the duration of $\tau = \tau_{NG} + \tau_{EL}$ was timed after the thorough manual stirring of the bath carried out for the purpose of ensuring that there were no lumps of the unmolten charge left on the bottom. The temperature of the liquid bath t_{MET} was also measured after the stirring.

The two-stage heats were conducted as follows. At the first stage, immediately after charging of the entire metal-charge, the shutter with the burner was placed in the door of the furnace, and the charge was heated up to the maximum temperature. Then the burner was switched off, the shutters were transposed, the furnace was filled with argon, the plasmatron was switched on, and then the melting and the subsequent heating of the liquid metal were conducted as per conventional procedure without any changes in the electrical regime.

The design of the shutter with the burner allowed to change the flame direction within wide limits in the course of heating the charge and, at the same time, practically completely eliminated the air infiltration into the furnace. With the small dimensions of the bath, this ensured the sufficiently uniform heating of the entire charge without the noticeable additional oxidation of iron and alloying elements.

The maximum power of the burner P_{NG} was 6 MW; the power density was more than 1 MW/t of metal-charge. Such power density of the oxy-gas flame during heating of scrap in the electric furnace has been achieved for the first time ever. It exceeded the achieved in practice maximum values of the density of actual electrical power of the furnaces of the same capacity. The average power of plasmatrons in the 6-ton furnaces during the period of melting, P_{EL} was equal to 2.2–2.4 MW. Since the power of the burners P_{NG}, in this case considerably exceeded P_{EL}, the two-stage process ensured not only the large electrical energy savings, but also a significant increase in the productivity due to shortening of the duration of melting τ, Sect. 7.4.1, formula (7.3).

The two-stage heats and the regular heats without the use of the burner alternated. This decreased the effect of the random factors on the results of comparison. For the remelting of clean materials in an atmosphere of argon, the insignificant heat evolution resulting from the chemical reactions of oxidation of the elements of the charge can be disregarded. The entire charge was thoroughly weighed. All this allowed to establish with high accuracy that the total oxidation of metal-charge in the two-stage process did not increase despite the presence in the charge of easily oxidized alloying elements. This result was obtained due to the uniform heating of charge by the rotary burner as well as due to practically complete absence of free oxygen in the freeboard of the furnace [4].

The tests conducted showed that the electrical energy efficiency coefficient of the plasmarc furnaces η_{EL} for smelting of different grades of steels and alloys varies within quite narrow limits and is equal on the average to 0.45. The energy efficiency coefficient of the gas η_{NG}, and the amount of useful heat absorbed by the scrap during the first stage of the process $E^*_{NG} = \eta_{NG} \cdot E_{NG}$, kWh, were determined by using the formula (7.4) and the known values of $\eta_{EL} = 0.45$, E_{NG}, and of the reduction in electrical energy consumption ΔE_{EL}. From the values of E^*_{NG} and the average heat capacity of metal-charge c, kWh/kg · °C, the medium mass temperature of the metal-charge heated by the burner t_{SCR} was determined according to the formula $t_{SCR} = E^*_{NG}/c, {}^\circ\mathrm{C}$.

For the two-stage heats, the power of the burner P_{NG} was varied within the limits of up to 5.5 MW, the burner operation time—up to 30 min, and the natural gas consumption—up to 43 m^3/t, or 400 kWh/t. Oxygen consumption in the burner was varied respectively within the limits of up to 86 m^3/t. Within these quite wide limits of gas consumption, the coefficient η_{NG} remained practically constant and was close to 0.5. In this particular case, it exceeded η_{EL} by 0.5 / 0.45 = 1.1 times.

With $\eta_{NG} = 0.5$, the gas consumption of 400 kWh/t corresponds to the enthalpy of scrap of $0.5 \cdot 400 = 200\,\mathrm{kWh/t}$ and to the medium mass temperature of the scrap t_{SCR} = 1050°C The enthalpy of the completely melted liquid metal, in this case, is equal to 380 kWh/t. Therefore, during the two-stage heats with the maximum gas consumption, approximately half of the total heat required for complete melting was obtained by the metal-charge from the oxy-gas flame.

For the conventional burners, the natural gas consumption for scrap heating does not exceed 6–7 m^3/t. This is explained by the rapid drop in the coefficient η_{NG} in the course of heating for the reasons discussed in detail in Sect. 7.2. The industrial trials conducted showed that the use of the HPR-burners for the high-temperature heating of scrap in the two-stage process ensures the sufficiently high constant value of η_{NG} for the gas consumption up to more than four times exceeding the maximum consumption in the conventional burners. With the constant η_{NG}, the electrical energy savings ΔE_{EL} grows directly proportional to an increase in the gas consumption, Fig. 7.6 [4].

With the natural gas consumption of 400 kWh/t (43 m^3/t), the average electrical energy consumption, in comparison with the regular heats, was reduced, on account of shortening of the melting period, from 840 to 380 kWh/t, i.e. by 460 kWh/t or 2.2 times, which sufficiently closely corresponds with the results of calculation according to the formula (7.4). The duration of two-stage heats with $P_{NG} \cdot \eta_{NG} / P_{EL} \cdot \eta_{EL} > 1$ decreases directly proportional to the duration of heating of scrap τ_{NG}, formula (7.3). With $P_{NG} = 5.5$ MW and $\tau_{NG} = 25$ min (0.42 hour), the duration of the power-on period of melting reduces from 117 to 52 min (by 2.3 times), and the tap-to-tap time—by 42 min (36%), which corresponds well with the results of calculation according to the formula (7.3). These results indicate one more potentiality of the two-stage process. With the sufficiently high actual power of the HPR-burners and for a certain limited level of productivity of a furnace,

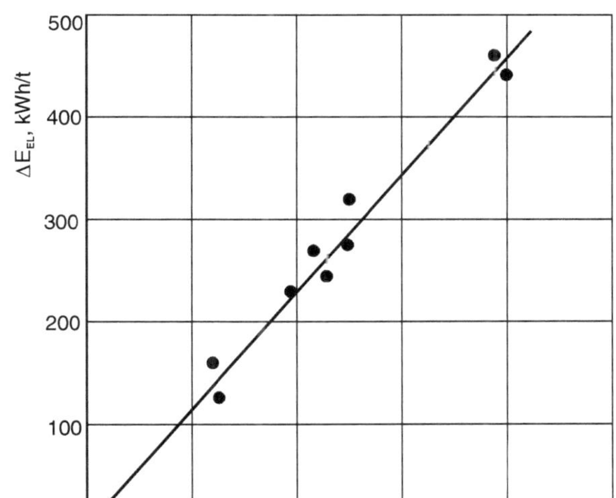

Fig. 7.6 Dependence of decrease in electrical energy consumption ΔE_{EL} on E_{NG} at the two-stage process in a 6-ton EAF

the power of the electrical system of a furnace can be considerably reduced. Under certain conditions, this direction can be found rational.

The tests conducted revealed another important feature of the two-stage process. Despite a very high unit power of oxy-gas burners which more than twice exceeded the power of plasmatrons, the temperature of the furnace lining only slightly exceeded the temperature of the surface of the scrap and was not higher than 1450–1550°C. With the arc heating, this temperature was approximately 200°C higher. This difference is explained by the fact that the temperature of the oxy-gas flame is considerably lower than the temperature of the arc plasma, as well as by the fundamental differences in the laws of heat transfer by radiation and by convection. If the two-stage process is implemented in the EAFs, this feature ensures the decrease in heat losses due to water cooling the wall and roof panels of the freeboard.

7.4.3 Process with Roof Burners in 100-ton and 200-ton Furnaces

Two-stage operating mode of the HPR-burners during the melting period was tested on an industrial scale in the furnaces at CMC and "Red October". Each portion of scrap was heated and settled down first by the burners without the use of electrical energy, and then by the burners together with the arcs or by the arcs only. Furthermore, the burners were used during the power-off breaks caused by the reasons of organizational nature, as well as during the electrodes replacement. During the simultaneous arcs and burners operation, the gas and oxygen consumption in the burners were being decreased.

The effectiveness of the burners operation was evaluated based on a change in the performance indices of the melting period, which, in comparison to the indices of the entire heat, are more closely correlated with the use of fuel for scrap heating. In the old furnaces which operated without secondary treatment of steel in ladles, the electrical energy consumption for the heat and tap-to-tap time strongly depended on the technological factors not associated with heating of scrap by burners. In accordance with common practice, the end of the melting period was determined by formation of the flat bath with the temperature of approximately 1560–1580°C.

The tests were conducted under the actual conditions of current production which are considerably less stable than the conditions in the plasmarc 6-ton furnaces. The heats with and without the use of burners alternated. However, with the small number of heats in both groups that were being compared, it was impossible to avoid random variations of some of the external factors, which affected the results of comparison. Thus, in the group of heats with the burners in the 100-ton furnaces, the yield reduced by 0.8%, and the chemical heat input increased due to the increased iron oxidation rate. In the same group of heats in the 200-ton furnaces, the yield increased by 1%, whereas the chemical heat input, which strongly affects the electrical energy consumption, decreased. In these same furnaces, during the heats with the use of burners, the power of arcs increased considerably. This increase to a great extent resulted from the increase in the stability of the arcing in the preheated scrap. Nevertheless, the detailed analysis of the obtained data unambiguously indicates the high effectiveness of the two-stage mode of melting scrap by the HPR-burners, Table 7.1.

The duration of burner operation without the arcs in the 100-ton furnaces was 29%, and in the 200-ton furnaces 36% of the total duration of the melting period. In the 100-ton furnaces, the average for the melting period power of the burners and that of the arcs were practically equal. In the 200-ton furnaces, the power of the burners was 16% lower than the power of the arcs. Despite that, the duration of melting was reduced in both cases. In the 200-ton furnaces the total duration of the melting period was shortened by 12.1% and the power-on time by 1.8 times; in the 100-ton furnaces those were shortened by 6.6% and 1.5 times respectively. During the melting period electrical energy consumption was reduced 1.5 times in the 200-ton furnaces, and that was reduced 1.7 times in the 100-ton ones. An increase in the effectiveness of the arcs operation in the preliminarily heated scrap contributed to the shortening of the melting period.

Table 7.1 Characteristics of furnaces for melting period with no use of fuel and using two-stage mode of HPR- burner operation

Characteristics	100-ton EAF, city of Chelyabinsk			200-ton EAF, plant of «Red October»		
	Without burners	With burners	Changing abs. /%	Without burners	With burners	Changing abs. /%
Gas consumption, m^3/t	–	17.8	+17.8	–	17.4	+17.4
kWh/t		164	+164		161	+161
Oxygen flow rate, m^3/t	–	36.0	+36.0	10.0	34.5	+24.5
Burner power, MW						
Without arcs	–	19.5	–	–	21[a]	–
With arcs					23	
Actual arc power, MW	21.8	19.5	–2.3	27.7	33.5	+5.8
P_{EL},			–10.6%			+20.9%
Duration of, min						
–Total melting	122	114	–8 –6.6%	165	145	–20 –12.1%
–Power-on	122	81	–41 –33.6%	165	92	–73 –44.2%
–Burners' operation without arcs	–	33	–	–	53	–
–Operation of burners with arcs	–	16	–	–	15	–
Electrical energy consumption during the melting period, kWh/ton	455	274	–181 –39.8	360	240	–120 –33.3%

[a] With taking into consideration of a slag door burner

For the 100-ton and 200-ton EAFs with HPR-burners, the coefficient η_{EL} can be assumed to be equal approximately to 0.75 for the melting period. Utilizing this value of η_{EL} and the formula (7.4), let us determine, in accordance with the data in the Table 7.1, the values of η_{NG}. For the 100-ton EAF, ΔE_{EL} = 181 kWh/t, E_{NG} = 164 kWh/t, and the ratio η_{NG}/η_{EL} = 181/164 = 1.10. However, taking into account the correction for the additional heat input resulted from a random increase in iron oxidation by 0.8%, this ratio should be decreased to 1.0. The value η_{NG} will be then equal to 0.75. For the 200-ton furnaces, the ratio η_{NG}/η_{EL} is equal to 120/161 = 0.75. Taking into account the correction for the decrease in iron oxidation by 1%, it should be assumed to be equal to 0.88. This value corresponds to η_{NG} = 0.88 · 0.75 = 0.66. Therefore, for the 100-ton and 200-ton furnaces, the average energy efficiency coefficient of the gas in the roof HPR-burners η_{NG} can be assumed to be equal to 0.7. This value only slightly differs from η_{EL}.

Knowing η_{NG}, let us find an increase in both the enthalpies of the scrap $\Delta E_{SCR} = E^*_{NG}$ and its medium mass temperature Δt_{SCR} obtained due to the burner operation. For the 100-ton EAF, ΔE_{SCR} = 164 · 0.75 = 123 kWh/t. This value of enthalpy corresponds to $\Delta t_{SCR} \cong 720$°C, Table 6.1, Chap. 6. For the 200-ton EAF, ΔE_{SCR} = 161·0.66 = 106 kWh/t and $\Delta t_{SCR} \cong 650$°C. For two furnaces, in the average $\Delta t_{SCR} \cong 685$°C.

The weighing of electrodes showed that their loss changed approximately proportional to the power-on furnace operation time. The average duration of the power-on period of melting in the 100-ton and 200-ton furnaces was reduced approximately by 40%, Table 7.1. This gives grounds to expect the corresponding reduction in electrode consumption for the whole heat as well. The durability of the lining of the 200-ton furnaces did not change during the trial period, but that of the 100-ton furnaces somewhat increased. This agrees well with the results obtained in the 6-ton plasmarc furnaces and is explained by the same factors.

The results of the industrial trials of the roof HPR-burners allow to draw the following main conclusions. If the power of the HPR-burners is the same as that of the electric arcs, then the burners have practically the same energy efficiency and rate of scrap heating as the arcs, at least in the examined range of powers and temperatures. However, it should be considered that these results have been obtained from the old furnaces with lining of sidewall and roof which operated with use of a charge containing predominately rolling-mill scrap. Such a charge is optimal for the burner operation. When using a regular light scrap some difficulties described above occur, Chap. 6, Sect. 6.1.5, and it should be expected that η_{NG} will decrease from 0.7 down to 0.6.

The quite high obtainable power of the HPR-burners approximated to that of the electric arcs makes it reasonable to examine the possibilities of the two-stage process in an EAF which allow heating the scrap during the first stage, without using the arcs, to a temperature of 800°C. In this process the share of electrical energy in the final enthalpy of the metal is approximately equal to that of fuel.

Such a process is no longer a strictly electric steelmelting process, and can be called fuel arc process. Appropriately, a furnace realizing this process can be named the fuel arc furnace (FAF).

7.5 Fuel Arc Furnaces (FAFs)

7.5.1 FAF with Scrap Heating in a Furnace Freeboard

Let us examine the basic calculated indices of the 120-ton FAF operating in the two-stage process mode. Similar to the Danarc Plus furnace, this furnace has the tall shell containing the whole amount of scrap (130 t) with the density of 0.7 t/m^3 charged in a single basket. The result of such charging is not only the shortening of the power-off time, but also a notable increase in the path length of the gases passing through the scrap from its lower layers reached by the penetrating flames of the HPR-burners, to the uppermost layers. This considerably increases the heat transfer from the gases to the scrap as well as the fuel efficiency coefficient η_{NG}.

The FAF is equipped with four sidewall HPR-burners with the power of 30 MW each and one oriel HPR-burner with the power of 20 MW. With the total duration of the heat equal to about 35 min the possible burner operation time is not more than 13 min (0.22 hour). The burners with the total power of 140 MW and $\eta_{NG} = 0.7$ operating in the power-off regime in the first stage of the process with a duration of 0.22 h can transfer to the scrap $140000 \cdot 0.7 \cdot 0.22/130 = 166$ kWh/t of useful heat. This value of enthalpy of the scrap corresponds to a medium mass temperature of 880°C, Chap. 6, Table 6.1. The maximum temperature of heating of the lower layers of scrap in this case will be equal to about 1050°C, and that of the upper layers ~700°C, which excludes the considerable iron oxidation. With $\eta_{NG} = 0.7$, the natural gas consumption is equal to $166 / 0.7 \cdot 9.5 = 25$ m^3/t (9.5 kWh/m^3 is the calorific value).

At the second stage, both the electric arcs and the heat of chemical reactions are the sources of energy. The heat of oxidation of iron and its alloys is being completely absorbed. The amount of this useful heat is about 100 kWh/t. The heat of oxidation of 17 kg/t of the coke charged with the scrap and injected into the bath gives approximately 25 kWh/t of useful heat. The enthalpy of the metal before tapping is $E_{MET} = 390$ kWh/t. Therefore, the amount of useful heat which must be introduced into the metal at the second stage of the process by the electric arcs is equal to $390-166-100-25 = 99$ kWh/t. With $\eta_{EL} = 0.7$, this value corresponds to the electrical energy consumption of $99/0.7 = 141$ kWh/t. With the duration of the second stage of 14 min (0.23 h), this value of consumption corresponds to the average actual power of the arcs equal to $141 \cdot 120/0.23 \cdot 10^3 = 73.5$ MW and transformer power of about 100 MVA. In accordance with the data on the most highly productive electric arc furnaces with capacity of 100–140 t, the duration of technological operations in the FAF conducted with both the burners and the arcs

Table 7.2 Energy characteristics of modern EAF and FAF

Characteristics	120-ton EAF	120-ton FAF
Consumption per 1*t* of liquid steel:		
Electrical energy, kWh/t	340	141
Natural gas, m^3/t	7	25
Oxygen, m^3/t:		
To burners	13	50
To the bath	34	30
Total	47	80
Coke, kg/t	17	17
Electrodes, kg/t	1.2	0.5
Actual electrical power consumed, MW	100 – 125	74
Total power of oxy-gas burners, MW	21	140

off, can be assumed to be equal to 8 min. Then, the tap-to-tap time for the FAF amounts to $13 + 14 + 8 = 35$ min.

Table 7.2 gives the basic energy indices of the 120-ton FAF in comparison to the best indices of the most modern EAFs of the same capacity and productivity in case of operating with light scrap. The energy differences between FAFs and modern EAFs are not quantitative, but rather qualitative, i.e. fundamental. The electric arcs are the main energy source for the EAF, whereas for the FAF the main source of energy is the flame of the burners. High productivity of more than 200 t/h is achieved in the EAF basically due to an increase in the electrical power, and that of the FAF—due to an increase in the power of the burners and fuel consumption. The short power-on time and relatively low electrical power of the FAF sharply decrease the electrode consumption. Stable arcing in the high temperature preheated scrap ensures a great reduction in the level of noise and flicker, which decreases the costs on the protection of the external networks against electrical interference. Furthermore, it is possible to expect a notable increase in the service life of the roof and wall panels as well as the central refractory part of the furnace roof, since, in contrast to the high-power electric arcs, the low-temperature flames of the HPR-burners do not produce intensive thermal effect on these elements. Reduction in the electrical power of the FAF must also be accompanied by a decrease of costs on electrical equipment. This reduction is of special importance for the regions which do not have sufficiently powerful grids.

However, development of high productivity FAFs on the base of EAF involves serious problems due to too short duration of burners operation. In this case, the HPR-burners operation control can be of great difficulty since it is necessary to provide an efficient adoption of vast input power of the flames under rapidly changing conditions of the heating of scrap in the course of the process and from heat to heat. In addition, HPR-burners are incompatible with the devices for blowing of oxygen and carbon into the bath which, just like the burners, are installed in water-cooled boxes. It is quite difficult and not always possible to install in a 120-ton furnace all of these devices plus five HPR-burners of total power of 140 MW. The control of furnace operation as well as its maintenance becomes considerably more complicated.

It is more feasible to realize a FAF on the base of a twin-shell furnace, Chap. 6, Sect. 6.2.3. The duration of scrap heating by the burners on such furnaces can be increased by approximately 1.3–1.4 times. The required total power of HPR-burners drops correspondingly while productivity of the furnace increases as compared to EAF. On the other hand, twin-shell furnaces are unpromising with the exception of some particular cases, because two separate furnaces make possible to achieve much higher productivity than one twin shell furnace. For this reason, renovations of existing twin-shell furnaces have already been carried out to modify them into two independently operating furnaces with separate transformers. For FAFs, the best conditions of scrap heating by the burners are provided in the external devices such as conveyors and shafts rather than in the furnace itself. Such versions of FAFs are reviewed below.

7.5.2 Conveyor FAFs with Continuous Scrap Charging into the Liquid Bath

As opposed to heating of scrap by off-gases, the powered burners can ensure heating of scrap to the high temperatures at a high velocity. This, in particular, is confirmed by the experience of preheating of scrap in the oxygen converters. In the past, such preheating was used for the purpose of increasing of the share of scrap in the converter charge. 100 tons of scrap could be heated by a burner with the power of 180 MW and the energy efficiency coefficient of the fuel 0.6 up to 750°C in approximately 7 min. It was established that the average mass temperature of scrap must not exceed 800°C. The efficiency of heating decreases sharply at a higher temperature.

In comparison with heating of the mass of scrap on the bottom of the EAFs or on that of the converters, the conveyor heating provides much better conditions for efficient use of the burners. These conditions are close to optimum due to the following features. The thickness of the layer of scrap on the conveyor is relatively small. As soon as the scrap on the conveyor reaches the maximum permissible temperature, it is taken out from the heating zone and charged into the bath. Such a process of heating allows keeping the power of burners at a constant maximum level corresponding to the speed of scrap charging, without overheating the scrap. Side by side with the continuous charging of scrap into the liquid bath, the second fundamental advantage of the Consteel process is the possibility of high-temperature heating of scrap by the burners. If the heating of scrap by the off-gases hinders the usage of this advantage, such a heating should be done away with altogether as it has already been mentioned.

Giving up the heating of scrap with the off-gases allows offering the most rational method of utilizing the burners for high-temperature heating of scrap on a conveyor of the Consteel type. The burners 1 are placed at a short chamber 2 adjacent to the furnace door for scrap charging, Fig. 7.7. The gases generated by

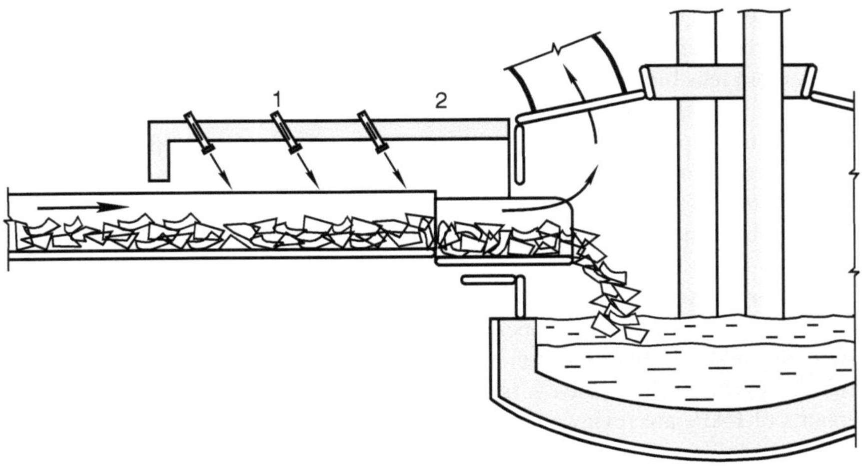

Fig. 7.7 Direct-flow FAF with conveyor charging of scrap into the liquid bath (designations are given in the text)

the burners move in the same direction with the scrap (the direct-flow method of heating) and enter the freeboard of the furnace. These gases along with the process gases are removed through the fourth opening in the roof, as it takes place in the conventional EAFs. This method excludes the need for additional consumption of energy for decomposing of dioxins and furans. Such consumption is inevitable when the burners along with the off-gases are being used for the heating of scrap.

The offered direct-flow method for heating of scrap by the burners allows developing a conveyor steelmaking furnace of the new type—FAF. The basic designed performances of such a 120-ton furnace with the heating of scrap up to the average mass temperature of 800°C are cited in Table 7.3.

The main energy feature of a new furnace is the fact that the arcs' power and the total burners' power are the same and equal to 75–80 MW. Such a furnace keeps all of the known advantages of the Consteel furnaces including environmental ones. In FAFs, the burners stabilize an off-gas flow which facilitates utilizing the heat from off-gases, for example, for producing electric power. At the same time, in comparison with the EAFs the direct-flow FAFs acquire some new advantages.

The electrical energy consumption is reduced by 1.8 times and the power of both the transformer and the arcs by 1.5 times at the same or even higher productivity. Electrode consumption is reduced. Considerable reduction in the electrical power allows using conveyor FAFs in the regions which do not have sufficiently powerful grids. Costs on electrical equipment and electric power supply are reduced.

The service life of the roof and sidewall panels as well as the central refractory part of the furnace roof notably increases, since the power of the arcs is sharply reduced whereas the high power burners are located outside of the furnace freeboard and have no effect on all these elements.

Table 7.3 Designed performances of a direct-flow conveyor FAF

1. Tapping weight	122 t
2. Tap-to-tap time	32 min
3. Productivity	230 t/h
4. Transformer power	100 MVA(80 MW)
5. Burners power	75 MW
6. Electrical energy cons.	200 kWh/t
7. Natural gas flow rate	20 m^3/t
8. Oxygen flow rate:	
for burners	40 m^3/t
for bath blowing	30 m^3/T

Air-gas burners are unsuitable for the heating of scrap in the conveyor direct-flow FAFs since they produce a lot of combustion products. Radically different requirements are imposed upon the oxy-gas burners for the high-temperature heating of scrap on the conveyor as compared to the sidewall burners in the EAFs. The flames of the burners must possess high kinetic energy and relatively low temperature to provide the intensive heating of scrap across the whole thickness of the layer and to avoid considerable oxidation and welding of separate lumps of the scrap. The required reduction in temperature can be achieved by slowing down the mixing of gas with oxygen, by recirculation of products of combustion, and by using gases with the lower calorific value, etc. For the uniform heating of scrap the burners should be installed across the entire width of the conveyor. For example, eight burners with the power of 9.4 MW each (the total power of 75 MW) may be used for a 120-ton FAF with a conveyor 2 m wide. Sections of the conveyor located in the chamber for the heating of scrap have to be provided by intensive water-cooling to ensure their required durability.

7.5.3 Shaft FAFs with Continuous Scrap Charging by a Pusher

In comparison with conveyor furnaces shaft furnaces create even more favourable possibilities for realization of high productive and economical direct-flow FAFs. Under conditions of shaft furnaces a required reduction in the temperature of oxy-gas flames can be provided by means of the most effective way using recirculation of combustion products which pass through the layer of scrap. The reduction in the combustion products temperature raises an achievable level of scrap heating temperatures since the heating becomes more uniform and soft. This allows expecting to obtain better main performance indices in the shaft FAFs in comparison with those of the conveyor ones, Table 7.3. It should be noted, that the charging of scrap with the pusher has proved its operability at temperatures of scrap close to 1000°C, Chap. 6, Sect. 6.2.4.

Design concept of the shaft FAF is shown in Fig. 7.8. Several lined combustion chambers (1) with oxy-gas burners (2) are located along the perimeter of the shaft

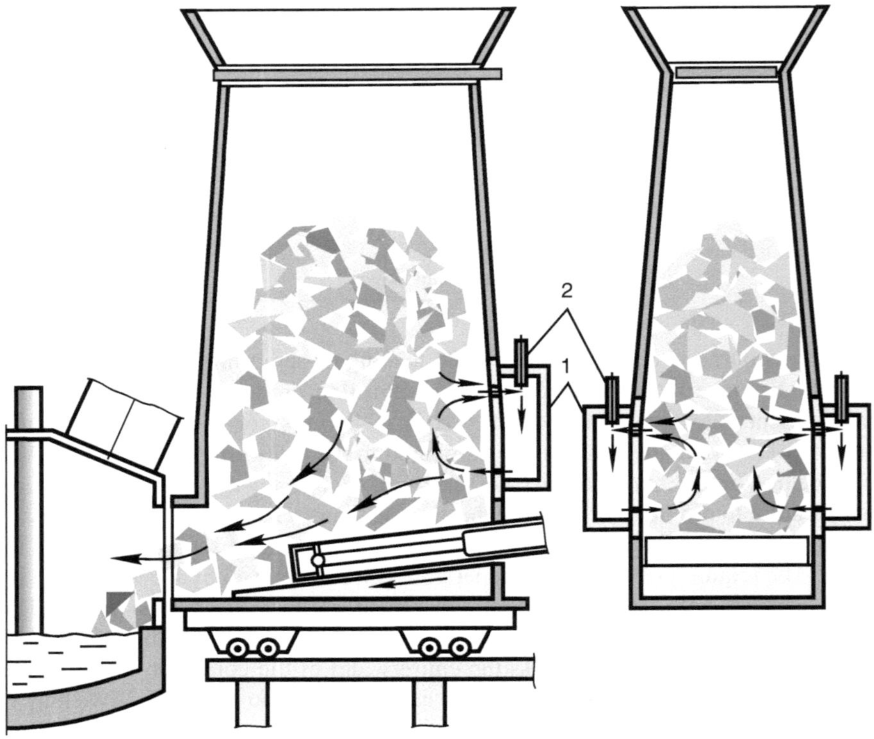

Fig. 7.8 Shaft FAF with continuous scrap charging by a pusher (designations are given in the text)

in its lower part. The burners operate with a minimal oxygen flow rate coefficient α approximating to 1.05. With the appropriate geometrical parameters of combustion chambers, high-speed flames of the burners create a considerable negative pressure in the upper part of the combustion chambers and positive pressure in their lower part. The shaft is air-tightly shut after being charged with the scrap.

The lower part of the combustion chambers has a row of openings of small diameter through which the hot high-speed gas jets penetrate into the scrap thickness. The upper part of the combustion chambers has a row of large openings through which the gases from the shaft are infiltrated into the combustion chambers where they mix with the natural gas combustion products. The gas flow from the lower openings splits due to the furnace draft on the one hand, and the negative pressure created by the burners in the upper part of the combustion chambers, on the other hand. The main volume of gases equal to the flow rate of combustion products of natural gas at $\alpha = 1.05$[1] goes into a furnace and heats the scrap on the way from the combustion chambers to the shaft exit. Thereat certain amount of

[1] When the power of the burners in 120-ton FAF is 75 MWt, the flow rate of combustion products of natural gas equals to approximately 5 m^3/s.

gases goes upwards and enters the combustion chambers through the upper openings. These gases continuously circulate through the combustion chambers. Their temperature drops since the gases go through the quite thick layer of scrap. When mixing these gases with the combustion products of natural gas they cool down the latter. Realization of such an aerodynamics requires the negative pressure in the upper part of the combustion chambers to be considerably higher than the negative pressure in the shaft exit, which is easily achievable. With a proper selection of both the height of the combustion chambers (a vertical distance between the upper and lower openings) and a value of the negative pressure in the zone of upper openings, the reduction in temperature of the gases which heat the scrap can be achieved to the required levels.

In order to increase the melting rate of scrap in the liquid bath of FAFs it is necessary not only to raise a temperature of scrap preheating but also increase the intensiveness of bath stirring in the scrap charging zone. The first one is achieved by replacing insufficient energy in off-gases with energy from natural gas. Multi-nozzles oxygen tuyeres submerged into the bath down to the slag-metal interface are required to improve the intensity of bath stirring. Long-term experience and fundamental studies on open-hearth furnaces have shown that such tuyeres, in comparison with a practice of bath blowing from above with separate oxygen jets which takes place at present, ensure considerably shortened tap-to-tap time due to the increased intensity of the bath stirring. The tuyeres have to be submerged into the melt close to the scrap charging zone.

During the melting the level of metal in the FAFs with continuous scrap charging rises rapidly. Therefore, the blowing at the slag-metal interface requires using sensors which fix the tuyere tip position at this interface. Such sensors are known; they were widely used in the open-hearth practice [6].

The offered both direct-flow conveyor and shaft fuel arc steelmaking furnaces with continuous scrap charging into the liquid bath are a new trend to a further development of EAFs. Such furnaces equipped with powerful oxy-gas burners and oxygen tuyeres submerged into the melt are able successfully compete with the state-of-the-art EAFs and replace them due to their high productivity, economization and environmental advantages.

7.6 Economy of Replacement of Electrical Energy with Fuel

The economic expediency of deep replacement of electrical energy with natural gas and of use of such steelmelting furnaces as FAFs can be evaluated by two methods: either on the scale of a plant as well as on the scale of the national economy of the entire country or its separate region. The results of estimation according to the first method are determined almost exclusively by the electrical energy and gas prices which can strongly vary over time and from one country to another.

Thus, for instance, in the USA in the period from 2000 to 2005, the price of electricity has barely changed, whereas the price of natural gas has approximately doubled. The cost of 1 KWh of electricity has practically equaled that of natural gas, which created unfavorable conditions for the furnaces of the FAF type. On the contrary, the calculations carried out with regard to the current prices in Russia, showed that the cost of the energy absorbed by the EAF bath is approximately 3 times higher in case of electrical energy usage (taking into account the electrodes consumption) than in case of usage of natural gas in the burners. The values assumed in these calculations were $\eta_{EL} = 0.75$ and $\eta_{NG} = 0.5$ [7]. Lately, the prices of gas in Canada and the USA sharply reduced due to development of new nature sources. A new economical situation contributes to developing and spreading of technologies such as FAFs which ensure a deep replacement of electrical energy with energy from natural gas.

Instability in gas prices and inadequacy of the price of the gas to its real cost makes it necessary to turn to the more objective method for evaluating the directions of the use of energy in EAF, namely, to the estimations on the scale of the country. This approach allows considering not only the prices of the energy-carriers, but also the fundamentally important interdependences between the two branches of the national economy, that of production of electrical energy and electric steelmelting. In the majority of the countries, in Russia and in the USA in particular, most of the electrical energy is produced in the thermal power stations (TPS) utilizing as a fuel not coal only, but natural gas as well. About 20% of electric power in the USA are produced by the gas TPS. In Russia this figure totals approximately 40%. The combustion of coal releases into the atmosphere considerably larger amount of CO_2, as well as sulfur oxides and other harmful emissions. As a result, substituting gas with coal in TPS would require quite significant additional environmental protection costs. Furthermore, transporting gas is considerably cheaper than transporting electricity.

Setting aside for a moment the vital ecological problems examined in Chap. 14, let us compare the effectiveness of the use of natural gas for production of electrical energy which later is used in EAF, with the effectiveness of the use of gas directly in the furnace. The overall efficiency coefficient of primary energy of natural gas in the chain of TPS – EAF, η_{TPS}^{EAF}, is determined by the expression:

$$\eta_{TPS}^{EAF} = \eta_{TPS} \cdot \eta_{EL.GR} \cdot \eta_{EL} \tag{7.5}$$

η_{EL} the efficiency coefficient of electrical energy in the furnace

$\eta_{EL.GR}$ the efficiency of electrical power grids taken with consideration for all energy losses due to voltage transformations

η_{TPS} the efficiency of the thermoelectric power stations operating on gas

Assuming, in accordance with the current data, $\eta_{EL} = 0.7$, $\eta_{EL.GR} = 0.92$ and $\eta_{TPS} = 0.41$, and using with the expression (7.5), we will obtain $\eta_{TPS}^{EAF} = 0.26$. Therefore, with the use of natural gas in TPS, only approximately one fourth of chemical energy of gas used in TPS is ultimately transferred to the metal-charge heated and melted by the electric arcs in EAF. For direct heating of scrap in a furnace by HPR-burners, the energy efficiency coefficient of the burners, η_{NG}, is considerably higher. Therefore, any expansion of the application of such burners must lead to in the savings on the overall natural gas consumption in TPS and in EAF (or in FAF) expressed in m^3/t of steel. The calculations of these savings must incorporate the energy consumption for production of oxygen. Utilizing the energy indices of modern EAFs and conveyor FAF, Table 7.3, let us carry out this calculation for the HPR-burners.

In comparison with EAF, the natural gas consumption in FAF increases by $20 - 5 = 15$ m^3/t, and the oxygen consumption by $70 - 37 = 33$ m^3/t. The electrical energy consumption for production of oxygen in the modern oxygen-compressor stations is approximately 0.55 kWh/m^3 of O_2. Production of additional 33 m^3/t of oxygen requires $33 \cdot 0.55 \cong 18.2$ kWh of electrical energy per 1 t of steel. With an increase in the gas consumption by 15 m^3/t and taking into account the electrical energy consumption for production of oxygen, the reduction in the electrical energy consumption in FAF will be equal to $340 - 200 + 18.2 \cong 158$ kWh/t or $158/15 = 10.5$ kWh/m^3 of natural gas.

In order to produce this amount of electrical energy in TPS ($\eta_{TPS} = 0.41$), it would be necessary to consume the gas with the calorific value equal to 9.5 kWh/m^3 in the amount of $10.5 / 0.41 \cdot 9.5 = 2.7$ m^3. Therefore, the application of the HPR-burners in the FAF saves for the national economy 1.7 m^3 ($2.7 - 1.0 = 1.7$) of gas per each 1 m^3 of gas consumed by the burners. These absolute savings are an evidence of an unquestionably high energy efficiency of replacement of electrical energy with gas by the means of the HPR-burners in FAF on the scale of the country. If we express these savings from a financial point of view, then they grow directly proportional to an increase in gas prices.

The realization of highly productive fuel arc furnaces with HPR-burners would allow to implement yet another option of effective replacement of electrical energy with fuel. We mean the option of combining FAF and oxygen converters at integrated metallurgical plants. Such plants, as a rule, have unutilized resources of coke gas which calorific value is equal to 4.1–5.3 kWh/m^3. This gas can be used in the HPR-burners in a mixture with natural gas.

As already mentioned, the fundamental feature of special oxy-gas burners ensuring their successful application in FAF, is the reduced temperature of the flame, in comparison with that of conventional oxy-gas burners. This reduction can be achieved in variety of ways including the use of a fuel with the lower calorific value. If the natural gas (9.5 kWh/m^3) is mixed with the coke gas (5 kWh/m^3, on average) in the proportion of three to two, then we will obtain a fuel with the calorific value equal to 7.7 kWh/m^3, which meets the requirements of the burners for FAF. At the same time, the fuel costs will be approximately by 40% lower, in comparison with the use of natural gas.

The use of fuel arc furnaces can give to the integrated plant yet another additional advantage. It will allow reducing the share of hot metal in the metal-charge processed at the plant due to the use in the FAF of 100% of the light, lowest-cost scrap. Usually, the share of scrap in the charge of the oxygen converters is 25%. If one fourth of steel at the plant is produced in the FAF, then the share of scrap in the entire metal-charge processed at the plant will increase to 44%, which can substantially reduce total production costs. As already noted, reduction in hot metal consumption at the steelmaking process provides a considerable decrease in emissions of CO_2 into the atmosphere.

References

1. Rudzki EM, Reinbold RI, Pease BK (1967) Oxy-fuel process of steelmaking. Iron and Steel 11:30–35
2. Kobayashi N (1999) Development of a new NSR process of oxygen scrap melting. Iron and Steel Engineer 61–66
3. Toulouevski YN (2000) High power oxygen-fuel burner for electric arc furnaces. MPT Int 6:66–69
4. Toulouevski YN, Mizin VG, Zinurov IY et al (1988) Flame-arc processes of electric melting. Stal 8:42–46
5. Toulouevski YN, Zinurov IY, Popov AN et al (1987) Electrical energy savings in arc steelmelting furnaces. Energoatomizdat, Moscow
6. Markov BL (1975) The methods of open-hearth bath blowing. Metallurgiya, Moscow
7. Kiselyov AD, Zinurov IY, Makarov DN et.al (2006) Effectiveness of oxy-gas burner applications in modern arc steelmaking furnaces. Metallurgy 10:60–62

Chapter 8
Basic Physical–Chemical Processes in Liquid Bath Blown with Oxygen: Process Mechanisms

8.1 Interaction of Oxygen Jets with the Bath: General Concepts

Until oxygen has become readily available on industrial scale, the source of oxygen intake into the bath of EAF used to be the atmosphere of the freeboard and the additions to the bath which contained ferrous oxides. Presently, only ferrous oxides introduced into the bath with the oxidized scrap must be considered as an additional source of oxygen. The furnace atmosphere has practically lost its significance as a source of oxygen. This has happened as a result of elimination of significant air infiltration into the furnace during the operation with closed slag door as well as sharp shortening of tap-to-tap time.

In the modern furnaces, practically all oxygen used during the heat is introduced into the bath by the high-velocity oxygen jets. These jets, depending on the amount of stored energy at the moment of impact with the bath, penetrate the bath to a greater or smaller depth, Fig. 8.1a, b. The hydrodynamic processes taking place during the oxygen blowing are quite complex and will be reviewed in detail in the following chapters. Here it is sufficient to consider only their certain most important features and point out that hydrodynamics of oxygen blowing greatly affects the physical–chemical processes in the bath, which, in turn, determine the final results of the heat. These processes include mainly oxidation of carbon as well as that of iron and its alloys, heating of the liquid metal by the external and internal energy sources, and melting scrap in the liquid metal.

When jet power is relatively low, the crater is formed on the surface of the bath. The reverse flow of oxygen blows splashes of metal and slag out of the crater, Fig. 8.1a. As the jet power rises the jet penetrates through the layer of slag and partially enters into the metal forming the so-called reaction zone. This case is mostly typical in practice. In the special literature, the space of jet immersion into the melt is divided sometimes into two reaction zones - primary and secondary. For the purposes of this book such detailing is deemed not necessary. Therefore, in the

Y. N. Toulouevski and I. Y. Zinurov, *Innovation in Electric Arc Furnaces*,
DOI: 10.1007/978-3-642-36273-6_8, © Springer-Verlag Berlin Heidelberg 2013

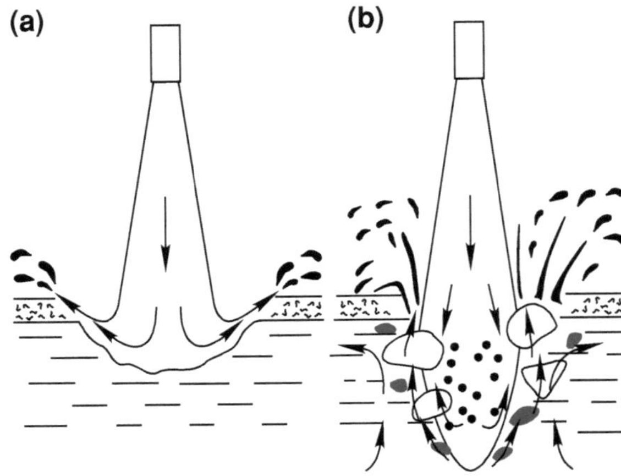

Fig. 8.1 Modes of interaction of the oxygen jet with the bath (elucidations are given in the text)

text below we consider the reaction zone as the entire, though quite limited in dimensions, space of jet immersion into the melt as opposed to the rest of the bath where the chemical transformations of the elements take place as well.

In the reaction zone, the metal and the slag are drawn into the oxygen jet, and both the jet and the melt are being atomized into separate gas volumes and droplets. When oxygen is in excess, some volumes of gas rise to the surface of the bath and increase splashing. Droplets of metal are oxidized in oxygen and are converted into ferrous oxides. These oxides are partially consumed for oxidizing other elements, carbon in particular, right in the reaction zone. They are partially dissolved in the metal, thus increasing oxygen content in it. A small fraction of the oxides rises forming the slag. The rest of ferrous oxides are spread all over the bath. This process is promoted by the intense turbulent pulsations in the reaction zone, which are transferred to the rest of the metal.

Spread of oxygen throughout the volume of the bath is also promoted by the high temperature of the reaction zone, which varies from 2200 to 2600°C depending on the periods of the heat. Solubility of oxygen in liquid iron increases sharply as temperature rises. In the reaction zone oxygen solubility is equal to 1.5–2.5%, and beyond its boundaries at the temperature about 1600°C its value is only 0.2%. Therefore, when the metal moves from the reaction zone to the bath areas with relatively low temperatures, oxygen evolves from the solution as finest particles of FeO with well-developed surface, which promotes the oxidation processes.

When oxygen is blown from below or from the side, the structure of the reaction zone is approximately the same as when oxygen is blown from above. However, when blowing from below is used, gravitation forces do not obstruct the spread of the oxygen jet as with blowing from above, but promote it. Therefore, the length of the reaction zone increases.

8.2 Oxidation of Carbon

According to the modern concepts, oxidation of carbon as well as of a number of other elements in the steel melting bath is a complex multistage process. Certain details of this process have been a subject of scientific discussions for a long time. Without going deep into details posing rather academic than practical interest, let us concentrate on the basic concepts, which have to be considered when resolving problems regarding intensification of decarburization processes.

As has already been mentioned above, in good accord with the laws of physical chemistry and industrial data, it is generally assumed that under conditions of steel melting bath carbon is oxidized only to CO. It is known that negligible portion of carbon can be oxidized to CO_2 as well, but from the practical point of view this process can be ignored.

The rate of chemical reactions increases greatly, as the temperature rises. At the temperatures of steel bath equal to 1480–1650°C the rate of carbon oxidation by oxygen reaches the values so high, that this rate itself does not restrict in any way the possibilities of intensification of decarburization process. However, in order to enable this reaction, carbon and oxygen must come into direct contact. For sustaining the process of decarburization, it is necessary to continuously deliver carbon and oxygen into the reaction zone and remove the product of reaction CO from there.

Rising to the surface and removal of gas bubbles of CO from the bath occurs quite intensely and does not hinder the process. Observations show that evolution of CO occurs quite uniformly across the entire surface of the bath without obvious dominance of reaction zones. Therefore, the area of oxidation of carbon is the entire bath where carbon is spread relatively uniformly. On the contrary, the sources of incoming oxygen are concentrated in quite limited in volume reaction zones. Such a depiction of the process is well in accordance with the simplified working model of the bath decarburization process, i.e., with so-called two-stage scheme of oxidation of carbon.

According to this scheme, the primary chemical act of blowing the bath with oxygen is oxidation of iron. This assumption is quite obvious since almost 100% of the bath consists of iron, and this element is the first to be oxidized. Then, the ferrous oxides formed in the reaction zones spread throughout the bath. They act as a transmitter of oxygen, which is consumed for oxidation of carbon and other elements, alloys of iron. This scheme agrees well with the depicted above scheme of interaction of oxygen jets with the bath. As applied to the EAFs and other hearth furnaces, this scheme is accepted by most metallurgists.

It is necessary to stress that the two-stage scheme does not exclude the possibility of direct oxidation of carbon in the reaction zones, but this process is given a secondary role. It should be mentioned as well that in oxygen converters, where relative intensity of oxygen blowing (m^3/ton min) is a few times higher than in the hearth furnaces, up to 70–80% of all the carbon can be oxidized in a reaction zone, according to some specialists.

Two-stage scheme explains quite obviously why the transfer of ferrous oxides from the reaction zones to the other areas of the bath is the stage of the decarburization process, which limits the rate of the whole process. If this transfer was just due to the gradient of concentrations and temperatures in the different areas of the bath, i.e., due to the quite slow process of molecular diffusion, the modern rates of heat in the EAFs would not be even remotely achievable. In reality, the bath is stirred intensely by the turbulent metal flows, which result mostly from the dynamic effects of the oxygen jets and CO bubbles on the bath. Under these conditions, the rate of oxygen transfer increases by many times.

Issues of the bath stirring will be reviewed in the following chapter. It should be added here that the role of the bath stirring increases even more when low-carbon steels are produced. When at the end of the heat carbon content drops below 0.2%, the rate of the further bath decarburization is limited to even greater extent by transfer of not only oxygen but carbon as well.

8.3 Melting of Scrap

The features of melting scrap in iron–carbon melt have been studied under laboratory conditions by melting samples in small induction furnaces as well as under industrial conditions in steelmaking units using the method of radioactive isotopes. The isotopes used were those which are practically not possible to oxidize in the steel melting bath, e.g., cobalt isotope. This isotope and other isotopes were introduced into liquid bath at the beginning of scrap melting; and then the radioactivity of metal samples was being measured during the heat. As mass of the liquid bath increased due to melting of scrap, the concentration of isotope in metal and the radioactivity of the samples decreased. This allowed estimating the intensity of the melting process.

Melting of scrap in a liquid bath is significantly different from melting by electric arcs in a freeboard. Rate of melting by the arcs is determined only by the processes of heat transfer from the arc plasma to the surface of scrap lumps. Melting in the liquid bath is complicated by the involvement of mass transfer processes. It has been determined that depending on the melt temperature and carbon concentration, as well as on the temperature of scrap preheating, it is possible to divide the overall duration of scrap melting at the liquid bath into three typical stages.

First stage: freezing of the melt. As cold or slightly heated scrap is immersed into the bath, there is a short period when the melt freezes over the surface of the lumps forming a solid crust. At the beginning, the thickness of this crust grows rapidly. Then, as the heating progresses and scrap temperature increases, this growth slows down due to reduced heat flux from the crust inside the lump. At the moment when this heat flux and heat flux from the melt toward the surface of the crust become equal, the growth of the crust stops and the crust thickness reaches its maximum. As the scrap temperature continues to rise, the crust starts to melt and then disappears. The smaller the mass of scrap lump and the initial temperature differential between the scrap and the melt, the smaller the thickness of the crust and the time it exists.

Second stage: diffusion melting. This process takes place during scrap melting in iron–carbon melt, for example in hot metal, when the temperature of the melt is significantly lower than the scrap melting temperature (1530–1540°C). In case of hot metal this difference is 200°C. The diffusion melting of scrap plays a significant role during the initial stage of heat in a converter. It can take place in the EAFs as well, especially in those which use hot metal in large amounts. Melting of scrap in hot metal would be impossible without diffusion of carbon from the melt into the surface layer of scrap lumps. As this layer is enriched with carbon, its melting temperature becomes lower than the temperature of the melt. The scrap, layer by layer, is converted to a liquid state and mixes with the bath.

The term "melting" itself does not correspond fully to the nature of the described process intimately associated with diffusion of carbon. On the other hand, the term "dissolution," which is also widely used for this process depiction, is not quite applicable since dissolution means mixing with the bath through diffusion without preliminary transition of iron into a liquid state. While further using the term "diffusion melting of scrap" it should be assumed the entire complex of thermal and diffusion processes.

Third stage: intensive melting. Such melting starts as soon as the melt temperature is significantly higher than the scrap melting temperature. Intense supply of heat to the surface of the scrap lumps resulted from the temperature difference and turbulent stirring of the bath assures greatly high rate of melting during this stage. It is worth noting that intensive stirring is a necessary condition required to reach such rates.

Powerful electric arcs and release of large amounts of chemical heat due to intensive oxygen use in modern furnaces assure a rapid growth of bath temperature. Calculations and special research data show that under these conditions, contrary to the heat in a converter, the stage of melt freezing is not quite explicit. This difference can be explained also by the fact that the EAFs use light scrap, while the converters use heavier scrap. Freezing of the melt over light scrap can take place, but it is very short-lived and happens only over some larger lumps. Therefore, it does not affect significantly the duration of scrap melting and electrical energy consumption.

The same can be said regarding the second stage—diffusion scrap melting. Diffusion of carbon from the melt to the surface layer of solid scrap is quite slow process compared to the rate of the bath temperature rising. Besides, unlike in converters, the carbon content in the EAF bath is not quite different from that of the scrap. The exceptions are the furnaces which use hot metal or reduced iron (DRI, HBI), where the carbon content could be 1–2%. In those furnaces, the conditions corresponding to the second stage can occur for quite brief periods of time.

Thus, in order to achieve high productivity, the scrap immersed into the liquid bath must be melted under the conditions of the third stage, i.e., when the melt temperatures greatly exceed the scrap melting temperature. Then, for a short tap-to-tap time, very fast heating of the liquid bath is required.

It is known, Chap. 1, Sect. 1.3.3, that at the present time most of furnaces operate with so-called "hot heel." Such a technology requires that the slag and approximately 15–20% of the metal from the previous heat are left on the furnace bottom. Presence of the hot heel significantly affects the process of melting of the scrap.

This effect is explained by the following. The light scrap, sheet bushelling for example, is placed in the bottom part of the charging baskets in order to avoid the damage of the bottom lining. As the first basket is charged, this scrap, which has a very large surface area, is immersed in the hot heel causing the temperature of the latter to drop by 100–200°C. Such an abrupt cooling of the hot heel with the low carbon content could slow down the heat process. But this is avoided by blowing oxygen into the melt. Blowing speeds up the melting of scrap by intensifying the melt stirring by O_2 jets and CO bubbles, amount of the latter is increased due to intensified oxidation of carbon. The early start of oxygen blowing is required when operating with the hot heel.

8.4 Heating of the Bath

Conditions of heat exchange determining the heating of the bath of the furnace are very complex. Heat from the external sources is delivered to the bath through its surface covered with slag as well as emitted inside the bath itself by internal energy sources. The external sources are electric arcs, oxy-fuel burners, and filling the furnace freeboard gases in case their temperature is significantly higher than the melt temperature. Such conditions are created by post-combustion of CO near the bath surface. The internal sources are chemical reactions of oxidation of carbon as well as iron and its alloys like Si, Mn, etc.

The bath loses heat through emission of CO bubbles, due to thermal conductivity through the lining of the bottom and the banks as well as due to radiation from the slag surface onto the water-cooled panels of the sidewalls and the roof. Great amount of heat obtained by the bath is consumed in heating and melting the scrap immersed in a bath. The main thermal losses of the bath are the losses through radiation and through emission of CO bubbles. The other losses can be ignored.

The main source of heat for the liquid bath is electric arcs, which heat it from above. During the last 30–40 years, electrical power of EAFs has more than quadrupled. On the 120-ton furnaces of a new generation, electrical power in the period of melting of scrap in the liquid bath reaches 120 MW. This power, at the efficiency coefficient of energy $\eta_{EL} = 0.70$, produces average specific heat flux of about 3000 kW/m^2 per entire surface of the bath 6.2 m in diameter. This is more than 10 times higher than the specific heat flux of the oxy-fuel flames in the most powerful open-hearth furnaces operated with intensive use of oxygen.

It is necessary to take into account that heat flux from the arcs is unevenly distributed over the surface of the bath. Only a small area in the middle of the bath surface is directly heated by the arcs. Therefore, the real densities of the heat fluxes incident onto the bath from the arcs are much higher than the average value shown above. For the three-electrode furnaces the local flux densities are several times higher than the average ones, and those of the single-electrode DC furnaces are ten times higher. Let us ask the following question: what is the mechanism of absorption by the bath of enormous amounts of external heat supplied to the bath in a way which does not promote absorption?

It is known that a rational way of heating liquid is heating from below, Chap. 2, Sect. 2.4. During heating, the liquid is stirred intensely, which shortens the duration of heating and reduces energy consumption by many times. The stirring occurs due to the difference in density of the heated and relatively cold volumes of liquid. The heated, less dense liquid due to gravitational forces rises up, while the cold, denser liquid drops down. This type of stirring is called gravitational. It occurs by itself and does not require consumption of external energy. Gravitational stirring is highly intense, and its intensity increases proportionally to the increase in the power of heat energy source. As a result, during the heating from below, the temperature field in the volume of liquid remains rather uniform regardless of increasing power of the heat source.

The bath in EAF is heated from above. Such a type of heating does not produce gravitational stirring. In comparison to non-metallic liquids, liquid steel has high thermal conductivity $\lambda = 23$ W/m × °C, Chap. 3, Sect. 3.2.2. Despite this, removal of heat from the arc zone only due to thermal conductivity of the bath would lead to absolutely unacceptable overheating of the upper layer of the bath. The fact that such a problem does not occur can be explained by intense stirring of the bath. Without intense stirring of the bath, the operation of the high-power EAF would be impossible.

The bath is stirred by the oxygen jets and the CO bubble rising to the surface. Electric arcs contribute to the stirring by exerting dynamic pressure on the bath surface. Lately, blowing the bath with inert gas is being widely used to further intensify the stirring. For this purpose, the special tuyeres are installed in the bottom of the furnace. For the high-capacity furnaces, the bottom blowing becomes mandatory.

Chapter 9
Bath Stirring and Splashing During Oxygen Blowing

9.1 Stirring Intensity: Methods and Results of Measurement

The matters discussed in the previous chapter show that the melt stirring intensity plays the decisive role in the physical–chemical processes of the bath. This parameter requires quantitative definition, since the development and the selection of innovations improving stirring must be based on quantitative data. The stirring intensity is usually evaluated by the time required for equalizing the bath composition throughout the entire bath volume after introducing a foreign material, a so-called tracer, into any small part of this volume. Under the industrial conditions, the additives of either copper or radioactive isotope of cobalt, which are not oxidized in the steel bath, are used as a tracer. The so-called bath uniform mixing time is determined by the method of sampling in the course of the heat and analysis of the metal samples from the different points of the bath. The bath uniform mixing time is characterized by practically identical content of the tracer in all of the samples.

Measuring the stirring intensity in the operating steelmelting units by the above-described method is a very complex and labor-intensive process. All the more attractive appears the option of studying the stirring processes by the methods of cold physical modeling. However, physical modeling gives reliable results only if the definite similarity conditions are complied with. These conditions imply quite strict requirements for the selection of both geometric parameters of a model and physical parameters of the liquids simulating metal and slag. These parameters must be governed by the requirements based upon the well-developed theory of similitude rather than by the ease of experiment. They cannot be selected at random and independently of each other. On the contrary, there must be a definite quantitative correlation between them.

Unfortunately, the requirements of the theory of similitude are either considered inadequately or even completely ignored in some studies. The liquid steel is simulated by the water, although the application of water interferes with developing a model in a required scale and does not allow to select a suitable liquid for

Y. N. Toulouevski and I. Y. Zinurov, *Innovation in Electric Arc Furnaces*, DOI: 10.1007/978-3-642-36273-6_9, © Springer-Verlag Berlin Heidelberg 2013

Table 9.1 The effect of bottom blowing on stirring intensity of oxygen converters[a]

Type of converter	Only with top blowing – LD	With combined blowing: top and bottom		Only with bottom blowing Q – BOP
Gas blown via the bottom	–	N_2 and Ar	O_2	O_2
Intensity of bottom blowing m^3/ton × min	–	0.15	0.9	5
Time of the complete bath stirring, s	100	40	20	4–5

[a] The average values are given [2]

the slag simulation [1]. The violation of similarity conditions depreciates the results of the studies of hydrodynamics of steelmelting baths with the use of models. Observed in a model picture of the processes of bath stirring, gas jets penetrating the bath, slag foaming, etc., might seem "resembling" the actual processes. However, this "resemblance" does not give grounds for the strong quantitative conclusions. The latter requires from the experiments in the models not "a resemblance," but the process similarity in a strict mathematical sense of this scientific term.

It should be noted that, using the cold physical modeling, it is impossible to ensure complete similarity to the real processes in a bath. Even in the most thoroughly set experiments similarity is not accurate. Therefore, preference should be given to those data on stirring, which are obtained by direct measurements in the steelmelting units. Unfortunately, systematic data are known only for the oxygen converters, Table 9.1 [2]. But they are of great interest for the EAFs as well.

Data in Table 9.1 are obtained with the use of copper additives as the tracers. These data show that, despite of the high intensity of oxygen top-blowing in the modern converters (4.5–6.0 m^3/ton × min), the additional injecting of even very small amount (0.15–0.9 m^3/ton × min) of gases through the bottom tuyeres increases the stirring intensity by 2.5–5 times, and, with replacing the top-blowing with bottom-blowing, approximately by 20–25 times. It does not mean that such replacement is necessarily expedient, since each of these methods has technological advantages and deficiencies, whereas their combination gives the optimum results.

9.2 Mechanisms of Bath Stirring

9.2.1 Stirring Through Circulation and Pulsation

Visual observations show that the bath of EAF is in constant motion. This visible motion is characterized by circulation in a horizontal plane. The linear velocity of this circulation is low: 1.0–1.5 m/s. However, because of the high density of the liquid metal (7830 kg/m^3), such a low velocity corresponds to quite high values of the criterion *Re* in a range of 10^5–10^6, which indicates the clearly expressed turbulent nature of motion, Chap. 3, Sect. 3.3.2. Besides the horizontal flows, there are vertical flows circulating in the bath, which are characterized by approximately the same velocity and level of turbulence. All these flows ensure the so-called circulation bath stirring.

The circulation of the melt occurs due to a number of factors. The high-speed jets of oxygen and the floating-up bubbles of CO are the key factors. The reverse flows of oxygen in the reaction zones, Chap. 8, Fig. 8.1, entrain the metal in the upward direction. The floating-up gas bubbles and, to a lesser degree, powerful electric arcs produce the same effect. The ascending flows of the metal cause its vertical circulation, which contributes to equalizing the temperature over the depth of the bath. The metal with the higher temperature flows over the bath surface in the direction from the electric arcs to the periphery and is replaced with the colder metal rising up from below.

The circulation stirring of the macro-volumes of the bath is of great importance. It ensures the continuous renewal of gas–metal interfaces in the reaction zones of oxygen blowing, the delivery of fresh metal into these zones, and dispersion of the iron oxides formed throughout the entire volume of the bath. This type of stirring plays the dominant role in the processes of bath heating and scrap melting in it.

The second type of stirring occurs due to the high-frequency turbulent pulsations of the micro-volumes of the bath, which are amplified with an increase in the flow velocity and in the criterion *Re*. This type of stirring is called pulsation stirring. Pulsation stirring ensures the continually renewed close contact of reagents required for chemical reactions to take place in the bath. In particular, it is the pulsation stirring that ultimately ensures the high decarburization rate of the bath required for the current level of intensification of the melting process in the EAFs.

The small-scale turbulent pulsations in the bath whose intensity is determined by the criterion *Re* are superimposed with the large-scale pulsations of metal caused by the motion of CO bubbles and by the processes in the reaction zones occurring due to penetration of oxygen jets into the bath. These additional pulsations substantially increase the intensity of bath stirring as well as considerably accelerate bath heating and scrap melting. Therefore, circulation and pulsation stirring supplement each other and thus enhance the total effect.

9.2.2 Stirring by Oxygen Jets and CO Bubbles

Earlier, the majority of metallurgists assumed that the CO bubbles played a dominant role in bath stirring, at least, in the open-hearth and EAFs. The effect of oxygen jets was considered to be secondary. However, recently, this opinion has changed due to a sharp increase in intensity of oxygen blowing.

It has been established that with an increase in the decarburization rate of the bath and, correspondingly, in the number of bubbles formed per unit time, the intensity of bath stirring by bubbles increases. However, this increase is characterized by a rapidly damped curve approaching a constant value. This is explained by the fact that an increase in the number of bubbles results in a decrease in the hydraulic resistance acting on the bubbles floating up in the bath. Consequently, the energy of the floating-up bubbles used for stirring decreases. Since the intensity of bath stirring by jets grows with an increase in consumption of oxygen, the relative effect of the jets on the stirring increases in the case of high blowing intensity.

It should be emphasized that the influence on the bath stirring of the oxygen jets and the CO bubbles is closely related. With a change in the intensity of the bath blowing, the decarburization rate of the bath changes as well. Therefore, it seems impossible to directly estimate the separate contribution to stirring of each of these factors. All speculations on the relative value of these contributions are based mainly on the results of theoretical calculations. Since the processes in the bath are extremely complex, these calculations inevitably include a whole series of assumptions and simplifications, which make their results insufficiently reliable for quantitative conclusions.

Leaving aside the questions that are primarily of theoretical interest, let us concentrate on the practical aspect of the matter, namely, on the actual bath stirring control capabilities. The analysis of this problem leads to the conclusion that, in fact, the only effective and accessible method of controlling, within the wide limits, the intensity of bath stirring is by varying the parameters of oxygen blowing. The effect of blowing is determined not as much by the oxygen consumption, as by the blowing methods, i.e., by design and operation modes of the blowing devices. With the identical oxygen consumptions, a change in the blowing method can produce many times more effective bath stirring, as it happens in the oxygen converters, Table 9.1. With an increase in oxygen consumption, the bath stirring intensity increases as well. However, this consumption can be increased only to a certain limit, which depends on a number of factors, but is ultimately determined by the technical characteristics of the blowing method.

9.3 Factors Limiting Intensity of Bath Oxygen Blowing in Electric Arc Furnaces

9.3.1 Iron Oxidation: Effect of Stirring

During dissolution in liquid iron, the oxygen is converted to atomic state. Concentration of dissolved oxygen [% O] is limited by metal temperature and by content of carbon [% C] in the metal. With an increase in temperature and a decrease in carbon content, the solubility of oxygen in iron increases, and concentration [% O] rises. According to the laws of physical chemistry, the reaction of carbon oxidation by oxygen dissolved in liquid iron, similar to oxidation reactions of other elements, cannot proceed to completion, i.e., to total consumption of one of the reagents. With a certain ratio between [% O] and [% C], the chemical equilibrium is achieved, which means that the concentrations of oxygen and carbon in the metal remain constant. At constant temperature and in the absence of external forces, this equilibrium can be maintained for arbitrarily long time.

At the temperatures of the bath of EAFs and oxygen converters and with the typical carbon content of the bath, the dependence between the equilibrium concentrations [% C] and [% O] is described by the known equation:

$$[\%\mathrm{C}] \times [\%\mathrm{O}] = 0.0025. \tag{9.1}$$

For [% C] $>$ 0.05, this hyperbolic dependence corresponds to relatively low concentrations of oxygen [% O], i.e., the degree of oxidation of the metal is low. With further reduction of carbon content, the equilibrium oxidation of the metal increases sharply, Table 9.2.

Continuous blowing of oxygen into the bath excludes the possibility of achieving the equilibrium between [% O] and [% C]. The metal in such a bath, to one degree or another, is always overoxidized. With high intensity of blowing and insufficiently intensive stirring, the oxygen content in the metal can be many times higher than its equilibrium concentration. At the same time, the non-uniformity of oxygen distribution throughout the volume of the bath increases. It is always higher than the non-uniformity of carbon distribution.

The highest degree of over-oxidation is observed in the reaction zones of the bath. The over-oxidation of the metal is reduced as the distance from these zones increases. The lower the carbon content, the greater volume of the bath is over-oxidized. As the intensity of oxygen blowing increases, the degree of over-oxidation of metal increases rapidly. This causes the excess oxygen evolution from the solution in the form of iron oxides, which partially pass into the slag, increasing the FeO content in the slag.

With the same oxygen consumption for blowing, the degree of over-oxidation of the bath to the great extent depends on the intensity of the bath stirring. In the oxygen converters, where the metal is stirred considerably more intensively than in EAFs, the metal is much less overoxidized and is closer to the equilibrium state despite the fact that the specific intensity of blowing in the converters (m^3/ton $\times$ min) is four to five times higher. The converter metal with the bottom blowing, which is characterized by the maximum stirring intensity, Table 9.1, has the lowest

Table 9.2 Concentrations of oxygen dissolved in metal in equilibrium with carbon content

[% C]	0.40	0.30	0.20	0.15	0.10	0.050	0.025
[% O]	0.0062	0.0083	0.0125	0.0167	0.0250	0.050	0.100

degree of oxidation. With the bottom blowing, the FeO content in the converter slag decreases by 2–4% in comparison with blowing from the top that considerably increases the yield.

As already mentioned, a tendency to shorten, in every possible way, the tap-to-tap time and to increase the productivity of EAFs forced the steelmakers to sharply increase the oxygen consumptions for the blowing of the bath. However, the special measures for increasing the intensity of the bath stirring were not being taken. As a result, the effectiveness of the use of oxygen reduced considerably. With exceeding the level of 25–30 m^3/ton of O_2, which corresponds to the specific blowing intensity equal to approximately 1 m^3/ton $\times$ min, each next step in shortening of tap-to-tap time and reduction of electrical energy consumption required a sharper increase in oxygen consumption. At the same time, the FeO content in the slag increased significantly and the yield reduced sharply, which indicated the increasing degree of bath over-oxidation. This deadlock has been resolved by an increase in the consumption of the carbon powder injected into the bath simultaneously with oxygen blowing.

Injecting the carbon into the bath. This technological operation solves two different problems, namely, the slag foaming and the reduction of iron oxides for the purpose of increasing the yield. The slag is foamed in order to immerse the electric arcs. The slag foaming requires relatively low carbon consumption, Chap. 1. A decrease in the FeO content in the slag and in the degree of over-oxidation of the bath caused by intensive oxygen blowing in combination with insufficient stirring, considerably increases the required carbon consumption. The carbon is usually injected into the slag. The reduction of FeO to iron occurs according to the reaction: FeO + C = Fe + CO. The advantages and disadvantages in this technological method are examined in Chap. 11 in connection with the design of the blowing devices.

9.3.2 Bath Splashing

Splashing of metal and slag during oxygen blowing of the bath has an essential effect on the furnaces' operation. Excessive splashing caused by the poorly designed tuyeres or too intensive blowing can make it necessary to reduce oxygen consumption. In the EAFs, the study of the processes of splashing of metal and slag has not been given an adequate consideration. The detailed studies have been conducted in the open-hearth and twin-shell furnaces [1].

These extensive studies are characterized by the wide use of methods of physical modeling of processes with the full compliance with the requirements of the theory of similitude. In these studies, liquid steel was simulated by a solution of iodide potassium (KJ), and slag was simulated by a solution of vacuum pump oil in kerosene. The ratio between the densities of both solutions complied with the similarity conditions. Concurrently, the industrial experiments were conducted on the furnaces equipped with different blowing devices. High-speed video recording was carried out. The magnitude of the entrainment of dust and splashes out of the freeboard as well as the granulometric and chemical composition of the entrainment were determined. Comparison of all these data allowed establishing the basic laws governing the splashing. The hydrodynamic processes in the baths of different hearth steelmelting furnaces are similar to each other, including the processes taking place during the oxygen blowing of the bath [3]. In view of that, the results of the work [1], with good reason, will be used below for describing the processes of bath splashing in the EAFs as well.

The granulometric composition of dust and splashes generated in the bath is quite non-uniform. The dust is formed not due to splashing only. The finest dust, or the so-called red fume, is generated mainly by evaporation of iron in the high-temperature reaction zones of the blowing. If metal and slag splashing is caused only by CO bubbles evolving from the bath, the size of the particles is very small and varies from 0.01 to 0.15 micrometers (1 μm = 10^{-3} mm). With the blowing, along with the small particles, the large particles with the size varying from a fraction of a millimeter to several millimeters appear. An amount of dust and splashes increases sharply. Without blowing, their overall mass is 2–3 g/m^3 of gases leaving the freeboard; with the moderate blowing, this number increases up to 10–15 g/m^3; whereas with the intensive blowing (about 1 m^3 of O_2/ton × min), it is equal to 60–70 g/m^3.

The relatively large splashes consist of metal and slag. They are ejected from the bath at a quite high speed. Some of them are ejected from the electrodes' gaps to the height of several meters. Predominantly, most of the splashes are not entrained by the off-gases, like the red fume, but settle onto the sidewall and roof panels, the roof elbow, and the stationary gas duct. The largest splashes fall back into the bath. In the furnace trials all these splashes are not being captured since dust collecting devices are installed outside of the freeboard. Therefore, the amount of splashing is usually strongly underestimated. With the intensive oxygen blowing the relatively large splashes can make approximately 20% of the overall mass of ejections.

The majority of splashes are formed in the zones of blowing. The splashing intensifies with an increase in the viscosity of slag due to reduction in its temperature. It also intensifies if the slag is inhomogeneous, especially when it contains the undissolved pieces of lime and other bulk materials. The splashing also increases with an increase in the amount of slag.

The intensity of splashing depends greatly on the design of oxygen tuyeres and their positioning relative to the bath surface. The single-nozzle and the triple-nozzle tuyeres with downward angle of 30° to the vertical were studied. In all cases, the single-nozzle tuyeres produce the maximum splashing, which leads to the increased wear of roof refractory. Because of that, the single-nozzle tuyeres have not found an application in the open-hearth furnaces.

As the triple-nozzle tuyere is approaching the bath surface and, after its submerging into the slag, the slag–metal interface, the following changes in the intensity of splashing are being observed, Fig. 9.1. With a large distance between the nozzles and the slag surface, the impact of the oxygen jets on the bath is minor and the splashing intensity is negligible. At a certain small distance from the slag

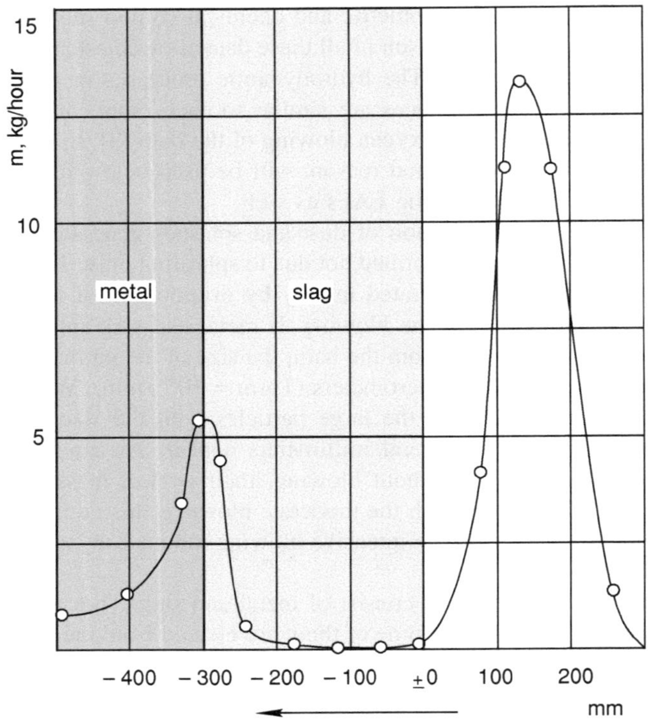

Fig. 9.1 Dependence of splashing intensity m, kg/h, on position of tuyere nozzles relative to the slag surface (results of physical simulation). The *arrow* indicates direction of motion of the tuyere

surface, the splashing intensity increases rapidly and reaches the maximum. At that moment the slag produces more splashes than the metal does. At even closer distance between the nozzles and the bath surface, the splashing intensity reduces sharply, and with the nozzles submerged into the slag, it reduces practically to zero. As the nozzles, after further submerging, get even closer to the slag–metal interface the splashing intensity increases again. The second maximum of splashing intensity appears, but its magnitude is much lower than that of the first maximum. The splashing stops altogether as soon as the tuyere submerges into the metal.

The presence of the second maximum of splashing intensity is explained by formation, at the slag–metal interface, of the cellular interlayer, which is formed only in the zones of blowing due to the mixing of metal and slag. This interlayer contains 65–85% of metal; the rest is slag and gas. The viscosity of this layer is considerably higher than that of its liquid components, which is the main cause of intensification of splashing. The thickness of this layer is proportional to the thickness of the slag. The presence of this layer has been detected both during the cold modeling of bath splashing and during the direct experiments not in the furnaces only, but in the oxygen converters as well.

The intense bath splashing is accompanied by a number of negative effects on the furnace operation. It increases the metal losses and decreases the yield; it leads to a sharp decrease in durability of the ceramic center section of the roof due to the interaction between refractory and slag splashes; it causes the excessive formation of the highly metallized skull on the water-cooled wall and roof panels. The latter hampers the operation of the burners and other devices located in the walls and in the roof.

The cases are known, when due to intense bath splashing, the roof was welded with sidewalls so that the swing of the roof involved some difficulties. This called for reduction in oxygen consumption for blowing. The excessive splashing leads to the fast clogging of the roof gas-exhaust elbow and the entrance section of the stationary gas duct with the slag–metal deposits, which increases the expenditures on their purging. With the high oxygen consumption, the relatively low level of splashing can be ensured by the dispersing of the blowing zones over the bath surface, as well as by submerging the oxygen tuyeres into the slag.

9.4 Oxygen Jets as a Key to Controlling Processes in the Bath

It was shown in Sect. 9.2.2 that varying the oxygen jets' parameters and methods of oxygen blowing into the bath are those basic controlling measures, which allow, in practice, to intensify melt stirring within the wide limits. All basic processes in the bath such as bath heating, carbon oxidation, and scrap melting are closely correlated with the stirring intensity. Therefore, the control of these processes is reduced, to a considerable extent, to the controlled change of the parameters and methods of jet blowing. The same relates to the metal and slag splashing.

In the previous chapters, the effect of oxygen blowing on the processes in the bath was examined without taking into consideration the characteristics of the jets themselves. This examination is insufficient. It should be supplemented by the analysis of those parameters of the jets, which have the greatest effect on their fundamental characteristics, such as long range and depth of penetration into melt.

The characteristics of the jets are determined by the design parameters of injectors, tuyeres, and other devices utilized for the blowing of the bath. These devices are being continuously improved. Chapter 11 is dedicated to the analysis of the existing designs and the directions of further development. The materials in this chapter assume a deep enough understanding of the fundamental principles of the jet streams. These principles are different for the sonic and supersonic jets, and they are very complex. They are examined in the books on gas aerodynamics with application, as a rule, of a large number of complex formulae, which make the physical essence of phenomena difficult to understand.

In order to help the readers, a brief summary of the laws of the jet streams is given in the following chapter at the elementary level with the use of simplified formulae which, on the other hand, do not compromise scientific accuracy of the concepts. Getting familiar with the materials in this chapter will allow the readers to be quite freely oriented in the field of jet streams to carry out the calculations of subsonic and supersonic jets and their nozzles. All this is necessary both for objective evaluation of proposed blowing devices and for development engineering.

References

1. Markov BL (1975) Methods of open-hearth bath blowing. Metallurgia, Moskow
2. Povolotski DI (2004) The fundamentals of steelmaking technology. Publishing house of SUSU, Chelyabinsk
3. Glinkov MA (1970) Thermal performance of steelmelting baths. Metallurgia, Moskow

Chapter 10
Jet Streams: Fundamental Laws and Calculation Formulae

10.1 Jet Momentum

Jetpenetration into the melt depends on the jet force acting on the bath surface. This force is unequivocally determined by the jet momentum I, which is expressed by the formula:

$$I = m \times v_1,\ \text{kgm/s}^2, \tag{10.1}$$

m – mass of the jet passing through the nozzle in unit time, kg/s
v_1 – initial velocity of the jet at the nozzle exit, m/s

The unit of measurement of the momentum corresponds to the unit of measurement of the force, which is Newton, 1 N = 1 kgm/s^2. Unfortunately, quite often the ability of jet to penetrate into the melt is characterized not by the momentum, but by its kinetic energy $mv^2/2$. When the comparison of jets is carried out, this mistake leads to the incorrect quantitative assessments of the relative effectiveness of the different jet devices. In formula (10.1), the jet momentum is equal to the amount of motion of jet mv. This is correct if the pressure of the jet at the nozzle exit section p_1 is equal to the pressure of the surrounding gaseous medium p_{MED} into which the jet is projected. If the pressure at the nozzle exit is higher than that of the surrounding gaseous medium, then the force of excessive pressure of the gas has to be added to its momentum mw. The jet momentum in this case is determined by the expression:

$$I = m \times v_1 + f_1(p_1 - p_{\text{MED}}), \tag{10.2}$$

f_1 – area of the nozzle exit section, m^2

Y. N. Toulouevski and I. Y. Zinurov, *Innovation in Electric Arc Furnaces*,
DOI: 10.1007/978-3-642-36273-6_10, © Springer-Verlag Berlin Heidelberg 2013

10.2 Flooded Free Turbulent Jet: Formation Mechanism and Basic Principles

This kind of jet is formed when a gas stream projects from a nozzle into a relatively large space filled with a gas of same or different density.[1] The jet is not constrained by the walls and interacts freely with surrounding relatively motionless gas, engaging it into its motion. The process mechanism is as follows.

Much as the gases and the liquids consist of the chaotically moving molecules, the turbulent flows consist of micro-volumes of gas or liquid, each of them being a small vortex. In a large free space or when the flows encounter obstacles, these vortices can grow to a macroscopic size. In the midst of the ordered motion of the flow, the vortices move chaotically in all directions stirring the flow intensively. The chaotic motion of the vortices has the character of high-frequency speed pulsations. At any point of a flow its velocity can be decomposed in two components: time-averaged velocity and pulsating velocity. Since the pulsations do not have predominant direction and are being summed as vectors, time-averaged values of the pulsating velocity components are equal to zero. However, due to these pulsations, the vortical jet volumes of high translational velocity are ejected from the jet into adjacent layer of gas. They slowed down there, transferring a portion of their momentum to the micro-volumes of gas and thus forcing gas to move with the jet.

The jet pulsations induce the pulsations of the surrounding gas as well. The vortical gas volumes enter the jet replacing those ejected from the jet. These volumes of a relatively low velocity are being accelerated in the jet, which is slowed down thereat. As a result of this turbulent mixing of the jet with the medium, a so-called boundary layer (1) is formed. This layer separates the motionless surrounding gas and jet core (2), Fig. 10.1. In the core, at a certain distance from the nozzle, the initial jet velocity v_1 remains unchanged. A considerable amount of energy is consumed for the turbulent mass and energy exchange between the jet and the surrounding gas. As a result, the average jet velocity and its kinetic energy decrease as the distance from the nozzle grows. However, in accordance with the mechanism of the jet formation, the momentum or the amount of motion remains constant and equal to the initial value:

$$I = mv = \text{const.} \tag{10.3}$$

The pressure in the free jet is also constant and equal to the pressure of the surrounding gaseous medium.

---

[1] The term "flooded jet" does not suggest that a jet is submerged into liquid. It refers to the case when a gas is projected into another gas. The density of one gas is only several times higher or lower than the density of another, whereas the densities of liquids exceed thousands times the densities of gases.

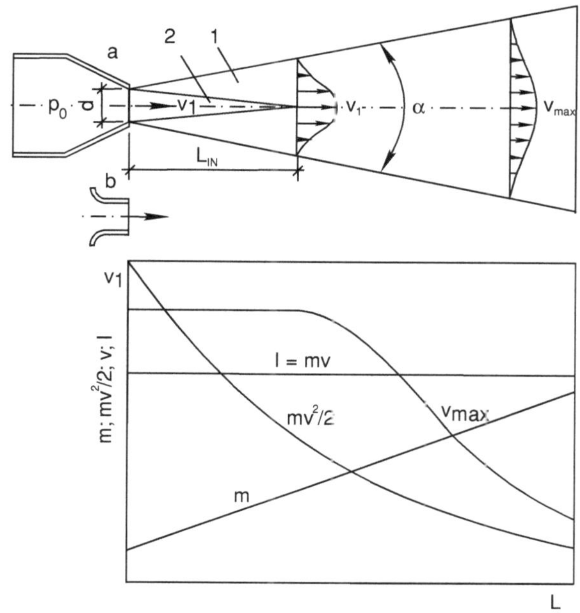

Fig. 10.1 Variation of free jet characteristics versus its length (**a**) tapering nozzle; (**b**) cylindrical (the rest of designation are given in the text)

The smallness of the time-averaged transverse components of the velocity in any cross-section of the jet in comparison with the longitudinal velocity is the characteristic property of the turbulent jets. In the engineering calculations of the jets, the transverse components of the velocity are not being considered because of their smallness. In the cross-sections of the jet the longitudinal velocity decreases away from the axis, where it has maximum value, toward the periphery, asymptotically approaching zero, Fig. 10.1. In the calculations, which do not require extremely high accuracy, the conical surface where the velocities along the axis of the jet are negligibly small is taken as a jet boundary. For instance, the peripheral jet areas where velocities do not exceed 1–3% of the velocities at the axis can be ignored. The expansion angle of the free subsonic jet is equal to 18–26° and increases with an increase in the degree of flow turbulence, the *Re* numbers, Chap. 3.

The initial and the main regions of the jet are distinguished. Within the limits of the initial region varying in length L_{IN} depending on *Re* from 4 to 8 diameters (calibers) of a nozzle *d*, the increasing thickness of the boundary layer gradually "eats away" the jet core. At the very end of the initial region, the jet core disappears, and the velocity v_1 remains the same at the axis only, Fig. 10.1.

The main region of the jet starts right after the initial region. In the main region, the velocity at the axis v_{MAX} as well as the average over cross-section jet velocity continuously decreases. It should be noted that the average velocity decreases within the limits of the initial region as well. The jet mass grows linearly along its length. The boundary layer thickness at the nozzle exit section is equal to zero;

after that it grows continuously. The jet core does not exist in the main region, and the turbulent boundary layer occupies the entire cross-section of the jet, Fig. 10.1.

10.3 Subsonic Jets: Cylindrical and Tapered Nozzles

The initial gas flow velocity at the nozzle exit v_1 depends on the following factors: the nozzle configuration, physical characteristics of the gas, the pressure of the gas before entering the nozzle p_0, and the pressure of the surrounding gaseous medium p_{MED} into which the jet is projected. The jets examined in this section are those projecting from the cylindrical and tapered nozzles, Fig. 10.1. For the sake of brevity, such nozzles will be called "simple." For small pressure differences $\Delta p = p_0 - p_{MED}$ and, correspondingly, small initial gas flow velocities at the nozzle exit v_1 not exceeding 150–200 m/s, the density of the gas, ρ kg/m^3, practically does not depend on the flow velocity, and the gas behaves as an incompressible liquid (water, for instance). In this range the velocity v_1, m/s, and the volumetric flow rate of gas V, m^3/s, through the nozzles in question are calculated using simple formulae (10.4) and (10.5) for the incompressible liquids:

$$v_1 = \varphi \times \sqrt{\frac{2(p_0 - p_{MED})}{\rho}}, \ \mathrm{m/s}, \tag{10.4}$$

φ – velocity coefficient taking into account the pressure losses due to hydraulic resistance of the nozzle
p_0 – pressure of the gas before the nozzle, Pa
p_{MED} – pressure in the jet at the nozzle exit which in this case is equal to the pressure of the surrounding medium, Pa
ρ– density of the gas before the nozzle, with due regard for its pressure and temperature, kg/m^3

With the rounded-off edges of the nozzle on the side of gas entry, Fig. 10.1, the smooth surface of the nozzle and its small length, the coefficient φ is equal to 0.95–0.97.

$$V = \mu \times f_1 \times v_1, \mathrm{m}^3/s, \tag{10.5}$$

f_1 – area of the nozzle exit cross-section, m^2
μ – coefficient taking into account certain contraction of the jet leaving the nozzle.

Depending on shape and surface finish of the nozzles, the value μ lies within the limits 0.90–0.95. Since the volume of gas depends on the pressure and the temperature, the volumetric flow rates of gases, formula (10.5), for the purpose of comparison, are usually expressed in the so-called normal (standard) conditions, i.e., the pressure equal to 760 mm of Hg (101 kPa) and the temperature equal to 0°C, Chap. 2, Sect. 2.1.

The mass flow rate of gas $m = \rho \times V$ kg/s is obviously equal to:

$$m = \mu \times f_1 \times v_1, \text{ kg/s} \tag{10.6}$$

At $v_1 > 200$ m/s, gases begin to compress noticeably, their density ρ grows as the velocity increases, and the use of formulae (10.4) and (10.5) leads to the increasing errors. If the simple nozzles are used, then, according to the laws of gas dynamics, v_1 cannot exceed a certain critical value v_1^* equal to the speed of sound a in gas forming a jet. Therefore, with an increase in p_0 and in the difference $p_0 - p_{MED}$, the velocity v_1 grows only to $v_1^* = a$, Fig. 10.2.

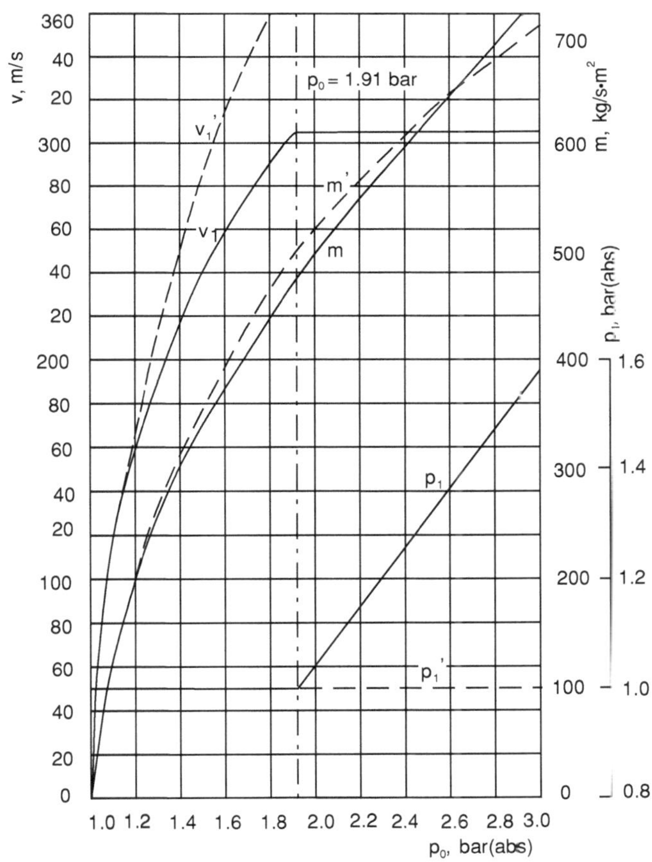

Fig. 10.2 Variation of air (oxygen) jet parameters at the exit of a simple nozzle depending on the pressure in front of the nozzle p_0 [1] (*dotted curves* – without taking into consideration of compressibility of gases)

The critical velocity v_1^* corresponds to the defined critical value of the ratio of absolute pressures p_{MED}/p_0^* in which p_0^* is the critical pressure of the gas before entering the nozzle, which corresponds to the critical velocity v_1^* as well. For the diatomic gases (oxygen, air), $p_{MED}/p_0^* = 0.528$. Hence, when the absolute pressure of the medium p_{MED} is equal to the normal atmospheric pressure 760 mm of Hg (101 kPa, or 1.01 bar), the critical pressure of oxygen (air) $p_0^* = 101{,}000/0.528 = 191{,}000$ Pa, or $p_0^* = 1.91$ bar abs (the absolute pressure is equal to the pressure on manometer plus 1.01 bar). With an increase in p_0 from $p_{MED} = 1.01$ to 1.91 bar (abs), the initial velocity of gas for the simple nozzles increases from zero to the critical velocity v_1^* equal to the speed of sound in gas, Fig. 10.2.

In the range of values of p_0 smaller or equal p_0^* ($p_0 \leqslant p_0^*$), the gas outlet pressure p_1 remains equal to p_{MED}. With further increase in the inlet pressure when p_0 begins to exceed p_0^*, the outlet pressure also increases and increasingly exceeds p_{MED}. The outlet gas velocity in this case does not change and remains equal to the critical velocity v_1^*. However, the mass m (kg/s $\times$ m^2) of nozzle cross-section and, therefore, the volumetric flow rate of gas corrected to the standard conditions V (m^3/s $\times$ m^2) grow, since the gas density increases directly proportional to the pressure p_1, Fig. 10.2.

The gas leaving the nozzle at the pressure exceeding the surrounding medium pressure continues to expand, and, at a small distance from the nozzle, the pressure in the jet equalizes with p_{MED}. Thus, one of the basic properties of the free flooded jet noted above, namely, the equality of the jet and medium pressures, holds true in this case as well. With the additional expansion of the gas outside of the nozzle, its velocity somewhat grows. However, in the engineering calculations this effect can be disregarded.

It needs to be explained why a gas jet ejected from a simple nozzle cannot exceed the speed of sound. The laws governing the high-speed motion of gases are quite complex and, in the majority of the cases, they are not easy to interpret illustratively. Nevertheless, it is possible to answer the presented question without resorting to the abstract language of formulae. The answer will not be comprehensive, but it will give some idea about the physical essence of the process.

The sound, which is detected by our auditory organs, is relatively small longitudinal vibrations of pressure in elastic medium, e.g., in air or water. Utilizing the rough but quite illustrative analogy, it is possible to say that these vibrations propagate from a small point-like source of sound (a bell, for example) in all directions, similar to circular waves spreading out from the point of impact created by a stone dropped in water. Only the acoustic vibrations propagate not over the surface of water but in the three-dimensional space (airspace, for example) and have not a circular but a spherical shape.

The propagation velocity of these relatively weak vibrations of pressure in gas dynamics is called the speed of sound. A remarkable feature of the speed of sound is that within quite wide limits, it does not depend on the magnitude of these vibrations, i.e., sound intensity, or their frequency, i.e., pitch. Gas dynamics is only interested in the fact that gases moving with velocities greater than the speed of sound behave completely differently from those moving with subsonic velocities. For this very reason, the speed of sound a is one of the basic characteristics in gas dynamics. Let us note that in case of large pressure changes (vibrations) associated with detonation, e.g., during explosions, the propagation velocity of the blast wave (sound of explosion), depending on conditions, can vary over wide limits and exceed many times the value a.

Examined in gas dynamics the speed of sound a depends only on the physical properties and the temperature of the medium in which the sound propagates. At 0°C, the speed of sound in oxygen is about 300 m/s, and in the medium humidity air approximately 330 m/s. These figures correspond to the speed of sound in the gas in a case when gas and sound source are not moving relative to each other, which means their own velocities are either equal to zero or identical. If the gas moves relative to the source of sound with a velocity v, then, relative to the external motionless observer and in accordance with the principle of addition of velocities, the speed of sound in the direction of motion of the gas (downstream) will be equal to $a + v$, and in the opposite direction (upstream) it will be $a - v$. If the gas velocity v relative to the sound source is equal to or greater than a, then the sound will not propagate in the direction opposite to the flow of the gas.

Let us examine the jet projecting from a simple nozzle at a constant pressure of the gas before the nozzle p_0. Let us start decreasing, step by step, the pressure of the medium outside the nozzle p_{MED}. With the subsonic initial velocities of the jet, the changes in the pressure of the medium p_{MED} propagating with the speed of sound will be able, when moving in the jet upstream, to reach the nozzle exit section. Due to this, the pressure at the nozzle exit section will decrease correspondingly and remain equal to p_{MED}. As a result of the increase in the pressure difference $p_0 - p_{MED}$, the velocity $v_1$and the mass flow rate of the gas m at the nozzle exit will increase with the decrease in p_{MED}. But as soon as the initial velocity of the jet v_1 becomes equal to the speed of sound, further decrease in the pressure of the medium p_{MED} will no longer be able to spread out against the gas flow to the nozzle exit section. Despite of further decrease in p_{MED}, the pressure at the nozzle exit p_1 will stop decreasing together with the medium pressure. A constant pressure difference $p_0 - p_1$ will be established. As this takes place, the initial velocity of the jet at the nozzle exit becomes critical and, with any further decrease in p_{MED} down to zero (projecting into vacuum), will remain constant and equal to a.

If thereafter p_0 is increased, this leads to an increase in the gas pressure p_1 and density ρ_1 at the nozzle exit and, therefore, to an increase in the mass flow rate of the gas m, Fig. 10.2. However, the gas projecting velocity v_1 in this case will not increase, since an increase in p_0 will be compensated by an increase in p_1 and ρ_1. Such is a physical picture of the processes, which gives an answer to the presented question.

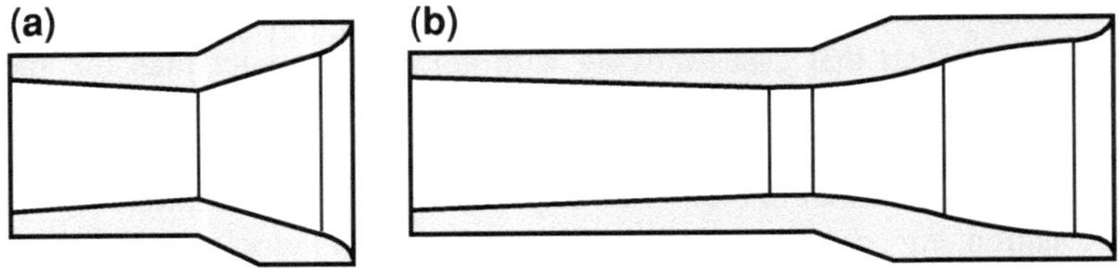

Fig. 10.3 A simple nozzle of de Laval (**a**) and coherent nozzle (**b**)

10.4 Supersonic Jets and Nozzles: Operation Modes

A depth of jet submersion into the melt and its ability to stir the bath are determined by the jet momentum, Sect. 10.1. The momentum grows proportional to the initial velocity of the jet, formulae (10.1) and (10.2). Therefore, in order to increase the effectiveness of the bath blowing, it is necessary to use oxygen jets with maximum possible supersonic velocities. Such velocities are reached in the supersonic nozzles.

Any supersonic nozzle consists of converging and diverging sections. In the nozzles of the simplest configuration which, for convenience, we will further call de Laval nozzles,[2] both sections have conical shape, Fig. 10.3a. The central convergence angle usually is 30–60°, while the central divergence angle is 8–16°. The transition between the cones and the nozzle inlet edge is rounded. The nozzles intended for increasing the jet length due to decreasing in the turbulence degree have more complex special configuration, Fig. 10.3b. Both the convergence and divergence angles are approximately halved which lengthens the nozzle significantly. Such nozzles, in accordance with accepted terminology, will be further called coherent. The manufacturing of such nozzles is complicated; therefore they are only used for special applications.

In the converging section of the nozzle operating similar to the simple nozzle, the gas is accelerated to the speed of sound, which is reached in the narrowest section of the nozzle. This section is called critical. The gas flow in the converging section is governed by the same laws as gas flow through the simple nozzle. The diverging section of the nozzle serves for further gas acceleration and for reaching the velocities exceeding the speed of sound. In this section of the nozzle in the direction of gas motion, the velocity of the gas increases and its pressure drops. The reasons dictating such a nozzle configuration are as follows.

As in the case of the motion of an incompressible liquid, during the subsonic motion of the gas, the velocity of the flow grows with the decrease in the passage cross-section and, on the contrary, drops with an increase in the cross-section. During the supersonic motion the gas behaves oppositely. Its velocity grows with an increase in the passage cross-section and decreases with the decrease in the

[2] Very often, all supersonic nozzles are called de Laval nozzles, though the latter initially had the form shown in Fig. 10.3a.

cross-section. This is related to the fact that at supersonic velocities the gas density rapidly drops with an increase in the velocity.

For the defined flow rate m of a certain gas (oxygen, for instance), its inlet pressure p_0 and pressure p_{MED} of the surrounding medium into which the gas projects, it is possible to select the areas of the critical section of a supersonic nozzle f_{CR} and its exit section f_1 which will bring the outlet pressure in jet p_1 to the level of p_{MED}. In this case, the potential energy of the compressed gas in front of the nozzle will be fully, to the utmost possible extent, converted into the kinetic energy of the jet, and the jet initial velocity will reach the maximum possible value for the given conditions.

The mode of nozzle operation at $p_1 = p_{MED}$ is optimal and is called the design mode. It is important to emphasize that, for the given values of m, p_0, and p_{MED}, there are the unique geometric parameters of the nozzle f_{CR} and f_1 for each gas ensuring the nozzle operation in the design mode.

Under the conditions of industrial practice, the pressure of the oxygen in front of the nozzles of the blowing devices varies due to a number of reasons. The oxygen flow rate varies as well. The latter can change also due to technological requirements. Therefore, the supersonic nozzles often operate in the modes deviating from the design modes. Such modes are called the off-design modes.

If the pressure p_0 is greater than the design pressure, then the outlet pressure at the nozzle exit exceeds p_{MED}. In this case, as already mentioned, the equalization of the pressure in the jet and the medium pressure occurs at a small distance from the nozzle, Sect. 10.3. In the supersonic jets this equalization occurs not smoothly, as in the subsonic jets, but by spasmodic fluctuations in pressure, which are accompanied by abrupt changes in gas density. The noticeable portion of the energy of jet is wasted on these fluctuations called a compression shock.

The nozzles operating in such off-design mode are called underexpanded, since the area of their exit section f_1 is smaller than that of required for the optimum nozzle for the given values of p_0 and p_{MED}. To achieve the same initial jet velocities, the higher pressures p_0 is required for the underexpanded nozzles, in comparison with the optimum nozzles operating in the design mode, due to the additional energy losses.

The nozzles with the area of the exit section larger than that of required for the given values of p_0 and p_{MED} are called overexpanded. In other words, nozzle is overexpanded, if the pressure p_0 drops below the design value. The primary purpose of the supersonic nozzle, i.e., acceleration of flow to the velocities exceeding the speed of sound, is not achieved for overexpanded nozzles. This happens due to the compression shocks appearing inside of the overexpanded nozzles, in which the gas density increases and its velocity drops. With significant over-expansion, a supersonic nozzle can prove to be less effective than a simple one. Therefore, in practice, the supersonic nozzle operation in the over-expansion mode should be avoided. For this purpose, the nozzle exit section is designed for the minimum or close to the minimum gas flow rate. With varying gas flow rates, such nozzle will work either in the design mode or the underexpanded mode, which is better than nozzle operation in the over-expansion mode.

10.5 Simplified Formulae for Calculations of High-Velocity Oxygen Jets and Supersonic Nozzles

The use of formulae (10.4), (10.5), and (10.6), not considering the influence of velocity on gas density ρ, leads to the significant errors in the case of gas projecting from the simple nozzles at velocities over 150–200 m/s. The higher the velocities are, the greater the errors. These errors are especially great in calculations of the velocity v_1, formula (10.4), Fig. 10.2. For the high subsonic and for the supersonic values of velocities, all calculations must be conducted according to the formulae which take into account the compressibility of gases. However, being presented in the general form, these formulae are bulky, inconvenient for calculations, and require the use of the tables of physical properties of gases. Such is the formula (10.7) used when not only the pressure of the gas entering the nozzle p_0 is known, but the pressure at the nozzle exit p_1, Pa, as well:

$$v_1 = \varphi \times \sqrt{2kRT_0/(k-1)[1-(p_1/p_0)^{(k-1)/k}]}, \tag{10.7}$$

k and R – adiabatic index and gas constant, J/K × kg of a given gas
T_0 – absolute temperature of gas in Kelvin, K
φ – velocity coefficient taking into account friction and deviations from the adiabatic process

Let us limit to the calculations of oxygen jets projected into the medium with the pressure p_{MED} close to atmospheric pressure (760 mm of Hg). This case corresponds to the operating conditions of oxy-gas burners and oxygen tuyeres for the bath blowing if they are not submerged into the melt. For oxygen, $k = 1.4$, and $R = 260$ J/K · kg. Let us assume the absolute temperature of oxygen is equal to 273 K or 0°C.

Let us also assume that the nozzle exit pressure p_1 is equal to p_{MED}. The conditions assumed make it possible to rewrite formula (10.7) in simplified form (10.8), which is more convenient for calculations:

$$v_1 = \varphi \times \frac{705 \times \sqrt{p_0^{0.286} - 26.9}}{p_0^{0.143}}. \tag{10.8}$$

This formula (10.8) can be used in two cases, which are quite important for practice: first, for the oxygen projecting from the simple nozzles, if the absolute inlet pressure p_0 does not exceed the critical pressure value equal to 1.91 bar and v_1 is lower than the speed of sound; and second, for the projecting from the supersonic nozzles operating in the design modes. Only in these two cases, as is shown above, the condition $p_1 = p_{MED}$ assumed in the derivation of formula (10.8) is fulfilled. Absolute pressure p_0 in formula (10.8) is expressed in the Pascals.

The density of oxygen ρ at p_1 = 760 mm Hg is equal to 1.43 kg/m^3. After determining v_1 from formula (10.8), knowing ρ and using formulae (10.5) and (10.6), it is possible to calculate the mass and volumetric oxygen flow rates for projection from the simple nozzles with known diameters and values of f_1. But if the oxygen flow rate is given, then, using the same formulae, it is possible to determine a required nozzle diameter.

For calculating the supersonic nozzle, let us use the dependence showing that the mass oxygen flow rate m, kg/s, through this nozzle is determined by only four parameters: the coefficient μ, the pressure p_0, and the temperature T_0 before entering the nozzle, and the area of the critical (the narrowest) section of the nozzle f_{CR}. At T_0 = 273 K (0°C), this dependence can be converted to the simple form convenient for calculations:

$$m = 0.00257 \times \mu \times f_{CR} \times p_0, \text{ kg/s}. \qquad (10.9)$$

Using this formula, it is possible to calculate m for the different p_0, if the diameter of the critical section of the nozzle (the throat) d_{CR} is known, or to calculate the throat diameter, if the flow rate m is given. The coefficient μ is usually taken equal to 0.92. The area of the nozzle exit section f_1 can be determined from the simplified formula:

$$f_1 = 19.1 \times m/v_1 \times p_0^{0.286}. \qquad (10.10)$$

In order to use the formula (10.9) not only for oxygen but also for other gases let us reduce it to next form:

$$m = G \times \mu \times f_{CR} \times p_0/T_0^{0.5} \qquad (10.9')$$

T_0 – absolute temperature of gas in front of the nozzle
G – coefficient depending on physical properties of gas only, Table 10.1

The formula (10.9′) can be used for simple nozzles as well if to substitute f, which is area of the nozzles exit cross-section, for f_{CR}. This makes sense at velocities of gas projecting from the nozzle $v_1 > 200$ m/s when the simplifier formulas (10.4) and (10.5) cannot be used.

As shown above, formulas (10.8), (10.9), and (10.10) can only be used for calculating the supersonic oxygen nozzles operation at the design modes, when the condition $p_1 = p_{MED}$ is fulfilled.

Table 10.1 Values of G-coefficient for various gases

Gas	H_2	He	CH_4	N_2	Air	O_2	CO_2	Ar
G	0.0107	0.0159	0.0294	0.0397	0.0404	0.0425	0.0477	0.0503

10.5.1 A Limiting Value of Jets' Velocity

Energy loss in the nozzle itself is quite small. If we disregard this loss, it may be shown then that the initial velocity of the gas, under any conditions, cannot be higher than a certain limiting value of v^*. The limiting velocity is reached when the enthalpy of the motionless gas before entering the nozzle I^* is fully converted into kinetic energy of the flow. In this hypothetical case, the enthalpy of the flow drops to zero, and v^* is determined by the formula:

$$v^* = \sqrt{2I^*}. \tag{10.11}$$

If the temperature of motionless air (oxygen) in front of the nozzle is close to 273 K (0°C), then the possible limiting velocity of gases amounts approximately to 740 m/s. This velocity can be increased only by raising the temperature of gases in front of the nozzle.

Let us explain the usage of the formulae given above by some examples.

Example 1. Oxygen passes through a supersonic nozzle of an injector. Diameter of the critical cross-section of the nozzle d_{CR} is equal to 18 mm. The temperature of oxygen in front of the nozzle is 0°C, the pressure $p_0 = 6 \times 10^5$ Pa (abs) (5 bar on a manometer). Oxygen exits into surrounding air with a pressure of $p_{MED} = 10^5$ Pa (abs). Jet velocity v_1 and oxygen mass flow rate m should be determined.

Formula (10.8) is used. At $\varphi = 0.95$: $v_1 = 0.95 \times [705\ (600{,}000^{0.286}–26.9)^{0.5} /600{,}000^{0.143}] = 424$ m/s. Let us find f_{CR}. When $d_{CR} = 0.018$ m, then $f_{CR} = 0.785 \times 0.018^2 = 0.000254$ m^2. According to formula (10.9), at $\mu = 0.92$, the oxygen mass flow rate $m = 0.00257 \times 0.92 \times 0.000254 \times 600{,}000 = 0.361$ kg/s. If the oxygen density $\rho = 1.43$ kg/m^3, then the oxygen volumetric flow rate amounts to $V = 0.361/1.43 = 0.252$ m^3/s or 908 m^3/h.

Example 2. Oxygen in the mass flow rate of $m = 36$ kg/s is supplied to a sidewall oxy-gas burner through a supersonic nozzle. The oxygen pressure in front of the nozzle $p_0 = 6 \times 10^5$ Pa (abs), the temperature is 0°C. The pressure of gases in a furnace freeboard p_{MED} is equal to 10^5 Pa. Main dimensions of the nozzle for its operation in the design mode should be determined.

Using formula (10.9) f_{CR} should be found: $f_{CR} = m/0.00257 \times \mu \times p_0 = 0.36 /0.00257 \times 0.92 \times 600{,}000 = 0.000253$ m^2. This area of the critical cross-section corresponds to $d_{CR} = 0.018$ m. When $p_0 = 6 \times 10^5$ Pa then the initial velocity v_1 amounts to 424 m/s (Example 1). According to formula (10.10) the area of the nozzle exit section $f_1 = 19.1 \times 0.36/424 \times 600{,}000^{0.286} = 0.000361$ m^2. A suitable diameter d_1 is equal to 21.4 mm.

10.6 Long Range of Jets

Quantitative estimation of this characteristic varies for the different processes. It depends on both specifics of the process and purposes of its analysis. For all cases, long range of the jet can be determined as maximum distance from the nozzle *L* where the impact of the jet on the process is still quite significant. For a number of processes, the velocity and kinetic energy of the jet are most important, whereas for the others the jet momentum is of great significance. The latter are the processes which depend on the jet action force on the attacked surface. This action is equal to the ratio of jet momentum to the cross-section of the jet at the point of contact with the surface.

For the processes of the first type, long range is assumed to be equal to a distance *L* where the value of the jet velocity is still high enough. For the processes of the second type, the same condition relates to the jet force action on the surface. Since long range is determined in different ways in these two cases, their values can differ significantly. We will return to the analysis of this topic in the next chapter while reviewing operation of the so-called jet modules.

Reference

1. Markov BL (1975) Methods of open-hearth bath blowing. Metallurgia, Moskow

Chapter 11
Devices for Blowing of Oxygen and Carbon into the Bath

11.1 Blowing by Consumable Pipes Submerged into Melt and by Mobile Water-Cooled Tuyeres

For decades, since the beginning of using oxygen on the industrial scale, the devices for oxygen blowing into the bath of EAFs, open-hearth furnaces, and oxygen converters have been continuously improved and modified. This equipment has been and still is attracting a great deal of attention due to the fact that the impact of this equipment on key performance indices of EAF is comparable with the impact of the devices for introducing electrical energy. The devices for blowing oxygen into the bath operate under very severe conditions. The requirements imposed upon these devices are stringent and diverse. These requirements include technological efficiency as well as high operational reliability and safety and reasonably low operational costs. These requirements are difficult to combine, and, as many years' experience shows, the devices ensuring higher performance indices are quite frequently preferred to the devices which are less technologically efficient but more reliable and simpler in maintenance.

In the different periods of time, various designs of devices for oxygen blowing were used in the EAFs. These devices differed not only in design features, but in fundamental features as well. These devices came and went, and the search for the most efficient versions continues. In this search, it is essential to rely on the enormous accumulated experience including related to the past experience gained over the years from oxygen blowing of the bath of the open-hearth furnaces.

As already mentioned, the hydrodynamic processes in the baths of EAFs and open-hearth furnaces are similar. Their key geometric parameters, i.e., ratio of the area of the heat-absorbing bath surface to the area of the internal surface of the bottom lining through which the heat is lost to the environment, are similar as well. The average value of this ratio for the electric arc furnaces is 0.84–0.86, and for the open-hearth furnaces 0.86–0.88. The values of their average bath depth are close as well [1]. The experience with oxygen blowing of the open-hearth baths gives many instructive examples for the current practice.

Y. N. Toulouevski and I. Y. Zinurov, *Innovation in Electric Arc Furnaces*,
DOI: 10.1007/978-3-642-36273-6_11,

The experience gained from operation of the basic types of blowing devices and the reasons for substituting one device with another are analyzed below. This analysis is not only useful, but essential for the selection of the promising trends of further innovations.

11.1.1 Manually Operated Blowing Through Consumable Pipes

It was this method which laid foundation for oxygen blowing of the bath in the EAFs. It gave quite satisfactory results when handled by experienced steelmakers. The pipes were used not only for blowing of liquid bath, but for scrap cutting in the zones of retarded melting as well. As a result, consumption of small amounts of oxygen allowed to achieve significant shortening of the melting period and reduction of electrical energy consumption. Unfortunately, the electrical energy consumption was reduced, essentially, due to oxidation of iron and significant reduction of yield, which, as is known, is not justifiable.

Since the diameter and the carrying capacity of the pipes manually handled by a steelmaker were quite limited, the method in question was not suitable for using oxygen in large quantities. Furthermore, it did not satisfy the health and safety regulations. Working for long term in front of the EAF's slag door is not only harmful for the human health, but very dangerous as well. For these reasons, this blowing method is no longer being used. However, the principle of the blowing of the bath by using the consumable pipes was recognized and implemented at the higher technical level in the manipulator of BSE Company.

11.1.2 BSE Manipulator

In the recent past, this manipulator has been widely used in a number of countries. At present, it is still being used at many plants.

All mechanisms of the manipulator controlling the consumable pipes, together with the safety devices, are mounted to the console (1), which is connected to the rotating supporting unit (2), Fig. 11.1, Chap. 7, Fig. 7.1. With the aid of this unit, located on the side of the furnace the manipulator is turned and brought into the operating position in front of the furnace slag door. During the off-period, the manipulator is turned away from the furnace to the initial position and does not interfere with the technological operations carried out in the furnace. The manipulator is equipped with oxygen pipes (3), located on both sides of the pipes for carbon powder injection (4). The length of the oxygen pipes is approximately 5.5 m. As they get burned, they are pushed forward from the movable boxes (5). The mechanisms of manipulator allow to change the pipe inclination in the vertical plane, to move the pipes in the forward–backward or left–right directions,

Fig. 11.1 BSE manipulator (designations are given in the text)

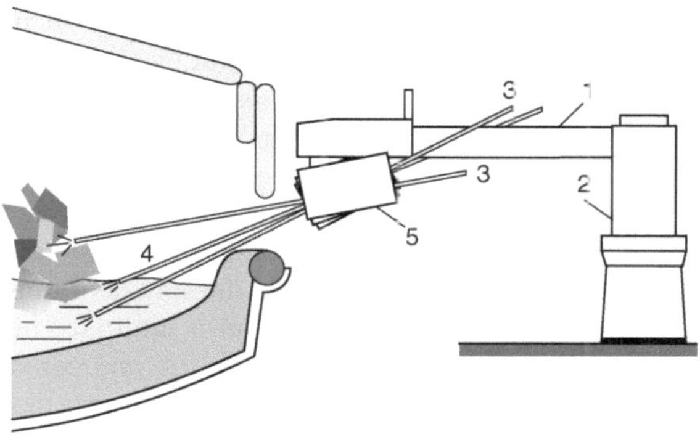

to submerge them in the melt or to direct them onto the scrap for cutting. The pipes are coated from the inside with a special coating preventing their rapid combustion. As the pipes get shorter, they are pushed forward from the boxes (5).

The possibility to control the direction of oxygen and carbon jets in such a flexible way, depending on the actual position of scrap in the furnace and the level of liquid bath, increases the effectiveness of the use of these energy carriers and is an advantage of the BSE manipulator. The possibility to submerge the pipes quite deeply into the melt should be specifically pointed out, because it promotes stirring of the bath and melting of the scrap submerged in the bath. However, to what extent these advantages are realized in practice depends in the decisive measure on the qualification of the operator controlling the manipulator. The operator has to monitor continuously through the half-open furnace slag door the rapidly changing situation in the freeboard and to use the capabilities of the manipulator in the best way possible depending on the specifics of the process in the course of each particular heat. With this kind of control, the experienced operator can ensure the sufficiently high efficiency of use of the BSE manipulator.

A strong dependence of the effectiveness of this device on the skills of the operator controlling it can be assessed differently. Some consider it to be an advantage, others a disadvantage. This question relates to the general problem of determining the role of operator in the automated control system of the steel-melting unit. This problem is discussed in Chap. 13.

The operator controls the manipulator while staying in the closed control cabin. The location of this cabin ensures the necessary field of view. This way, the health and safety regulations are completely satisfied. When the pipes installed on the manipulator are consumed, they have to be manually replaced by the operator. Thus, the need for physical labor is not completely excluded, not to mention the quite considerable expenditures on pipes and maintenance of all the mechanisms of the manipulator.

A serious shortcoming of the BSE manipulator is that its blowing zone is limited by the section of the bath near the furnace slag door. This shortcoming impedes the effective use of oxygen in large quantities and is especially noticeable in the big furnaces. In such furnaces with the oxygen consumption for blowing higher than 25–30 m^3/ton, it is necessary to disperse the blowing devices along the perimeter of the bath. Otherwise, with increase in the oxygen flow rates, the effectiveness of the use of oxygen drops, the bath overoxidation and splashing increase, which leads to the yield reduction. When carbon powder is injected in the area near the slag door, the substantial portion of this expensive material is lost with the foamed slag overflowing the sill before carbon could react with iron oxides. With the significant flow rates, the injection of carbon into the bath should also be dispersed.

During manipulator operation, the slag door must be at least half way open. In this case, a large quantity of cold air is infiltrated into the furnace, which increases the scrap oxidation, decreases the yield, and considerably increases the costs on off-gas evacuation and its cleaning. At present, the majority of the modern EAFs operate with the closed slag door. This mode of operation is incompatible with the use of the BSE manipulators.

The discussed above as well as some other shortcomings necessitated the substitution of manipulators with the consumable pipes with more advanced blowing devices. For a long period of time, at many plants, the mobile water-cooled tuyeres were used as an alternative.

11.1.3 Mobile Water-Cooled Tuyeres

These tuyeres are inserted into the freeboard with the help of the manipulators not only through the slag door, but through the openings in the sidewalls and roofs of the furnaces. The vertical roof tuyeres were used during the initial stage of development of similar devices in the Soviet Union and a number of other countries. This method was borrowed from the open-hearth furnace practice. However, due to the operational problems, its application in the EAFs has soon ceased.

The water-cooled tuyeres, Fig. 11.2, developed by Berry Metal Company, USA, became widespread. These tuyeres characterized by large length were inserted into the furnace through the openings in the sidewall panels at an angle to the bath surface of about 40°. At the end of the blowing, the tuyeres were removed from the freeboard and lowered back into the horizontal position. Usually, two tuyeres were installed in the furnace. The tips of the tuyeres could have two oxygen supersonic nozzles and one channel for the carbon injection positioned parallel to the axis of the tuyere, Chap. 12, Fig. 12.5. With respect to this axis, the oxygen nozzles were positioned so that the oxygen jets attacked the bath at an angle of about 30° to the vertical.

Let us note that the inclination of the oxygen nozzles in these tuyeres was practically the same as the inclination of the oxygen tuyeres in the open-hearth furnaces. In both cases the inclinations were chosen so that they could in the best way fulfill two contradictory requirements, namely, maximum possible penetration of the oxygen jets into the melt and, at the same time, minimum possible splashing, Chap. 9, Sect. 9.3.2. The close agreement between the optimum directions of the oxygen jets attacking the melt is yet another evidence of

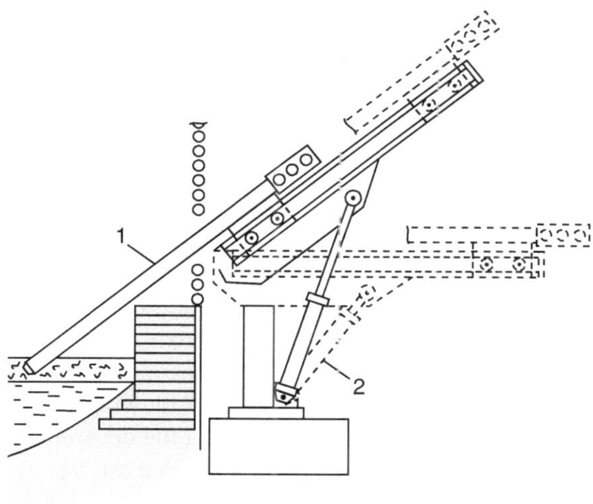

Fig. 11.2 Mobile water-cooled tuyere 1 – tuyere; 2 – hydro-cylinder

hydrodynamic similarity of the baths of the open-hearth and electric arc furnaces with respect to the blowing processes.

The dependence of the effectiveness of blowing on the position of oxygen tuyeres relative to the surface of the liquid bath is of great practical importance. As early as in the initial stage of introduction of oxygen blowing in the open-hearth furnaces, it has been clearly established that the effectiveness of blowing increases sharply with submerging the tuyeres into the bath. This has led to substitution of the method of blowing by the tuyeres positioned above the slag, which was used earlier because of the fear of the tuyeres burn-back and water leakage, with the method of submerged blowing. This substitution has been promoted by the improvement of the design of the open-hearth tuyeres, which ensured essential increase in their durability.

It has been established later that the maximum effectiveness is achieved when the nozzles of tuyeres are submerged to the slag–metal boundary [2]. In comparison with positioning the tuyeres above the slag, in this case, even with the same or even lower oxygen flow rates, the effectiveness estimated based on the rate of decarburization and heating of the metal has increased by several tens of percent. It seems obvious that this has occurred due to a sharp increase in the intensity of stirring in the bath.

In the Soviet Union, for practical realization of optimum blowing regime, the electrical sensors for detecting the tuyeres position relative to the slag–metal boundary have been developed. Once, these sensors have found widespread use in open-hearth furnaces of a number of plants. These sensors were indicating, with high accuracy, when the tip of tuyere was passing from the gas phase to the slag and from the slag into the metal. In many furnaces, the position of tuyeres at the slag–metal boundary was kept by automatic regulators with the help of these sensors with accuracy of 20–30 mm [2].

Comparative tests carried out in furnaces equipped with this automation system have shown that blowing into the slag is considerably less technologically effective in comparison with blowing into the slag–metal interface. With the tuyeres raised 150–200 mm above this boundary (without withdrawing the tuyeres from the slag), the tap-to-tap time increased considerably, the specific oxygen flow rate and the oxidation of slag increased, and the yield reduced.

The effectiveness of blowing by the tuyeres submerged 100–200 mm below the metal surface has been studied as well. By that time, the open-hearth tuyeres had been perfected to such a degree that their durability allowed to carry out such industrial experiments for the duration of many tens and even hundreds of heats, with the duration of blowing approximately 2.0–3.0 h per heat in 300–600-ton furnaces. Carried out in the furnaces of a number of plants, these tests have not revealed any obvious advantages of blowing with submerging of tuyeres into the metal. On the contrary, even though the decarburization rate increased, the rate of metal heating dropped, which increased the tap-to-tap time. This is explained by the fact that the tuyeres cooled the liquid metal. Calculations show that the cooling effect of three tuyeres submerged into the metal 150–200 mm below its surface is equivalent to a double increase in heat loss through the bottom in a bath of a 600-ton furnace. There is every reason to believe that established for the open-hearth baths' dependencies of the effectiveness of the blowing of the bath by the oxygen water-cooled tuyeres on their position relative to the slag–metal boundary are accurate for the EAFs as well.

Let us return to the analysis of the practical experience of the use of the mobile water-cooled oxygen tuyeres in the electric arc furnaces. These tuyeres have been developed for the submerged blowing, whose advantages were already well known from the operational experience gained from the open-hearth furnaces. However, the tuyeres in the EAF could not withstand such a mode of operation. Submerging of the tuyeres into the slag caused so frequent burn-backs of the tuyeres that, according to the report of the Nucor Yamato Steel Company (USA), the costs on their tips replacement for two furnaces amounted to $ 400,000 a year [3].

Because of low durability of tuyeres, the operators avoided submerging the tuyeres into the melt and kept the tuyeres mostly above the slag during the blowing. This mode of use, to a considerable extent, has eliminated the potential advantages of the mobile water-cooled tuyeres and, in the majority of cases, their application has ceased. This case shows that selection of the directions of steel-melting process intensification depends on resolution of the problem of durability of the water-cooled elements, blowing devices in particular. Problems of cooling of the EAF components are examined in detail in Chap. 12.

Difficulties in operating the manipulators with submerging into the melt consumable pipes and water-cooled tuyeres, low durability of tuyeres, high operational expenditures, strong dependence of effectiveness of these devices on operator's qualification and quality of work, impossibility of automating without losing most of their technological advantages – all of these called for the search of the new, free from the enumerated deficiencies, directions in the oxygen blowing of the bath of the electric arc furnaces. This search has led to the development of a new type of water-cooled blowing devices, the so-called jet modules. These new devices use the long-range supersonic oxygen jets with special characteristics, which are called coherent jets.

11.2 Jet Modules: Design, Operating Modes, Reliability

These are the multifunctional devices. They fulfill functions of the oxy-gas burners used for scrap heating, the tuyeres used for blowing the bath with oxygen, the injectors used for blowing carbon into the melt for the purpose of slag foaming and FeO reduction. All structural elements of the modules are usually placed in water-cooled boxes protecting these elements from high temperatures as well as from damage during scrap charging. The boxes are inserted into the furnace through the openings in the sidewall panels, which considerably decreases the distances from the nozzles of the burners and from injectors to the bath surface. There is a wide variety of design versions of the jet modules. The advent and development of this direction is associated with the PTI Company (USA) and with the name of V. Shver. Let us examine the arrangement of the module by the example of a typical design of PTI. Compared to other modules this module can be installed closer to the sill level. Thus, a distance from the oxygen burner nozzle to metal surface does not exceed of 700 mm. Reducing oxygen jet length improves oxygen efficiency. This is a substantial advantage of the PTI module [4]. Further, this design and the similar ones have gained wide acceptance in many countries around the world.

The PTI module contains the water-cooled copper block (1) in which the oxy-gas burner (2) with water-cooled combustion chamber (3) and the pipe (4) for the injection of carbon powder are located, Fig. 11.3. The burner (2) has two operating modes. In the first mode, it is used as a burner for heating of scrap and operates at its maximum power of 3.5–4.0 MW. The gas mixes with oxygen and partially burns within the combustion chamber (3). At the exit from the chamber, the high-temperature flame is formed, which heats and settles down intensively the scrap in front of the burner. The combustion chamber protects, to a considerable extent, the burner nozzles from the clogging by splashes of metal and slag.

In the second operating mode, the burner is mainly used as a device for blowing of the bath. The gas flow rate considerably decreases, and the oxygen flow rate sharply increases. In this case a long-range supersonic oxygen jet is formed. In this mode, the function of the burner alters. It is reduced to the maintenance of the low-power pilot flame. This flame shrouds the oxygen jet increasing its long range, prevents flowing of the foamed slag into the combustion chamber, and protects the nozzle of the burner from clogging as well.

The burner is controlled by a computer, which switches its operating modes in accordance with the preset program. Immediately after scrap charging, the first mode is switched on. In several minutes, it is switched to the second mode. The highly heated scrap can be cut by oxygen considerably easier than cold scrap.

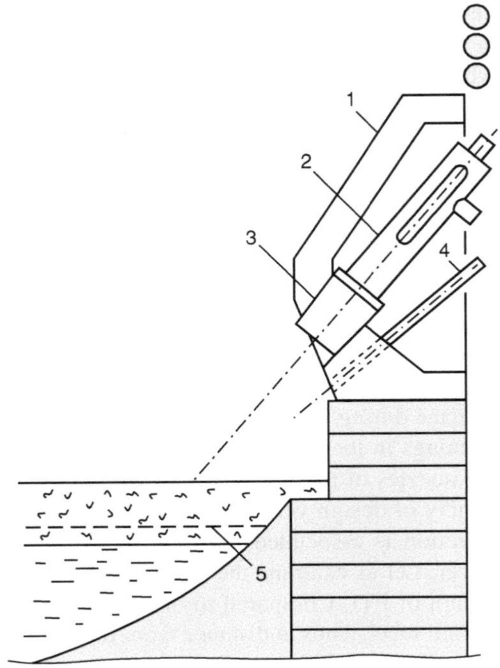

Fig. 11.3 PTI module 5 – furnace sill level (the rest of designations are given in the text)

Therefore, the preliminary operation of the burner in the first mode greatly facilitates penetration of the supersonic oxygen jet through the layer of scrap to the hot heel on the bottom. This ensures the early initiation of the blowing of the liquid metal with oxygen, which is the necessary condition for achievement of high productivity of the furnaces. While the upper layers of scrap continue to descend to the level of the burner, the alternation of the operating modes is carried on and is repeated after charging of the next portion of scrap. This considerably increases the effectiveness of the use of oxygen in the initial period of the heat before the formation of the flat bath.

The module operating reliability in a decisive measure depends on durability of the protective boxes and wall panels in the zone of action of the burner. These water-cooled elements operate under super severe conditions. Moreover, the closer to the bath surface, the more severe the conditions. The blow-back of the oxygen jets reflected from the scrap lumps are the main cause of damage of the boxes and panels in the burner zone. Alternating operating modes of the burner reduces this problem, but does not eliminate it completely. In order to increase the durability of the water-cooled elements of the modules, some companies prefer to install them at a greater height, even though this installation increases considerably the length of oxygen jets required for the blowing of the bath and calls for an increase in their long range.

11.2.1 Increase in Oxygen Jets Long Range: Coherent Jets

In practice, the problem of increase in the long range of the jets is solved by two methods, namely, by the use of a pilot flame and the so-called coherent supersonic nozzles.

The first method consists in supplementing the oxygen jet with the annular co-flowing pilot flame of the burner. The gases in the flame have low density due to the high temperature. In this case, the known effect of increase in the long range of the jet flowing into the less-dense ambient medium is used. The following figures give an idea of the quantitative aspect of the effect in question. In the laboratory furnace, with the increase in the temperature from 260°C (533 K) to 1630°C (1903 K) and corresponding decrease in air density by 1903/533 = 3.57 times, the long range of supersonic air jet has doubled (from 0.8 to 1.6 m) [5].

Let us also take into consideration a fact that, even without the supplementing pilot flame, the oxygen jets propagate not in the cold air, but in the EAF freeboard filled with gases with the temperature in the order of 1500–1600°C and higher. These gases contain a large amount of CO. Taking into account post-combustion of CO in the O_2 jets, the density of the gases is not much different from the gas density in the pilot flame. Therefore, based on the figures given above, the additional increase in the long range of oxygen jets resulting from supplementing the jets with the pilot flame is relatively small.

The general use of the pilot flame in the modules, which requires significant additional consumption of natural gas, is required not so much for increasing the long range of oxygen jets, as for its protective functions as well as for its ability to facilitate penetration of oxygen jet to the melt through the layer of scrap. At the same time, the probability of reflection of oxygen jets from the scrap lumps to the water-cooled elements decreases.

The second method of increasing the jets' long range consists in configuring the optimum profile of the supersonic nozzle, which ensures the minimum turbulence of the flow, Fig. 10.3b, Chap. 10. By analogy with the lasers, these nozzles and the supersonic jets formed by them are called coherent. Due to the quite low turbulence, the coherent jets involve into their motion several times smaller mass of the ambient gas, expand considerably slower, and maintain initial velocity at considerably greater distance from the nozzle, in comparison with the jets flowing from the simple de Laval nozzles, Fig. 10.3b.

In the tests conducted in the laboratories on the testing ground, the achieved increase in the length of the initial region of the coherent jets was up to 37 calibers, which approximately 5 times exceeds the average length of the initial region of the jets flowing from the simple de Laval nozzles [6]. Let us recall that initial region is the region of the jet where its axial velocity does not decrease and remains equal to the initial velocity, Chap. 10, Sect. 10.2, Fig. 10.1.

Such results of the stand tests can make a false impression regarding the possibility of quite significant increase in the distances from the jet modules to the bath surface. It is necessary to understand that the conditions of the coherent nozzles operation in the EAF differ considerably from the laboratory conditions. Even the simple de Laval nozzles are quite sensitive to the most insignificant deviations of the oxygen parameters from the design parameters. To an even greater degree, this relates to the coherent nozzles. It is noted that small deviations from the design operating conditions eliminate the advantages of these nozzles, whereas under the production conditions, such deviations occur constantly. Furthermore, even quite small deposits of droplets of metal and slag inside the nozzles, which cannot be avoided in practice, cause strong turbulence of the flow disrupting its coherence. It is also necessary to note that there are no any direct or indirect data obtained under the actual conditions in the furnaces, which could confirm the results of the stand tests.

However, when analyzing the capabilities of jet modules with coherent oxygen nozzles, the following main aspect should not be neglected. As noted in Chap. 10, Sect. 10.1, the impact force of the jet on the bath, which determines submerging of the jet into the melt and stirring of the melt by the jet, depends neither on the jet velocity nor on its kinetic energy. This force is determined by the ratio between the momentum of the jet and the area of its cross-section at the bath level, i.e., by the specific pressure of the jet on the bath surface measured in kg/m^2. The momentum of any free jets, regardless of their velocities and nozzle configurations, does not change over the length of the jet and remains equal to the initial momentum, Sect. 10.2, formula (10.3). In case of identical oxygen flow rates and initial velocities at the nozzle exit, the momenta of the free jets are also identical; moreover, the lower velocity can be compensated by a proportional increase in the flow rate and vice versa.

The actual advantage of coherent jets in the processes of their impact on the bath consists in their smaller expansion, i.e., in the smaller central expansion angle α, Chap. 10, Fig. 10.1. In the electric arc furnaces, the oxygen jets attack the bath not at a right angle, but at a sharp angle β, which is close to 45°. However, this is not essential for further analysis. Without significant error, it is possible to assume that the area S of a surface of the liquid bath, over which the jet exerts pressure, is related to the length of the jet L and to the angle α by the following formula:

$$S \cong \pi\,(L \times \tan\alpha)^2. \tag{11.1}$$

The values of α of the supersonic jets do not exceed 10°. In this range, tanα changes directly proportional to the angle α. Therefore, the same directly proportional dependence is valid for the area S and the value of $(L \times \alpha)^2$.

According to the data obtained from the ground tests, the angle α in supersonic coherent jets decreases approximately by 1.5 times in comparison with regular jets from the simple de Laval nozzles. As a result, the area S of the coherent jet, in case of identical distance from the nozzles to the bath surface measured along the jet axis L, decreases by $1.5^2 = 2.25$ times in comparison with the regular supersonic jet. Accordingly, in case of identical momentum, the pressure on the bath surface (kg/m^2) in the coherent jet is 2.25 times higher. If we assume identical values of area S and pressure, then, from the elementary geometric relationships, it follows that for the identical angles β the length of the coherent jet L and the height from its nozzle to the bath can be increased not more than 1.5 times. Such is an actual increase in the long range of the coherent oxygen jets in the processes of their interaction with the liquid bath. This increase is considerably lower than that obtained in the ground trials when determining the long range by jet velocity.

The advantages of the coherent jets must manifest themselves to a fuller extent in the processes of scrap cutting, which require the high degree of purity of oxygen. As the distance from the nozzle increases, the initial purity of oxygen and the initial velocity of jets are reduced practically to the same extent. Therefore, the coherent oxygen jet is capable of cutting scrap effectively at the distances, which are several times greater than those reached by the ordinary supersonic jet.

11.2.2 Effectiveness of Use of Oxygen, Carbon, and Natural Gas in the Modules

When installing in the furnaces, the modules are dispersed along the perimeter of the freeboard with due consideration of its temperature asymmetry and other characteristics of a particular furnace, Fig. 11.4. The possibility of the optimum spacing of the points of injection of oxygen and carbon appears to be the fundamental advantage of modules in comparison with the use of the manipulators with consumable pipes and the water-cooled tuyeres. Such spacing allows to considerably increase the total oxygen and carbon consumptions per heat without reducing the effectiveness of their use.

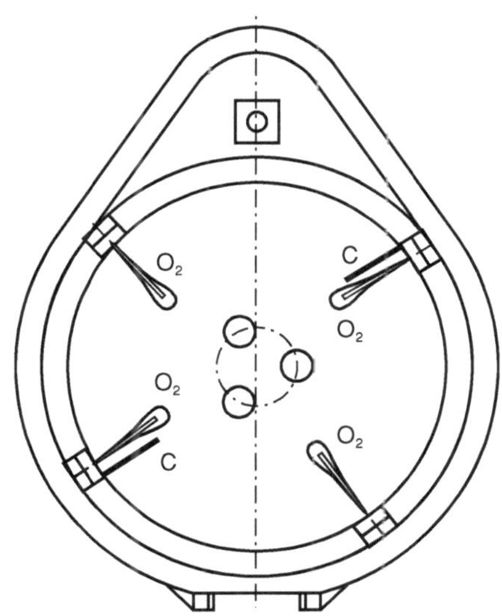

Fig 11.4 An example of modules dispersal along the furnace perimeter

In the furnaces of average and large capacity, four or five or even larger number of modules are installed. Quite often, separate modules are used for injecting carbon, Fig. 11.4. It should be noted that the advantages of coherent nozzles and those of jets with the long range higher than usual can become apparent when the modules are installed at a relatively great distances from the bath. In the PTI modules, located close to the bath, with the length of oxygen jets of approximately 700 mm, Fig. 11.3, the substitution of simple de Laval nozzles with coherent nozzles did not lead to any apparent improvements in the furnaces' operation.

The essential drawback of all modules is a quite limited use of natural gas for heating of scrap. In the EAFs equipped with the modules, the consumptions of natural gas are usually not higher than 6–8 m^3/ton, i.e., are at the level which, many years ago, was typical for the old furnaces with three sidewall burners. This is basically explained by the low power of the modular oxy-gas burners which, like in the seventies, is equal to 3.5–4.5 MW.

The concept of the modules gives only the supporting functions to the oxy-gas burners, which help increase the effectiveness of the use of the blown oxygen and the reliability of the devices' operation. These functions include protecting the oxygen nozzles and nozzles of the burner itself from clogging with metal and slag by means of maintaining the pilot light, enveloping the blowing oxygen jets with the pilot light for the purpose of increasing their long range, facilitating the scrap cutting with oxygen in front of the burner. Approximately 30% of the natural gas consumed is used for maintaining of the pilot light. The task of the high-temperature heating of the large masses of scrap for the purpose of increasing the productivity and maximum possible substituting of electric power with natural gas is not being set. This task and the means of achieving these objectives, namely the HPR-burners, Chap. 7, Sect. 7.3, are incompatible with the concept of the modules.

The technological effectiveness of the modules as a whole can be reliably inferred only from a change in the corresponding performance indices of the furnaces. Unfortunately, the published data are insufficient for drawing the unambiguous conclusions. In the majority of the cases, substituting one blowing device with another has been carried out gradually, and the old and the new devices have worked simultaneously for quite a long period of time. Besides, in the majority of case of such substitution, a significant increase in oxygen and carbon consumption has occurred, and it remains unclear to what extent the improvement of the performance indices was due to the new design of the blowing device, and to what extent due to the increase in consumption of the energy carriers which could take place in case of using the old equipment as well.

On the average, the obtained results are considerably lower than those which could be expected based on the increase in the long range of both coherent oxygen and enveloped by the pilot flame jets. The increase in the productivity of the furnaces and reduction of the electrical energy consumption due to installation of the modules do not exceed several percent (with a parallel increase in the oxygen and carbon consumption). This leads one to assume that the depth of penetration into the melt of the oxygen jets generated by the modules and stirring of the bath by the jets are nevertheless insufficient and do not properly compensate for the fact that these most important characteristics are worse than those of the oxygen blowing by the tuyeres submerged into the bath.

At present, the jet modules gradually replace all other types of blowing devices in the EAFs. However, this is explained not so much by improvement in the performance indices of the electric arc furnaces, which do not always appear sufficiently convincing, as by increase in the operational reliability of the new devices, as well as by the reduction of the operational expenditures and the possibility of almost complete automation of their operation. It can be assumed that the jet modules will in turn give place for more effective blowing devices the usage of which will ensure the fundamental advantages of the submerged blowing without reduction in the reliability and increase in the operational expenditures.

11.3 Blowing by Tuyeres Installed in the Bottom Lining

The stationary tuyeres for the submerged blowing can be installed in an inclined position in the lining of the banks or in the bottom of a furnace. They are usually dispersed along the perimeter of the furnace or the bottom area. The operation of these devices can be automated in the same way as the operation of the jet modules. The main problem of this method of blowing is durability of the tuyeres themselves and of the adjacent parts of the lining. Installation of the water-cooled tuyeres in the lining at the level lower than the level of the liquid bath is considered absolutely unacceptable because it creates an explosion hazard in case of water leakage. As a result, the tuyeres cooled by other methods are used for operating under these conditions. Among these are the converter-type tuyeres for the bottom blowing and the tuyeres with evaporative cooling.

11.3.1 Converter-Type Non-Water-Cooled Tuyeres

Using the experience of bottom blowing accumulated in the oxygen converters, Klöcker Company (Germany) in the 1990s has developed the technology of EAF's steel melting with blowing of the bath from below, which has been called the K-ES (ES stands for electric steel). In order to implement this technology, several blowing tuyeres of the pipe-in-pipe type are installed in the furnace bottom. The design of these tuyeres is analogous to that of the tuyeres for the bottom blowing in the oxygen converters, Fig. 11.5. Oxygen is delivered through the inner pipe, while natural gas or other hydrocarbon fuel is delivered through the annular gap. The oxygen jet entering the liquid metal is enveloped by gas, which prevents direct contact of oxygen with the bottom lining. Hydrocarbon gas entering the high-temperature zone undergoes carbonization and forms at the tuyere exit a porous carbonaceous build-up ("a mushroom"), which protects the refractory lining against intensive wear, Fig. 11.5.

Simultaneously with the bottom oxygen blowing, powder carbon is injected into the bath, in some way or other. A portion of carbon in the form of coke is charged into the furnace together with the scrap. Carbon monoxide CO evolving from the bath in large quantities is post-combusted above the bath with the help of the oxygen tuyeres installed in the sidewall panels along the entire perimeter of the furnace. The wall oxy-gas burners are used for the scrap preheating. In the liquid bath stage, more oxygen is added for the post-combustion of CO. With the use of the K-ES technology, the total oxygen flow rate for blowing of the bath and post-combustion reaches 55 m^3/ton, and carbon consumption is 27 kg/ton. The preponderant amount of oxygen is used for post-combustion of CO, and the share of carbon injected into the bath is equal to about 50% of its total consumption.

The use of this technology and of some of its elements in a number of plants in Italy and other countries has ensured a significant decrease in tap-to-tap time, as well as in electrical energy consumption. As in the oxygen converters, the bottom blowing has improved the bath stirring and facilitated its attaining a state of equilibrium. The content of oxygen in the metal and of FeO in the slag has decreased, and the yield has increased. However, due to fundamental design differences between oxygen converters and EAFs, especially due to relatively small depth of the furnace bath, the intensity of the bottom oxygen blowing in the K-ES process could not be brought up to the desired level. It remained relatively low despite increasing the number of tuyeres. Therefore, the obtained results have not been substantially different from those achieved by other methods without serious problems due to intensive wear of bottom tuyeres and bottom lining. This wear, in comparison with regular operation of furnaces with top oxygen blowing, has required significant additional efforts and operational expenditures for bottom maintenance. As a result, the bottom oxygen blowing in the EAF has not become widespread. In the furnaces with bottom tuyeres, in the majority of cases, these tuyeres are used not for the oxygen blowing, but for injection of inert gases,

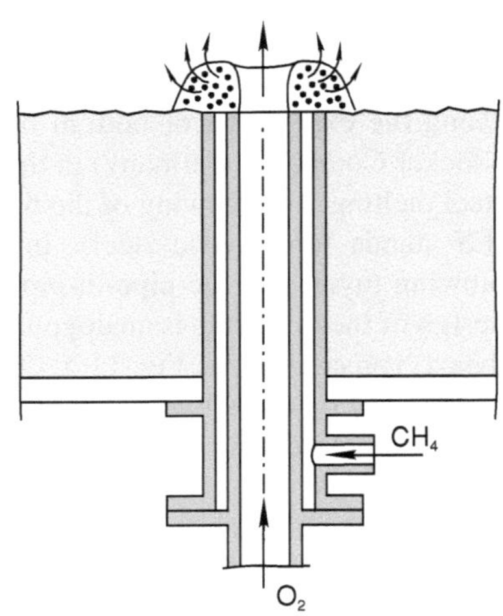

Fig. 11.5 Tuyere for bottom blowing

namely, argon, nitrogen, or their mixtures. Although their blowing is carried out with low intensity of the order of 0.1 m^3/ton × min, it appears to be quite effective, because it considerably improves the bath stirring and facilitates its homogenization. The bottom tuyeres become essential for furnaces with a large capacity.

11.3.2 Tuyeres Cooled by Evaporation of Atomized Water

In the second-half of the 1990s, understanding the fundamental advantages of the submerged blowing resulted in the development by the KT-Köster Company (Germany) of the blowing tuyeres of a new type, the so-called KT-tuyeres. These tuyeres are installed in the lining of the banks of bottom of the furnace. According to the initial concept, the oxygen KT-tuyeres had to be installed at the slag level, and the KT-tuyere for the carbon injection even lower, i.e., near the slag–metal boundary. The developers of this system believe that the explosion-proof nature of the KT-tuyeres is totally ensured by the original method of evaporative cooling. The cooling water is used in the mixture with the compressed air. The air supply rate per tuyere is equal to 50 m^3/h. The water is atomized in the heat-stressed frontal zone near the tuyere head. The small drops of water evaporate on the extended surface of the head. There is no water in the tuyere head itself. In case of the tuyere burn-back, only a small amount of highly atomized water can get into the liquid metal.

In this system, the heat is removed not so much due to heating of water as in the usual water-cooled devices, as due to its evaporation. It is known that the quantity of heat required to heat water from 0°C to the boiling point of 100°C is approximately 5 times less in comparison with the quantity of heat required for its total evaporation at this temperature. This allows reducing water consumption for cooling of one KT-tuyere with the diameter of 110 mm to approximately 2 m^3/h. The mixture of air, residual portion of atomized water, and water vapor is sucked off from the tuyere by the vacuum pump. As a result, there is no excess pressure in the tuyere. This increases its safety even more. After exiting the tuyere, the air separates from the water and exits into the atmosphere, the vapor condenses, and water is returned into the cooling system of the furnace.

The industrial trials of the KT-tuyeres have shown that the cooling system utilized in them does not allow the contact of the tuyere head with the liquid metal. Therefore, at a later stage, not only oxygen KT-tuyeres have been installed at the slag level at the sufficiently great distance from the slag–metal boundary, but carbon tuyeres as well. As the slag foaming takes place, the heads of the KT-tuyeres are submerged into the slag to a significant depth. It is impossible to install the stationary blowing devices of the module type at that height.

Additional compressed air is used for injecting of carbon powder into the slag in the amount of 300 kg/h. Up to 1500 m^3/h of oxygen and from 50 to 100 m^3/h of natural gas are delivered to the oxygen KT-tuyeres. This gas, like the gas in the bottom tuyeres, is introduced through the annular gap surrounding the oxygen jet.

The flow of gas protects the refractory from the contact with oxygen and, thus, prevents rapid wear of the lining. Compressed air and oxygen provide protection from the slag flowing in the KT-tuyeres. Similar to jet modules, oxygen KT-tuyeres are used for blowing of the bath during the liquid bath stage, and used as burners for scrap heating at the beginning of the heat. Oxygen and carbon KT-tuyeres are installed in pairs, next to each other. These pairs are dispersed along the entire perimeter of the bath. The oxygen, carbon, natural gas and compressed air supply to the tuyeres is fully automated.

KT-tuyeres have been become widespread. The obtained results confirm a sharp increase in the effectiveness of blowing when the tuyeres are submerged deep into the slag. In this respect, the most demonstrative are the data obtained from 100-ton DC EAF operating on a charge material containing from 80 to 90% of metallized pellets [7]. The pellets are charged into the bath by a conveyor through the opening in the furnace roof.

In 2003, manipulators with consumable pipes were replaced with the KT system. Four oxygen and two carbon KT-tuyeres were installed. Before the installation of these tuyeres, it took 60 min to charge 65 ton of pellets. It was impossible to charge the pellets faster, because this led to the formation of large “icebergs” from the unmelted pellets floating on the bath surface. When working with the KT system, 74 tons of pellets are charged in 33 min. In this case, the furnace productivity with regard to the pellets’ melting process increased from 65/60 = 1.08 ton/min to 74/33 = 2.24 ton/min, i.e., more than doubled.

Though an increase in the oxygen flow from 30 to 45 m^3/ton and in the electrical power of the furnace by 18% have taken place, such a sharp rise in productivity cannot be explained by these factors. These results could be obtained only in case of radical increase in the intensity of bath stirring by oxygen jets due to submerging of the heads of the KT-tuyeres into the slag. The same can be said about the results obtained from the 100-ton EAF of another plant operating on scrap. Installation of the KT-tuyeres has led to a sharp acceleration of melting of the large lumps of scrap submerged in the liquid metal. This has allowed to charge into the furnace considerably larger lumps of scrap and, thus, to decrease the cost of scrap preparation for melting without decreasing the productivity. The maximum permissible size of the lumps has been increased from 0.5 to 2 m [8].

In addition to advantages resulted from the stirring intensification, the submerged blowing of the bath by the KT-tuyeres also facilitates the slag foaming and increases the durability of the central refractory part of the furnace roof. Facilitating and improving the slag foaming allows to increase the electrical power of the furnace [6] and to use cheaper powders with the reduced carbon content [7]. Increase in durability of the central part of the roof by 2–3 times reduces the consumption of refractory materials and repair costs. It is known that the durability of the roof refractories is directly related to intensity of bath splashing. The studies of splashing conducted by the methods of physical modeling and under the industrial conditions have shown that submerging tuyeres into slag decreases splashing by many times, Chap. 9, Sect. 9.3.2. Application of KT-tuyeres in practice completely confirms the results of these studies, which proves, once more, the similarity between blowing processes in the baths of open-hearth and EAFs.

Unfortunately, even in case of the KT-tuyeres installed not lower than the slag level, the evaporative cooling does not ensure their sufficiently high reliability and durability. Although the design of KT-tuyeres allows for the possibility of replacing the worn part of the head without dismantling of the entire tuyere, their maintenance costs, in comparison with the jet modules are quite significant. This limits their wide use. Further increase in the effectiveness of the submerged blowing requires the development of more durable stationary tuyeres capable of operating reliably and safely at the slag–metal boundary level and even lower. Such tuyeres can be developed only based on their intensive cooling with water and on condition that complete safety of their use is provided.

11.3.3 Explosion-Proof Highly Durable Water-Cooled Tuyeres for Deep Blowing

Production safety is one of the most important and most complex problems. Two possible approaches can be taken when selecting the innovations related to this problem. In each specific case, there is a struggle between these opposing approaches. In the first approach, the principle of prohibition prevails. A new, potentially dangerous element of technological process or equipment is simply ruled out on the basis of the prevailing stereotypical opinions about degree of its hazard. The second approach allows to accept such innovations, if the conditions of their safe operation can be provided for and assured.

The history of technics shows that the first, prohibiting approach has no prospects. If new technics ensure major advantages and if it is feasible to make it practically safe, it will necessarily be accepted and implemented. Steelmaking, like other technologies, is developed by means of application of more and more effective, but in the same time more complex and, in principle, far more unsafe equipment. However, due to improvements and mastering the equipment, the modern steelmaking shops are not more, but less dangerous than those 50 or more years ago. This results from implementation of new technical solutions, as well as from the fact that new equipment is operated and maintained by well trained, highly skilled, experienced, and disciplined personnel. If these requirements for the personnel were not met, the modern steelmaking shop would continually present danger to life and health of the personnel. Thus, despite all technical achievements, which reduce the danger to the minimum due to the means of informatics and automation among other factors, the human factor remains the key element of safety assurance.

Let us illustrate the aforesaid with a typical example. From the very beginning of development of EAFs with the high-power transformers (in the 1960s), it became obvious, that the potential advantages of these furnaces could not be realized without substituting the quickly fracturing lining with the wall and roof water-cooled panels. However, at that time, one of the inventors of the high-power electric arc furnaces, W.E. Schwabe, strongly opposed the water-cooled panels. He was concerned with the experience obtained by Japan, which was the first to start the widespread application of these panels. The explosions due to the water leaks resulting from the burn-backs occurred repeatedly at the Japanese plants during the implementation stage of the panels. Later, the new technology was mastered, received ample recognition, and became widespread.

Unfortunately, although extremely rare, explosions due to burn-backs of the panels occur at the present time as well. However, this is always a result of major violation of regulations for manufacturing, installation, or operation of the panels, as well as of improper actions of the poorly trained operators who lacked the experience of dealing with water leaking into the furnace. But at the present day, no one will ever consider getting back to the refractory lining. Each of these accidents leads to analysis and elimination of its causes and reinforces preventive measures in order to ensure that such an accident does not occur in the future.

Summing up the aforesaid, we may stipulate three basic conditions which must be observed when implementing new, potentially dangerous innovations in the technological process. First, such innovations must quite considerably improve the basic performance indices of technological process, as well as its productivity, efficiency, etc. Second, innovations must satisfy the following unconditional requirement, namely that compliance with clear, simple, and easy to follow rules must completely ensure the operational safety of innovation. Third, the innovations must be put in operation only after the operators and maintenance personnel get proper training on working with new equipment.

We will further distinguish the submerged blowing into the slag and the deep blowing into the slag–metal boundary or even lower, i.e., into the liquid metal. Does the deep blowing satisfy the first condition which requires a sharp increase in the effectiveness in comparison with the existing blowing methods? The data already given above show that, with an increase in the depth of tuyere submersion into the melt, the intensity of stirring and the effectiveness of blowing grow quite significantly. Yet it is necessary to give one more example, namely the results obtained from testing of the so-called SIP process in the open-hearth furnaces. These results must be taken into consideration, since, as was already shown, the baths of open-hearth and EAFs, with respect to the processes of the oxygen blowing, are similar.

The SIP process is an impressive example of effectiveness of deep oxygen blowing of the bath of the hearth furnaces by lateral single-nozzle tuyeres. In the beginning of 1980s, this process was tested in Hungary. Later, over a period of several years, the extended industrial trials were conducted in the Soviet Union, mainly in the 600-ton open-hearth furnaces of the Magnitogorsk Metallurgical Combine which operated on the charge containing 65% hot metal and 35% scrap. The tuyeres of the converter type for bottom blowing were installed in one row on the hearth bank on the rear wall side of the furnace at the level of the slag–metal boundary or somewhat lower. Oxygen was blown through the internal pipes of the tuyeres, while natural gas was blown through the annular gaps between the pipes. In comparison with blowing of the bath by the usual roof six-nozzle oxygen tuyeres which (and this should be emphasized) were also submerged in the bath to the level of slag–metal boundary, the following results were obtained.

With the identical, or even somewhat smaller, oxygen flow rates, the duration of the periods of melting and refining (from hot metal charging to tapping) reduced by 30–40%, the rates of carbon decarburization and bath heating increased by 1.5–2 times, the yield increased by 1%. The heat power of the furnaces remained at the same level. These results can only be explained by the quite sharp increase in bath stirring intensity.

At the same time, the wear of the lining of the bottom banks in the tuyere zones reached an unacceptable rate. The studies by the method of physical modeling showed that this wear was mainly caused by the turbulent pulsations of the bath formed by the oxygen jets nearby nozzles. Unlike in the converter practice, the annular gas jets enveloping the oxygen jets directed horizontally or inclined did not prevent the rapid wear of the lining. Mushroom-shaped build-up protecting refractory, Fig. 11.5, was not formed. Installation of special high-density and high-fireproof blocks made of expensive ceramic materials in the tuyere zones did not solve the problem either. Their durability under these conditions was also unacceptably low. As a result, despite persistent efforts to overcome this obstacle, the attempts to master the SIP process eventually had to be abandoned.

Taken altogether, the data examined do not leave any doubt that the deep blowing could radically increase productivity and operating economy of the EAF. However, this process requires the application of highly durable water-cooled tuyeres. When examining the possibilities of its realization, it is first necessary to clarify what exactly the danger of blowing by these tuyeres is. The water cooling of the submerged in the bath blowing tuyeres does not exclude the possibility of water entering the bath in case of burn-back. Does it mean that blowing by these tuyeres is always unacceptable due to safety conditions? With respect to the mobile tuyeres, the long-standing practical experience accumulated at the open-hearth and EAFs gives an unambiguously negative answer to this question.

Roof oxygen tuyeres have been used in the Soviet Union in more than 100 open-hearth furnaces (2–3 tuyeres in a furnace) for more than 30 years. The durability of these tuyeres has amounted on the average to 30–40 heats prior to their burn-back. They were replaced only after burn-backs and, as a rule, not during the heat, but after tapping. A burnt tuyere was not raised, and blowing was carried out until the completion of heat. A tip of tuyere was always submerged in the slag as deeply as possible and kept as near the slag–metal boundary as possible. It happened quite often that a water jet was spurting out from a burnt tip for several hours under the pressure of several bars, hitting under the layer of slag whose thickness varied from 150 to 300 mm. In the course of more than 30 years of operation, such situations have happened hundreds of thousands of times and have been considered usual. There have been no explosions in the process.

For a long time, the open-hearth and the twin-shell furnaces at a number of plants have been operating with submerging tuyeres into the metal at the depth of 100–150 mm. The tuyeres in the metal have burnt back more frequently, but even in case of this kind of blowing the safety-related problems have not occurred. The experience of operation of the mobile water-cooled tuyeres in the electric arc furnaces is quite similar in this regard. During the extended period of usage of these tuyeres, the situations have happened thousands of times when, in case of tuyeres burn-back, the water jets have been blown under the pressure in the bath. The use of these tuyeres is being gradually abandoned, and not due to safety-related issues, but because of their low durability and high operating costs, Sect. 11.1.3.

In contrast to the mobile tuyeres, there are reasons to regard as quite dangerous the installation of stationary water-cooled tuyeres and other water-cooled devices in the lining of the bottom banks of the EAF even higher than at the level of the sill of the slag door. The accumulated practical experience shows that, in this case, a real danger is posed not by the noticeable tuyeres burn-backs, because they are easily and rapidly detected due to a number of signs indicating the burn-backs, which allows to stop the water leakage in a timely manner by shutting down the burnt tuyere. A real danger is posed by the small concealed leakages, which are difficult to detect early enough. Such leaks damage the inner, hidden from observation, layers of lining, which leads to severe accidents with explosions and metal escape. Therefore, these installations must be used only with the application of highly reliable technical equipment which guarantees practically inertialess detection of any smallest hidden water leaks.

The schematic diagram of installation of water-cooled tuyere in the lining of the bottom banks of the EAF completely satisfying this requirement is shown in Fig. 11.6. A complete safety of installation is ensured by the system in which the tuyere (1) is cooled by water circulating in a closed circuit (2), which includes a pump (3), a heat exchanger (4), and a water pressure sensor (5). Physically and chemically prepared water circulating in the circuit is cooled in the heat exchanger (4) by regular water.

An amount of water circulating in the closed circuit is very small, i.e., several liters in all. Water is a practically incompressible fluid. Therefore, the occurrence of even very minor water leak from the tuyere is detected practically inertialess by the sensor (5) due to a sharp drop in the pressure of water in the circuit. As this takes place, in response to a signal from the sensor (5), the valves (6) are shut down cutting off the water supply into the tuyere, and an injector (7) starts working, which creates rarefaction in the tuyere and thus sucks water out. The amount of water which in this case can get into the furnace and the bottom lining is negligibly small and cannot cause any negative consequences. The installation shown in Fig. 11.6 is quite compact and can serve both for deep oxygen blowing and for deep injection of carbon powder into the bath. It should be emphasized that unlike vertical bottom tuyeres, Fig. 11.5, the inclined position of tuyeres in bottom banks allows to carry out blowing of the bath with a high intensity.

The main problem with these installations is durability of the tuyeres which can come into contact with liquid metal and, furthermore, undergo the blow-back of the oxygen jets reflected from the unmelted scrap lumps. The resulting heat flows impacting on the tuyeres are extremely high and, in case of usual methods of

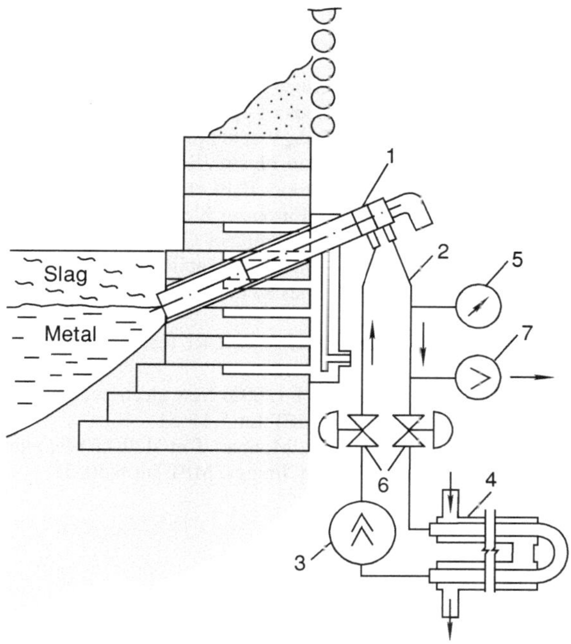

Fig. 11.6 A safety water-cooled oxygen tuyere installed in the bottom lining for intensive deep blowing of the bath (designations are given in the text)

cooling of the tuyeres, cause the burn-backs. Cooling of the tuyeres in the circulation circuit with individual low productivity, but high pressure pumps allows to use new hydraulic diagrams capable of reliably diverting any extremely high heat flows occurring in practice. These problems are examined in detail in the following chapter.

Ensuring the durability of the lining adjacent to the tuyere is no less important problem. Cooling the lining by the tuyere itself is insufficient. Therefore, installation of an additional copper ribbed cooler, Fig. 11.6, whose water-cooled part is placed outside of the bottom, is provided for in the tuyere zone. Such coolers are capable of sharp enhancing the durability of the lining. This is ensured by an increase in the density of heat flux removed out of the lining surface approximately six times [9].

Mechanism of this process is as follows. With an increase in the heat flux the temperature gradient in the vicinity of the lining surface rises. The lining layer thickness having high temperatures decreases. The impregnation pace of the lining with oxides of iron and other elements, which sharply deteriorates refractory properties, reduces correspondingly. Deceleration of this process provides with significant increase in the lining durability.

The described above installation solves the specific technical problem and must not be regarded as universal. It is yet another example which shows that further increase in the effectiveness of devices for blowing the bath with oxygen and carbon requires new, nontraditional approaches to this problem.

References

1. Glinkov MA (1970) Thermal performance of steelmelting of baths. Metallurgia, Moskow
2. Markov BL (1975) Methods of open-hearth bath blowing. Metallurgia, Moskow
3. Pujadas A, McCauly J, Tada Y et al (2002) Electric Arc Furnace energy optimization at Nucor Yamato Steel. 7th European Electric Steelmaking Conference, Venice
4. Jaroslav B, Shver V, Akemore R Bl et al (2002) An improved method of applying chemical energy into the EAF. 7th European Electric Steelmaking Conference, Venice
5. Allemand B, Bruchet P, Champinot C et al (2000) Theoretical and experimental study of supersonic oxygen jets – industrial application in EAF. Process Technology Conference
6. Sarma B, Mathur PC, Selines RJ et al (1998) Fundamental aspects of coherent gas jets. Process Technology Conference
7. Malek AO (2004) DC EAF with DRI feeding rates through multipoint injection. MPT Int 2:58–67
8. Rondi M, Bosi P, Memoli F (2002) New electrical and chemical technologies implemented in the Dalmine steel plant. MPT Int 5:44–51
9. Kirschen M, Kronthaler A, Molinari T et al (2007) Economical aspects of using water-cooled copper blocks in refractory linings. MPT Int 6:30–31

Chapter 12
Water-Cooled Furnace Elements

12.1 Preliminary Considerations

A significant number of water-cooled elements are used in the design of the modern electric arc furnaces. This chapter reviews only those elements which greatly affect productivity and cost effectiveness of the furnaces, as well as the selection of directions of intensification of the heat. These elements include, first of all, tuyeres for injection of oxygen and carbon into the liquid bath, protective boxes of the jet modules, and wall and roof panels. These elements are periodically exposed to the very high heat fluxes which occur in the bath reaction zones during oxygen blowing, as well as in the case of blow-back of oxygen jets reflected from the scrap lumps and under the conditions of radiation produced by the open powerful electric arcs.

Insufficient durability of these elements requires limiting the electrical furnace power and the power of oxy-fuel burners, as well as reducing oxygen consumption and even changing the bath blowing method as, for example, substituting the immersed blowing by mobile tuyeres. Chap. 11, Sect. 11.1.4, with less efficient blowing from above. Increasing durability of water-cooled elements is one of the most important tasks related to the further improvement of EAF's productivity. This chapter contains information necessary for understanding the causes of the water-cooled elements failures as well as recommendations for calculation and design of these elements. Particular attention is paid to the principles of development of the water-cooled elements capable of operating reliably under the harshest conditions. To understand the information below it is recommended to refer when necessary to the relevant sections and formulae of Chap. 3. The references to those formulae are given.

12.2 Thermal Performance of Elements: Basic Laws

Let us review the scheme of the simplest water-cooled element, Fig. 12.1a. The flat uniform wall of the element receives the heat flux of the power Q_1, kW, from the external surface side, and transfers the heat flux Q_{WAT} to water from the

Y. N. Toulouevski and I. Y. Zinurov, *Innovation in Electric Arc Furnaces*,
DOI: 10.1007/978-3-642-36273-6_12, © Springer-Verlag Berlin Heidelberg 2013

internal surface side. The wall thickness δ, m, is small in comparison to its length and width, therefore the heat flux can be considered one dimensional. The wall is made of metal with high thermal conductivity coefficient λ (copper or low-carbon steel, for instance). Due to this fact as well as due to the small wall thickness, the thermal inertia of the wall with changing of Q_1 can be ignored assuming that the element is in a stationary state at any given moment of time.

In this state, the enthalpy of the wall does not change and the wall heat balance equation $Q_1 = Q_{WAT}$, kW, is observed. This condition is a basis for determining the required water flow rate. If the areas of external and internal surfaces are the same and equal to S, m^2, then the heat flux densities q_1 and q_{WAT} passing through both these surfaces will also be the same ($q_1 = q_{WAT}$) and equal to Q/S. In the stationary state at $q_1 = q_{WAT}$ the temperatures of the external surface t_1 and of the internal surface t_2 are constant and related to q through Eq. (12.1), Chap. 3, formulae (3.1) and (3.30).

$$q = \lambda\,(t_1 - t_2)/\delta = \alpha(t_2 - t_{WAT}), \tag{12.1}$$

α – coefficient of convective heat transfer from wall to water, kW/m^2 × °C
t_{WAT} – water temperature, °C

Eq. (12.1) is valid only when $q_1 = q_{WAT}$. But the equality of these specific heat fluxes occurs only in some particular cases. In the pipe-shaped element with d_2 much smaller than d_1 the external surface area S_1 is significantly larger than the internal surface area S_2, and q_{WAT} can significantly exceed q_1, Fig. 12.1b. In such cases, which are common in practice, the heat flux concentration on the water-cooled surface should be considered. The degree of concentration is characterized by a coefficient K, which is equal to S_1/S_2 or q_{WAT}/q_1. When $q_{WAT} > q_1$, $K > 1$. If, for instance, due to its ribbing (as shown in Fig. 12.1c), the water-cooled surface exceeds the external surface area of the element, then $q_{WAT} < q_1$ and $K < 1$. In this case, instead of the heat flux concentration, the heat flux dissipation on the water-cooled surface takes place.

Different types of wear of water-cooled element occur in practice. Usually the intensive wear is caused by the combination of heating the element's external surface to the high temperature t_1 and inevitable under the heat conditions abrupt

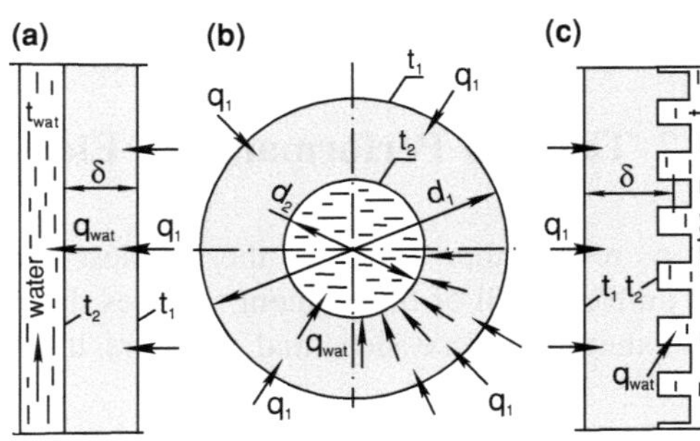

Fig. 12.1 (**a**, **b**, **c**). Schematic diagrams of water-cooled elements

cooling. Under the conditions of high t_1 and sharp temperature fluctuations and due to gases and slag attack, the wall material composition and structure changes, and the material corrodes and flakes off. The cracks develop, and their propagation leads to water leaks. Under these conditions, with quite high t_1, the cracks develop even in such material as copper, which hardens during its service life and looses its ductility.

When a maximum critical temperature $t_{1\max}$ for the given material is exceeded, the wear is accelerated and in a short period of time can lead to a burn-back. For the copper wall $t_{1\max}$ is approximately 300°C, whereas for the low-carbon steel wall $t_{1\max}$ is approximately 400°C.

There is one more type of wear. It can occur even at the relatively low temperatures t_1, if the internal wall surface temperature t_2 is close to the water boiling temperature t_S at the given pressure or somewhat exceeds it. In this case, the wall wear can start from the internal surface and is directed outward, therefore, the element looks undamaged until the burn-back occurs. This type of wear is caused by the thermo-chemical corrosion, which develops only when cooling water is of poor quality and contains a lot of dissolved oxygen and chemicals forming the deposits on the wall when heating water. In the modern steelmaking shop water treatment is given appropriate attention. Therefore, the wear through the internal thermo-chemical corrosion in the water-cooled elements is not typical and is not discussed any further.

The most reliable indicator of the potential durability of the water-cooled element is its external surface temperature t_1. At the low t_1 the element's service duration can be measured in years. The rapid wear of the element is usually related to the high surface temperatures. Let us review, which factors affect this parameter.

Let us rearrange the left part of Eq. (12.1) as follows:

$$t_1 = t_2 + q\delta/\lambda. \tag{12.2}$$

It is easy to see that the value $q\delta/\lambda$ is the temperature gradient in the wall $\Delta t = t_1 - t_2$. This value grows directly proportional to q and the thermal resistance of the wall δ/λ. The heat flux density q_1 is determined by the external factors, which are not controlled by a water-cooled element designer. In order to obtain low temperature t_1 in the case of high q_1, it is necessary to use the materials with the highest thermal conductivity and to reduce the wall thickness δ to the value acceptable under the given strength conditions. It is unreasonable to increase the wall thickness to create a wear reserve because the wear acceleration surpasses this reserve.

Assuring the low temperature level of the internal surface t_2 is of great importance. This value is not only one of the components of t_1 value, formula (12.2). It affects the very mechanism of heat removal from the wall to water. The change of this mechanism can lead to abrupt wear acceleration and rapid burn-back.

From formula (12.1) the temperature t_2 can be expressed as follows:

$$t_2 = t_{WAT} + q_{WAT}/\alpha, \tag{12.3}$$

t_{WAT} – water temperature in the channel, °C; usually the mean value is used
q_{WAT} – heat flux density through the waterside surface, kW/m^2

The coefficient of convective heat transfer from the wall to water α depends mostly on the water velocity in channel v, m/s, and significantly increases as the velocity rises. When α and t_{WAT} are constant, temperature t_2 grows linearly as q_{WAT} increases, and the lower the coefficient α, the quicker this increase is, formula (12.3), Fig. 12.2. However, such growth is observed only to a certain limit. As soon as t_2 exceeds the water boiling temperature t_S at the given pressure, Table 12.1, by a few degrees, the further growth of t_2 slows down sharply.

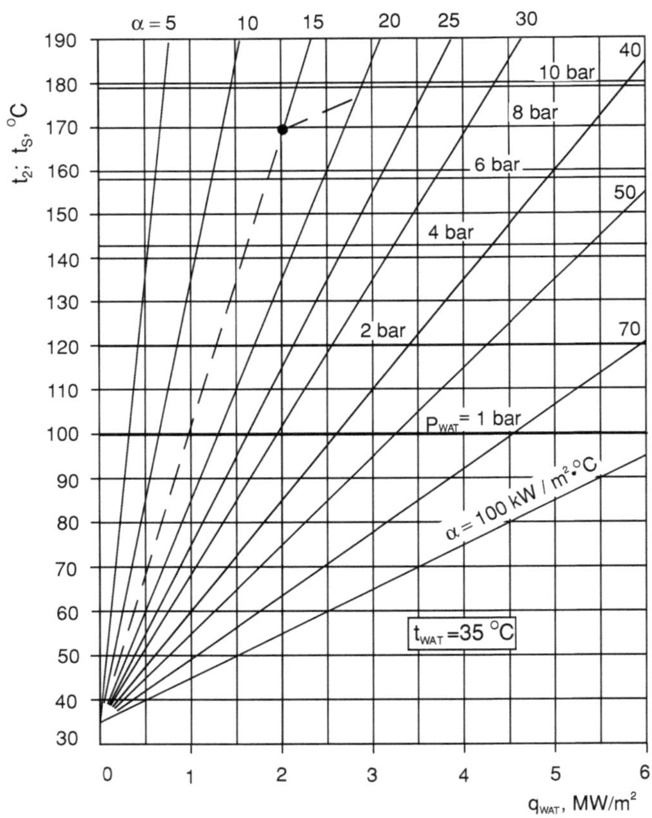

Fig. 12.2 Dependence of temperature of a water-side channel wall t_2 on q_{WAT}, α, and p_{WAT} (abs). *Horizontal straight lines* are the water boiling temperatures t_S at various values of p_{WAT}. The average water temperature is 35°C

Table 12.1 Water boiling temperature t_s depending on pressure p, abs

p, bar	1	2	3	4	5	6	7	8	9	10
t_S, °C	99.5	120.4	133.7	143.7	151.9	158.9	165.1	170.5	175.4	179.9

For example, when $\alpha = 15$ kW/m^2 × °C and water pressure is $p_{WAT} = 8$ bar (abs), reduction of t_2 growth rate will take place at the temperature close to the boiling point $t_S = 170.5$°C at this pressure (dotted broken line in Fig. 12.2). This can be explained by occurrence and development of a new process, the so-called local or surface bubble boiling, which significantly intensifies the heat transfer from the wall.

The mechanism of this process is as follows. When $t_2 > t_S$, despite the fact that water flow in general remains cold, the vapor bubbles are formed on the wall. When they detach from the wall, they pass through the boundary layer of water and condense when entering the cold core of the flow. These bubbles turbulize the laminar sub-layer of the boundary layer; its thermal resistance to the heat flux drops sharply, and the coefficient α increases, respectively. As q_{WAT} grows further, the number of bubbles formed per unit time grows, and α continues to increase.

However, increase in α continues only to a certain level critical for the given conditions, q_{CR}. When the heat flux exceeds this level, the number of bubbles becomes so huge that they coalesce into a solid–vapor film, which has negligible thermal conductivity and completely insulates the wall from water. If the vapor film is stable, this leads to the abrupt growth of t_2 and t_1 to such high values that heat transfer occurs through radiation. In this case, the wall burns through immediately. The calculation of the wall temperature t_2 in the case of local boiling is discussed further down, Sect. 12.3.4.

12.3 Principles of Calculation and Design of Water-Cooled Elements

12.3.1 Determining of Heat Flux Rates

Heat flux densities q_1, kW/m^2, received by the water-cooled elements in EAF depend on many factors and vary within the wide limits. Only in some simplest cases these flux densities can be determined theoretically with acceptable accuracy by the analysis of heat exchange conditions in the freeboard and in the liquid bath. Therefore, the experimental data obtained as a result of direct measurements should be used in calculations.

The data provided by the different authors differ significantly. Besides, the conditions under which the measurements were made can differ from the actual working conditions of that particular element. All of the above brings substantial inaccuracies in determining the heat flux densities q_1.

Heat flux rates are the most important input data for water-cooled elements' calculations. Obviously, the potential accuracy of the final results of these calculations cannot exceed the accuracy of determining q_1 values. Under these conditions, simplification of the calculation methods is quite justified since in reality the calculations using the graphs and the simplified formulae by the accuracy are not inferior to the calculations carried out using complex mathematical models.

The thermal and temperature operating conditions of water-cooled elements in the steelmaking units change sharply in the course of the heat. When designing the elements, not the time-averaged heat flux densities must be taken into consideration, but their maximum or peak values q_{1max}. These fluxes act for short periods but periodically reoccur. These very fluxes determine the elements durability and are responsible for the elements premature failures. The peak heat fluxes can reach the values so high that they are able to strip the skull from the surface of the element in a few minutes, melt down the surface, and even cause burn-back. Therefore, the cooling system of the element has to be designed specifically for the peak values q_{1max}. The typical maximum values of heat flux densities for various water-cooled elements under EAF's conditions are shown in Table 12.2, which is compiled using the published experimental data from a number of authors.

For the panels without the skull build-up, the values given in Table 12.2 agree with the density of the black body radiation of highly dust-laden gases in the freeboard, which is equal to 680 kW/m^2 at 1700°C. Occasionally, much higher heat fluxes may appear in the wall panels as a result of unstable burning of the electric arcs. With the aid of high-speed filming, it has been determined that under certain conditions the arcs could be very frequently blown from underneath the electrodes toward the furnace walls. When this happens, in the certain small areas of the wall panels the heat fluxes increase by several times for a short period of time. On the other hand, the values of q_{1max} drop down to 200–300 kW/m^2 for the panels, which are heavily encrusted with slag.

The values of the heat fluxes on the copper heads of oxygen tuyeres immersed in the slag and in the metal, Table 12.2, have been determined with certainty in the open-hearth furnaces at the bath temperatures of about 1650°C. These fluxes were determined by the difference between the total flux onto the tuyere and the flux onto its side surface. The former was determined by the flow rate and the temperature of the cooling water, whereas the latter one was determined by measuring

Table 12.2 Maximal heat flux densities for EAF's water-cooled elements

Elements	Operating conditions	$q_{1.max}$, kW/m^2
Wall and roof panels	Depending on the arc shielding extent	600–700
Heads of mobile	At submerging into slag	1300–1700
Oxygen tuyeres	At submerging into metal	4000–6000
External body pipes of mobile oxygen tuyeres	Above the bath	700–900
Boxes of jet modules and wall panels	At blow-backs of oxygen jets	4000–6000

the elongation of the external pipe of the tuyere [1]. According to the other data, in the case of head immersion into the metal, even with the skull build-up, the heat flux can reach 10 MW/m^2. Such fluxes have been measured in the experiments in a small induction furnace at the metal temperature of 1900°C [2]. The heat exchange coefficient from the liquid metal to the surface of the skull on the head was approximately 25 $kW/m^2 \times$ °C. This value can vary significantly depending on the intensity of metal stirring in the reaction zone [2].

12.3.2 Minimum Necessary Water Flow Rate

The heat flux densities in the different sections of the external surface of the elements are usually various. The total maximum heat flux Q_{max} received by the element is determined by adding the fluxes through all of its sections during the most thermal stressed for this element period of the heat.

$$Q = q_1 \times S_1 + q_2 \times S_2 + \ldots + q_n \times S_n, \text{ kW}, \tag{12.4}$$

S_n – external surface area of the nth section, m^2.

As it follows from the heat balance of the element, the total heat flux Q is used on water heating. The dependence between the mass M or volumetric V water flow rate and Q is determined through the following expressions:

$$M = Q/c\,(t_{WAT2} - t_{WAT1}), \text{ kg/h}, \tag{12.5}$$

c – heat capacity per unit mass of water, kWh/kg × °C
t_{WAT2} – final water temperature at the element outlet, °C
t_{WAT1} – initial water temperature at the inlet, °C
t_{WAT2}–t_{WAT1} = Δt_{WAT} – increase in water temperature after passing through the element, °C

$$V = Q/\rho \times c\,(t_{WAT2} - t_{WAT1}), \text{ m}^3/\text{h}, \tag{12.6}$$

ρ – density of water, kg/m^3

It is obvious that

$$V = M/\rho, \text{ m}^3/\text{h}. \tag{12.7}$$

Since water density ρ and its thermal capacity c vary insignificantly in the temperature range from 20 to 60°C, with an inaccuracy of less than 1% the dependence (12.6) can be expressed by a simplified formula:

$$V = Q/1.15\,(t_{WAT2} - t_{WAT1}), \text{ m}^3/\text{h}. \tag{12.8}$$

In order to prevent salt deposits on the channel walls, for chemically untreated water of low quality, the temperature gradient value Δt_{WAT} should be chosen so that the final water temperature t_{WAT2} will not exceed 40–50°C. In any case, it should be taken into account that the increase in the final and therefore average water temperature in the element causes increase in the temperature of the internal t_2 and external t_1 surface of the element, formulae (12.3) and (12.2). By the same reason, even for the chemically treated water the temperatures exceeding 70°C are undesirable. If in expressions (12.5) and (12.6) the maximum allowed temperature gradients $\Delta t_{WAT.max}$ are involved, then the obtained M and V values correspond to the minimum required water flow rate.

The minimum flow rates correspond to the minimum water velocities in the element channels in which flow sections are often determined by independent design considerations. If such velocities turn out to be insufficient due to the low coefficient of convective heat transfer α or other considerations, then the water flow rates are increased. In most cases, the water flow rates are determined not only by the maximum allowable values of t_{WAT2}, but, in the first place, by the conditions of heat transfer which directly affect the elements' durability.

12.3.3 Critical Zone of the Element

When designing the element, the most attention must be given to the so-called critical zone of the element where its external surface temperature t_1 and internal temperature t_2 can reach maximum values. The calculations must determine such parameters of water-cooling system, which allow to reduce t_1 и t_2 in the critical zone to the desired level. It should be taken into consideration that the rapid wear even of small area of critical zone of the element will sharply shorten the durability of the entire element.

In most cases, the location of the critical zone is known in advance. The critical zone is usually the section of the external surface of water-cooled element in which cooling runs into the greatest difficulties. These difficulties can occur either due to concentration of the highest heat fluxes in the critical zone or due to the design difficulties when it is impossible to supply water to the zone with sufficient velocity or in sufficient amount. The worst case is when both factors have an effect in the same zone. The heat transfer calculations should be started from determining temperatures t_1 и t_2 in the critical zone.

12.3.4 Temperature of Water-Cooled Surfaces

This temperature t_2 is determined by expression (12.3). When the local (surface) boiling is absent, the value of α in expression (12.3) is a coefficient of convective heat transfer in the case of forced turbulent flow, which depends mostly on the water velocity. It can be calculated by simplified formula from 3.36 from Chap. 3. Formula (3.36) includes the value of the hydraulic channel diameter d_h. For the round channels, $d_h = d$. For the channels of rectangular shape, $d_h = 4S/2(a + b)$, where S is cross-section area of the channel, a is the channel width, b is its height, and $2(a + b)$ is perimeter of the channel. For the relatively narrow ring and rectangular channels (slots) with the width of a, $d_h = 2a$.

When local boiling appears, the value of coefficient of convective heat transfer increases to α^*. The latter depends on two basic factors, namely the water flow velocity in the channel v and the intensity of turbulization of laminar sub-layer by vapor bubbles. The effects of both these factors are added together. The higher the water velocity and the more the intense boiling, the higher is α^*.

This mixed process of heat exchange is very complex. Various criterion formulae for determining α^* are known. These formulae summarize the results of vast experimental research. They are complex, contain the parameters which are hard to determine, and are not suitable for simplified calculations of the water-cooled elements. For such calculations, it is recommended to determine the coefficient α^* depending on the ratio of the following two coefficients of heat transfer: coefficient α_q for boiling in large volume without forced water flow and purely convective coefficient α without local boiling, which depends on flow velocity formulae (3.35) and (3.36), Chap. 3. It is recommended to assume $\alpha^* = \alpha$ for $\alpha_q < 0.5\alpha$, $\alpha^* = \alpha_q$ for $\alpha_q > 2\alpha$, and it is recommended to use the interpolation dependence for the intermediate range $0.5\alpha < \alpha_q < 2\alpha$ [3].

$$\alpha^* = \alpha \times \frac{4\alpha + \alpha_q}{5\alpha - \alpha_q}. \tag{12.9}$$

To determine α_q, the well-known simplified formula for water boiling in a large volume can be used:

$$\alpha_q = 3.14 \times q^{0.7} \times p^{0.15}\,\mathrm{W/m^2}, \tag{12.10}$$

q – heat flux density, W/m^2
p – absolute water pressure, bar

Let us illustrate the above considerations by an example.

An example. The water-cooled element contains a channel of diameter $d = 40$ mm. Water velocity in the channel is $v = 3$ m/s, water temperature is $t_{WAT} = 30$°C, and pressure is $p = 3$ bar (abs). Density of heat flux toward water is $q_{WAT} = 2000$ kW/m^2. The temperature of the channel surface t_2°C must be determined.

First step. Let us determine whether heat removal without local boiling is possible at the given parameters. From formula (3.36) and Table 3.4 at $v = 3$ m/s, $d_h = 40$ mm, and $t_{WAT} = 30$°C we find $\alpha = 12.1$ kW/m^2 × °C. In the case without local boiling, from formula (12.3) $t_2' = 30 + 2000/12.1 = 195$°C. Using Table 12.1 we find that for $p = 3$ bar (abs) the boiling temperature of water is $t_S = 133.7$°C. Since $t_2' > t_S$, we prove that the local boiling occurs.

Second step. In the case of local boiling, the temperature t_2 differs from t_2' and is determined by the coefficient of heat transfer α^* rather than α. To determine α^*, let us first calculate α_q, formula (12.10):

$$\alpha_q = 3.14\,(2000 \times 10^3)^{0.7} \times 3^{0.15} = 95,397\,\mathrm{W/\,m^2 \times {}^\circ C}$$
$$\text{or}\ \ 95.4\,\mathrm{kW/\,m^2 \times {}^\circ C}.$$

Since $\alpha_q > 2\alpha$, then we assume $\alpha^* = \alpha_q = 95.4$ kW/m^2 × °C.
Third step. Let us find the actual temperature t_2 in the case of local boiling. In this case, formula (12.3) is transformed to the form:

$$t_2 = t_S + q/\alpha^*, \tag{12.3'}$$

$$t_2 = 133.7 + 2000/95.4 = 154.7\,^\circ\mathrm{C};\ t_2 \cong 155^\circ\mathrm{C}.$$

In practice, the case of $\alpha_q < 2\alpha$ can be encountered only when water flow velocities v are very high. In most cases with local boiling, $\alpha^* = \alpha_q$ can be assumed.

Since $\alpha^* > \alpha$, it can seen that the cooling elements in the local boiling mode can be effectively used for removal of large heat fluxes. However, in reality, this mode has significant drawbacks and cannot be recommended.

First, when the local boiling occurs, the temperature t_2 cannot be below boiling point of water t_S, which exceeds 140–150°C at the excessive water pressure in the element of 3–4 bar, Table 12.1. When the initial water temperature is about 25°C, the convective cooling mode allows to assure much lower t_2 and, respectively, improve the element durability, since the reduction of t_2 leads to the equal reduction of the external surface temperature t_1, formula (12.2).

Second, the local boiling mode is potentially dangerous due to the fact that under certain unfavourable conditions the bubble boiling in some local areas of the element can change to the film boiling which would immediately lead to burn-back. One should not count on the fact that when water velocity is quite high, about 3–5 m/s, the steady film boiling requires such high heat fluxes, which are not achievable in the practical operation of EAFs. The research carried out on the transparent elements with the aid of high-speed filming has shown that when configuration of the channels is complex, the high-frequency water pulsations occur in these channels. The water velocity can sometimes for the short periods of time drop almost to zero in the poorly flown-around sections of the channels in turbulent zones behind the sharp corners, where the water flow detaches from the walls. In such zones the bubble boiling and the film boiling can continuously take turns. This does not lead to an immediate burn-back, but sharply increases the temperatures t_2 and t_1 and the wear rate. The specifics discussed above lead to the conclusion that in order to assure reliability and durability of the water-cooling elements, the modes of operation with local boiling should be avoided. The cooling system should be designed in such a way that the temperature of the wall surface flown around by water t_2 is significantly lower than the boiling point t_S.

12.3.5 Temperature of External Surfaces

When the temperature t_2 is known, the temperature t_1 is determined by expression (12.11), which follows from formula (12.2), as well as by expression (12.12):

$$t_1 = t_2 + \Delta t, \tag{12.11}$$

$$\Delta t = t_1 - t_2 = q_{AVE} \times r, \tag{12.12}$$

$q_{AVE} = (q_1 + q_{WAT}) \times 0.5$ – the average density of heat flux over the thickness of the wall δ, kW/m^2
$r = \delta/\lambda$ – thermal resistance of the single-layer wall, m^2 × °C/W

The water-cooled elements of the EAFs are usually made of copper or low-carbon steel. Sometimes, the pipes made of alloy Cu – 68% and Ni – 30% are used. This alloy assures good bonding with copper when the pipes are used in cast elements. The coefficients of thermal conductivity of these materials significantly depend on temperature. When determining the average across the wall values of λ the graphs, Fig. 12.3 should be used. When the heat fluxes are high, even small changes of λ have a significant effect. The data from Fig. 12.3 refer to rolled high-purity electrolytic copper. Insufficiently deoxidized and insufficiently dense cast copper has λ values lower by 15–20%. Addition of 0.4% chrome to copper to improve heat endurance reduces λ by 1.5–2 times.

In the simple water-cooled elements, similar to the one shown in Fig. 12.1, the temperature fields and heat fluxes are one dimensional or close to those. In the

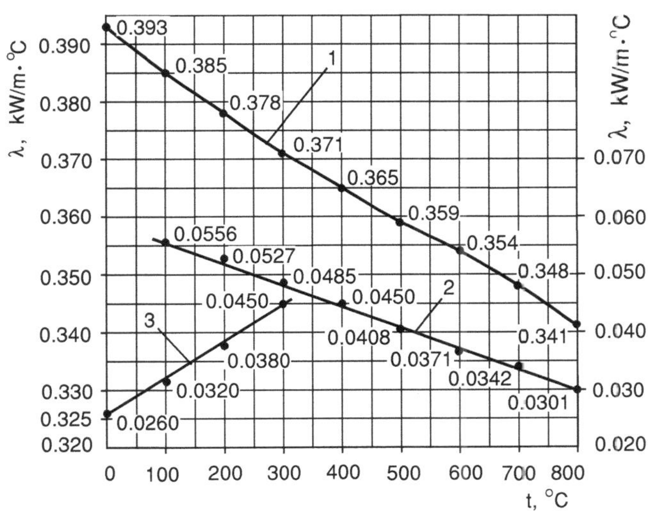

Fig. 12.3 Dependence of thermal conductivity λ on temperature 1 – rolled copper (Cu 99.9%); 2 – low-carbon steel; 3 – alloy Cu 68%, Ni 30%

elements of more complex geometry we should consider changes of these variables in two or three space coordinates rather than one. In these cases, instead of analytical formulae the numerical methods which were introduced in Chap. 3 Sect. 3.2.8 should be used. These methods are quite labor-consuming but allow obtaining all necessary values, such as t_1, with required accuracy.

For approximate calculations, the problem is simplified by substituting two- or three-dimensional temperature field with the equivalent one-dimensional one. The equivalency of the fields is determined approximately from the geometrical considerations. Using this method, it is possible to determine the thermal resistance of the complex wall r and determine t_1 with the practically acceptable accuracy. Using such simplified methods of calculation is illustrated below in Sect. 12.4.2 on the actual example.

12.3.6 General Diagram of Element Calculation

The temperatures t_1 и t_2 in acting elements where all geometrical parameters, thermal and physical properties of the materials, water flow rates, and the water temperatures are known can be calculated using the formulae given above and do not present any difficulties. The data obtained as a result of such verifying calculations allow to uncover the causes of insufficient durability of the element and to establish the ways to improve it. Let us remind that the calculation should be done, in the first place, for the critical zone of the element, which is found during the operation by the most significant wear.

The calculations of the thermal performance of the new element, which are the intrinsic part of its design, are much more complex. The sequence of such calculations is schematically shown in Fig. 12.4. They cannot be completely formalized and in each practical case require a creative approach. The calculation is comprised of a number of consecutive stages. If the intermediate result obtained at the certain stage is unacceptable this requires a return to the previous stages in order to change the design and mode parameters and repeated calculations. For example, for a given design option at the chosen water flow velocity (stage 5) the heat transfer calculations (stage 6) lead to the satisfactory values of t_1 и t_2 (stage 7). However, the verifying calculation of the hydraulic resistance of the element Δp_{h} (stage 8) showed that at the chosen water velocity it is unacceptably high. In this case, it is necessary to reduce Δp_{h} by changing the parameter of water circuit and to repeat stages 5–8 of the calculation.

12.3.7 Hydraulic Resistance of Elements

During the design of a new element and improving of existing elements their hydraulic resistance Δp_{h} is an important limiting factor, which was not previously reviewed. When a fluid flows the resistances always exist, which obstruct its

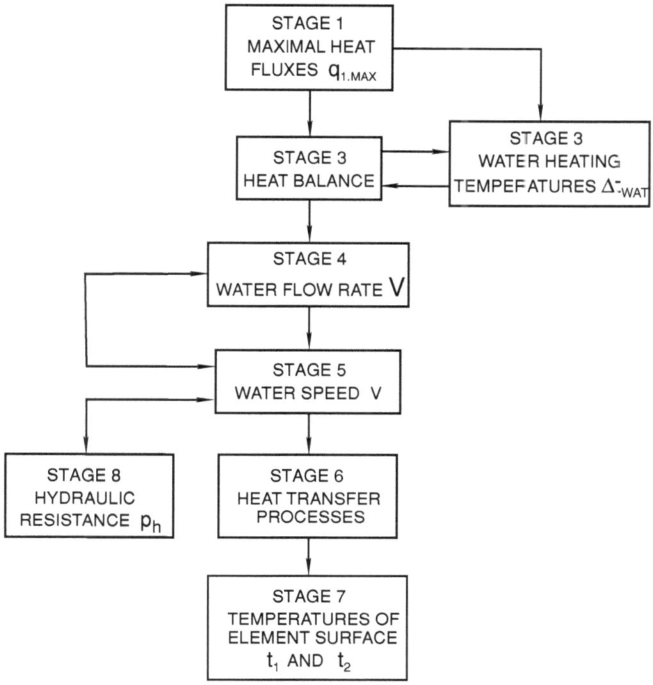

Fig. 12.4 Water-cooled elements calculation sequence

movement. Mechanical energy is spent to overcome these resistances. This energy is usually expressed as a gradient of pressure required to create a flow of fluid. Pressure is measured in bars (bar). Hydraulic resistance of an element Δp_h at the given water flow rate V, m^3/s, equals to the difference of pressures at the inlet p_1 and at the outlet from the element p_2:

$$\Delta p_h = p_1 - p_2, \text{ Pa (or bar)}. \tag{12.13}$$

Maximum value of Δp_h is always strictly limited by the water pressure difference, which practically exists in the furnace between its inlet collector and outlet collector. Usually this difference does not exceed 1.5–2.5 bars depending on the local conditions. The water circuit of the element must be calculated in such way that at the certain water pressure determined by the local conditions it must assure not only the necessary flow rate V, m^3/s, but also sufficiently high flow velocities v, m/s.

High velocities of water v are necessary to intensify the heat transfer to water in order to reduce the wall temperature t_2. However, as the velocity increases, so does the hydraulic resistance Δp_h, and usually directly proportional with the square of the velocity v^2. In the necessary cases, in order to overcome this conflict and assure the intense heat transfer without exceeding the maximum allowable level of Δp_h, the water velocities must be correctly distributed along the different sections of the element. The maximum velocities must be concentrated in the critical zone where they are required to prolong the durability of the element. At the other portions of the element the reduced velocities are used in a way which allows to compensate high hydraulic resistance of the section in the critical zone.

All hydraulic resistances are divided into two groups based on their origin: friction resistances and local resistances. The friction resistance Δp_{FR} is caused by the fluid's viscosity and is exhibited during the flow without detachment along the solid wall, e.g., in the straight pipes and channels. The resistance of such sections of the elements is calculated by the formula:

$$\Delta p_{FR} = (\zeta_{FR} \times L/d_h) \times \rho \times v^2/2, \text{ Pa}, \tag{12.14}$$

ζ_{FR} – coefficient of friction resistance
L – length of tube or channel, m
d_h – hydraulic diameter of channel, m
ρ – fluid (water) density at the given temperature, kg/m^3
v – velocity of water, m/s

Coefficient ζ_{FR} depends mostly on wall surface roughness and the degree of flow turbulence, which is defined by the criterion *Re*. As the roughness of metallic pipes increases due to their oxidation, corrosion, and deposits, the coefficient ζ_{FR} increases significantly. Under service conditions of copper and steel water-cooled elements of EAF, $\zeta_{FR} = 0.03$ can be assumed. Formula (12.14) is applicable to the flows with fully developed turbulence at $Re \geq 10^4$. Usually, in the practical applications of water-cooled elements in EAF, much higher values of *Re* are typical. If the channel consists of portions with different d_h and v, then the calculation of Δp_{FR} is carried out by separate portions and the obtained values are summed.

The local resistances are caused by the formation of vortices at the points of passing of different obstacles by the flow. Such obstacles are as follows: channel inlet and outlet, narrowing and widening of the channel, turns, etc. These resistances are calculated by the formula:

$$\Delta P_{LOC} = \zeta_{LOC} \times \rho \times v^2/2, \tag{12.15}$$

ζ_{LOC} – coefficient of local resistance is determined by the obstacle shape and correlation of its geometrical dimensions. The comprehensive data on various ζ_{LOC}, as well as additional information on ζ_{FR}, can be found in the handbook [4]. All local resistances are summed. The total hydraulic resistance of the element is assumed to be equal to the sum of local resistances and friction resistance.

This method of calculation is based on the assumption that the resistance of consecutively linked sections of the water circuit is equal to the sum of its separate resistances. In the reality, it is not true. The resistance of each section depends on the type of water motion in the previous sections. For example, the resistance of the straight section beyond the turn is significantly higher than the resistance of the same straight section before the turn. This can be explained by the deformation of the velocity field of the flow and formation of the additional vortices at the turn. Due to the same reason, the resistance of bent pipes is higher than the straight ones.

All mentioned above leads to a conclusion that usually the calculations of the hydraulic resistances of the elements are relatively not accurate. Therefore, the calculations must be carried out with a safety margin. In all cases where possible the calculation results must be verified on the manufactured elements or their prototypes. However, we have to stress that at the design stage, the calculations of the hydraulic resistance are absolutely necessary. The corresponding examples will be given in the next section during the analysis of thermal performance of the elements working under conditions of high heat fluxes. The purpose of these examples is to introduce reasonable design of the elements as well as the designs suitable for the demonstration and familiarization of the calculation methods.

12.4 Examples of Calculation Analysis of Thermal Performance of Elements

12.4.1 Mobile Oxygen Tuyere

The general view of the tuyere is shown in Fig. 11.2 of Chap. 11 and the tuyere head is shown in Fig. 12.5.

Input Data

In accordance with the actual data of tuyere operation in one of 120-ton EAFs we assume the tuyere diameter is equal to 219 mm, water flow rate V = 22 m^3/h (0.0061 m^3/s), and water temperature at the tuyere inlet t_{WAT1} = 18°C. The critical zone of the tuyere is the head copper face (1) with the two oxygen nozzles, Fig. 12.5, and the copper external pipe (2) near the head. The cross-section area of annular channel (3) $S = 0.0059$ m^2, hence water velocity in this channel $v = 0.0061/0.0059 \cong 1$ m/s. Considering the head channels' configuration we assume that the tuyere face is washed with water, which has approximately the same velocity.

Two tuyere positions are reviewed, Fig. 11.2, Chap. 11. At the first position the tuyere head is in contact with the slag surface. The tuyere length being in the freeboard amounts to 2.2 m and the surface area absorbing heat $F = 1.53$ m^2. At the second position the tuyere head is immersed in foamed slag for a length of 0.3 m. The area of the tuyere surface being in the slag F_{SL} = 0.25 m^2. Above the bath q_1= 700 kW/m^2, and in slag it is 1400 kW/m^2, Table 12.2. Concentration of heat fluxes on the water-cooled surfaces of the critical zone is insignificant and can be ignored.

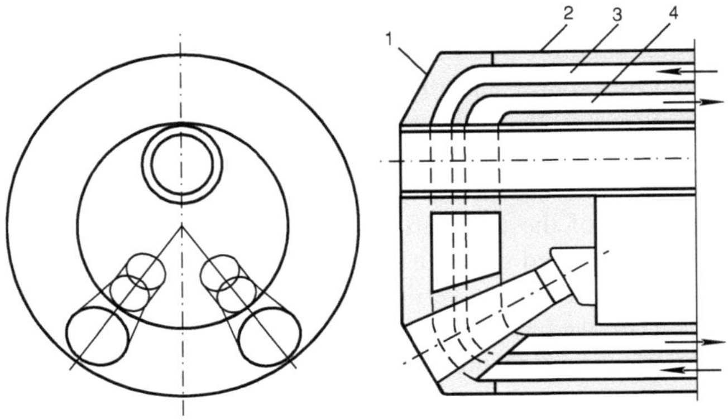

Fig. 12.5 The head of a mobile oxygen tuyere (designations are given in the text)

Calculation of Basic Parameters of Thermal Performance of Tuyere

The water in the external annular channel (3) of width 9.7 mm moves toward the head where water reaches the maximum temperature t_{WAT2}. During the water flow in the opposite direction in the internal annular channel (4) it is not heated. In order to determine t_{WAT2} let us convert formula (12.8) to look as follows: $t_{WAT2} = t_{WAT1} + Q/1.15\ V$.

At the first position $Q = 700 \times 1.53 = 1071$ kW and $t_{WAT2} = 18 + 1071/1.15 \times 22 \cong 60°C$.
At the second position $Q = 1071 + 1400 \times 0.25 = 1421$ kW; $t_{WAT2} = 18 + 1421/1.15 \times 22 \cong 74°C$. Using formula (3.36) and Table 3.4 from Chap. 3 we will find the convective coefficient of heat transfer from a wall to water α.

At the first position if $v = 1$ m/s, then we find $\alpha = 2.83 \times 1.0^{0.8}/(2 \times 0.0097)^{0.2} = 6.2\ \text{kW/m}^2 \times °C$. Assuming that the local boiling is absent, let us find temperature of the wall t_2' at the first approximation using formula (12.3): $t_2' = 60 + 700/6.2 = 173°C$. When water pressure in the head $p_{WAT} = 3.5$ bar (abs) then the water boiling point $t_S = 138°C$. Since t_2' is much higher than t_S, cooling of the critical zone of the tuyere, even without submerging it into slag, occurs actually in the mode of developed local boiling. In this mode, when submerging the tuyere into slag, we find α_q using formula (12.10).

$$\begin{aligned}\alpha_q &= 3.14 \times (1400 \times 10^3)^{0.7} \times 3.5^{0.15} \\ &= 76{,}208\,\text{W/m}^2 \times {}^\circ\text{C or } 76.2\,\text{kW/m}^2 \times {}^\circ\text{C}.\end{aligned}$$

Since α_q is much more than α, let us assume α_q to calculate the actual value of t_2. According to formula (12.3′): $t_2 = 138 + 1400/76.2 = 156°C$.

With the aid of formula (12.2) we will find temperature of the external head face wall surface t_1. The thickness of the copper wall $\delta = 19$ mm. In formula (12.2) the thermal conductivity coefficient for copper λ is determined by average temperature of the wall, which depends on an unknown value of t_1. Therefore, in principle, there is a need to use a method of the successive approximations. However, in the temperature range from 100 to 400°C the thermal conductivity coefficient of copper λ_{Cu} varies by 5% only. Thus, for regular calculations even the first approximation leads to sufficiently accurate result. This takes place in the given case as well.

Let us assume that average temperature of the wall is equal to 300°C. This temperature corresponds to $\lambda' = 0.371$ kW/m × °C, graph (1), Fig. 12.3.

According to formula (12.2): $t_1' = 156 + 1400 \times 0.019/0.371 \cong 228$°C. At the second approximation when the average temperature of the wall is equal to (156 + 228) × 0.5 = 192°C and $\lambda'' = 0.379$ kW/m × °C the repeated calculation gives $t_1'' = 226$°C. Thus, the second approximation was not required despite the first rough evaluation of the average wall temperature. The value of t_1 obtained is quite acceptable.

The calculation results lead to a conclusion that the main disadvantage of the tuyere is the cooling of the critical zone in the mode of developed local boiling of water. In this case, low reliability of this cooling method is enhanced by a high level of heat fluxes (up to 1400 kW/m^2), quite slow water velocity (about 1 m/s), and intolerably high temperature of its heating (up to 74°C). The latter is caused by unreasonable choice of water flow direction. Water enters the head already very heated during passing through the tuyere stem. Under such conditions, in some sections of the critical zone the bubble boiling can sporadically switch to the film boiling back and forth. These factors are those explain low durability of similar tuyeres and operator's refusal to immerse them into slag, Chap. 11, Sect. 11.1.3.

In order to increase durability of the tuyere, it is necessary to raise water flow rate at least 1.5–2 times which would reduce its heating temperature to the acceptable level. At the same time, in the critical zone v and α would increase. Even bigger rise of v and α can be achieved by the installation of helical insert into the annular gap, which allows to reduce the cross-section of the channel for water passage. In similar cases, this method is quite effective since at the same flow velocity in the channel twisting of the flow increases α by additional 50–80%. The measures pointed above would require increasing water pressure at the tuyere inlet since its hydraulic resistance grows. As far as change of the direction of water flow in the tuyere is concerned, it requires the radical re-design.

12.4.2 Elements with Pipes Cast into Copper Body and with Channels

Elements with pipes, Fig. 12.6a, find extensive application. Choice of the pipe material, along with other requirements is dictated by a need in tight and strong bond with the copper box. After cooling of the casting even the tiniest gaps are not

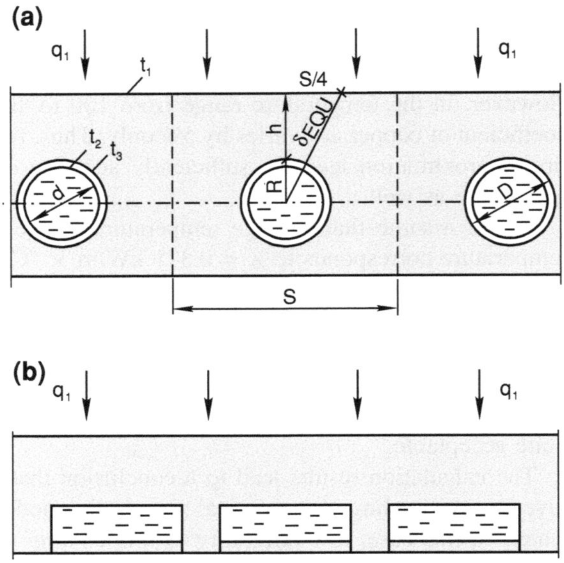

Fig. 12.6 Water-cooled elements: **a** – with pipes cast in the body; **b** – with channels

permissible between the casting and the pipes. Such gaps must not appear also during the operation, after numerous and abrupt box temperature fluctuations. This requirement can be met, to some extent, by the pipes of copper–nickel alloy of the following composition: Cu – 68% and Ni – 30%. The significant disadvantage of this material is quite low thermal conductivity in comparison with copper, Fig. 12.3.

Input Data

Geometrical and other parameters of the element, Fig. 12.6a, have been borrowed from the practical design of protective boxes of jet modules. The element thickness of 100 mm and the minimum distance from the pipes to the external surface h = 36 mm is dictated by the casting technology as well as assuring of the required element strength. It has to withstand the impacts of falling heavy scrap lumps during the furnace charging.

The water velocity in the pipes is v = 5 m/s, $t_{WAT.AVE}$ = 30°C. The pitch between the pipes S varies within quite wide limits. The calculations are carried out for S = 120 mm. Flow q_1 is assumed to be equally distributed across the width and along the length of the element.

Calculation of the Basic Parameters of Thermal Performance of the Element

The calculation is carried out for the section of the element with the width $S = 0.12$ m and the length of 1.0 m; D = 48 mm, d = 40 mm. The area of external heat-absorbing surface $F_1 = 0.12 \times 1.0 = 0.12$ m^2. The surface area flown around by water:

$F_2 = \pi \times 0.040 \times 1.0 = 0.126\,\text{m}^2$. Since $F_1 \cong F_2, q_1 = q_{WAT}$ can be assumed. When the critical zone of the box is lowered into the foamed slag but the oxygen jet blow-backs are absent, $q_1 = 1400\ \text{kW/m}^2$, Table 12.2. If $d_h = 40$ mm, $t_{WAT.AVE} = 30°\text{C}$, and $v = 5.0$ m/s let us find $\alpha = 18.1\ \text{kW/m}^2 \times °\text{C}$, formula (3.36), Table 3.4, Chap. 3.

Let us determine the temperature t_2, Fig. 12.6a. Assuming that the local boiling is absent, we use formula (12.3): $t_2 = 30 + 1400/18.1 = 107°\text{C}$. The absence of the local boiling is confirmed.

The ratio of $D/d = 1.2$. When the diameters' ratio is small enough like this, the pipe wall can be assumed flat, Chap. 3, Sect. 3.2.7.

Let us find the temperature t_3. Converting formula (12.2) to the given case, Chap. 3, Sect. 3.2.3, we find: $t_3 = t_2 + q \times \delta/\lambda$; $\delta = 0.004$ m. After iterations we find that for the pipe wall thickness $\lambda = 0.038\ \text{kW/m} \times °\text{C}$, Fig. 12.3, curve 3:

$$t_3 = 107 + 1400 \times 0.004/0.038 = 254°\text{C}.$$

Due to intense stirring of the turbulent water flow, the temperature t_3 can be assumed uniform across the entire pipe perimeter. The temperature gradient in the pipe wall Δt, despite its small thickness was 253–107 = 147°C. This is related to the high density of heat flux and relatively low thermal conductivity of the wall. High temperature gradient Δt is a serious drawback of the element since t_1 increases by Δt.

To calculate t_1 it is necessary to determine thermal resistance of the copper layer located in the path of heat flux from the external surface of the element to the pipe surface. In all cases, the thermal resistance is proportional to a certain equivalent length of this path, or what is the same, equivalent layer thickness δ_{EQU}. Since the discussed element has relatively complex configuration, δ_{EQU} can be calculated with any required accuracy using one of the numerical methods, Chap. 3, Sect. 3.2.8. Instead of labor-consuming programming, for the practical calculations with acceptable accuracy the simple formula can be used Fig. 12.6:

$$\delta_{EQU} = \sqrt{(S/4)^2 + (R+h)^2}. \tag{12.16}$$

This formula provides the value δ_{EQU} approximately equal to the average path of heat flux. In this case, $\delta_{EQU} = 0.067$ m. Assuming that for the cast copper after iterations $\lambda = 0.8 \times 0.365 = 0.292\ \text{kW/m} \times °\text{C}$, let us find t_1 using formula (12.2): $t_1 = 254 + 1400 \times 0.067/0.292 = 575°\text{C}$.

Such a value of t_1 considerably exceeds the permissible level for copper (300°C). This explains why the elements similar to those presented in Fig. 12.6a cannot work reliably at the high densities of heat fluxes close to 1400 kW/m^2. The cracks form inside them and, in some cases, the copper body burns back up to the very pipes. The latter phenomenon is seemingly related to the partial detachment of the pipes during the service.

The possibility of the detachment of the pipes is a main and dangerous drawback of the discussed design. The reliability of the thermal contact of the pipes with the element body is hard to control after casting. To eliminate detachment during the service, it is necessary to sharply reduce maximum temperatures of the element and therefore, the temperature gradients during the hot and cold periods of the heat. The temperatures t_1 are possible to reduce by using copper pipes and by bringing them closer to the external surface. However, it must be taken into consideration that the copper pipes detachment is even harder to eliminate than the copper–nickel ones.

To reduce heat fluxes absorbed by the element, its surface is made with the grooves designed to retain the skull. However, the practice shows that on the furnaces operating at high temperatures the skull in the critical zones occasionally melts and flows down, baring the surface of the element. It is easy to prove that reducing pipe pitch compared to S = 120 mm, Fig. 12.6a, does not provide noticeable results as well.

The calculations similar to the above have shown that the relatively acceptable maximum surface temperatures of the elements with pipes, which do not exceed 300°C, can be reliably assured only if the heat flux densities q_1 are significantly lower than 1400 kW/m^2.

The channels for water in the cast elements can be fabricated using not only pipes but also special casting methods as well as by drilling and milling of the castings. In the elements made of rolled copper plates the milling, Fig. 12.6b, allows to optimize the shape of the channels and their layout on the water-cooled surfaces of various configuration. By using this method, at the different surface sections, the channel cross-section and water velocity are possible to change, as well as to use consecutive and parallel flow circuits. Other methods do not provide such wide opportunities.

Since the pipes are not used, this method does not have disadvantages associated with their additional thermal resistance as well as the possibility of detachment. This improved the operational reliability of the elements. However, in the channels, regardless their formation method, practically achievable values of α are maximum 25–28 kW/m^2 × °C. Even higher α requires the water velocity in the channels to be so high that it leads to unacceptably high hydraulic resistances of the elements.

Due to this reason, the elements with channels cannot resist action of blow-back of oxygen jets and immersion in liquid metal when the heat flux densities q_1 reach 4000–6000 kW/m^2, Table 12.2. At these q_1 and the values of α less than 30 kW/m^2 × °C, the convective mode of cooling is substituted with the local boiling, which sharply reduces the operational reliability of the elements and their durability. The simple calculation proves that the local boiling is unavoidable, formula (12.3): $t_2 = 25 + 5000/30 = 192$°C.

12.4.3 Jet Cooling of the Elements

As mensioned above, the development of water-cooled elements capable of working under the harshest heat conditions is quite pressing problem. Solving this problem requires increasing α to the values close to 100–150 kW/m^2 × °C. In the metallurgical practice, the most available method to achieve such α is using of so-called jet cooling.

The sharp increase in α when the jet cooling is used can be explained by the fact that the water jets, which attack the cooled surface under the angle close to 90°, highly turbulize the boundary layer. As known, the laminar sub-layer contains the main thermal resistance to the heat transfer process from the wall to water, Chap. 3, Sect. 3.3.3. When water jets are used, this sub-layer is destroyed. An additional important advantage of jet cooling is that this method allows to quite effectively cool down such element surface sections to which it is impossible to bring water channels in close proximity due to design features.

The results of the experiments conducted at the institute "Energostal," Kharkov, Ukraine, have shown wide possibilities of jet cooling [5]. The work was conducted in order to improve durability of the blasting tuyeres of the blast furnaces. Their durability sharply drops if they come in contact with liquid iron when the iron level rises due the abnormal furnace operation.

During the experiment, hot metal with the temperature 1360°C was poured from the height 0.4–0.5 m onto the copper plates 110 × 110 mm of various thickness. The water jets attacked the plate from the distance of 20 mm from below at the angle of 90°. The water nozzle diameter was 5 mm; the initial jet velocity was 20 m/s; and the number of jets were 25. The heat flux density q_1 was 9–15 MW/m^2. The coefficients of convective heat transfer from the water side α approached 100 kW/m^2 × °C. The plates with the thickness of 9 mm did not melt partially, and when the thickness was 20 mm they melted partially but did not burn-back.

12.4.4 Oxygen Tuyere for Deep Blowing of the Bath

The effectiveness of jet cooling can be greatly improved if each water jet could be enclosed into a separate copper cylindrical heat-removing cell, Fig. 12.7a. This can be explained by the intense heat removal along the sidewall of the cell and the additional turbulization of the flow due to its 180° turn. The experimental research of such systems has shown that at the defined dimensions of cells heat fluxes up to 10 MW/m^2 can be removed from their face heat-absorbing surface without the local water boiling. The intensity of resulting heat transfer in the cell is described by exteremely high values of α, which grow when criterion *Re* increases. *Re* is calculated by the velocity of water flowing out from the internal pipe. When $Re > 40 \times 10^3$, the coefficient α can reach 150 kW/m^2 × °C.

The oxygen tuyere for deep blowing of the bath, which is water-cooled by the discussed method, is shown in Fig. 12.7. The diagram of the tuyere installation in the bottom bank lining of the EAF is shown in Fig. 11.6 of Chap. 11. It is totally explosion-proof, which is assured by the water-feeding system in the closed circuit. The face of the tuyere must reliably withstand both the blow-back of oxygen jets and the recurrent immersions of the tuyere into liquid metal when the

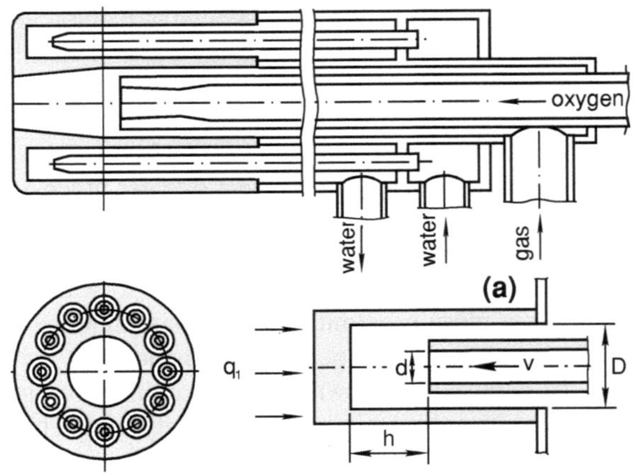

Fig. 12.7 High-durable oxygen tuyere for deep blowing of the bath: (**a**) jet cell

unavoidable in practice fluctuation of the bath level occurs. The calculation results presented below show that when $\alpha = 150$ kW/m^2 $\times$ °C the high durability of the tuyere under these conditions is quite achievable.

The tuyere head is made of rolled high-purity copper, Fig. 12.7. It contains 12 cylindrical cells with water-feeding pipes. All pipes are fed water in parallel scheme. The supersonic nozzle with the pipe supplying it oxygen is located at the tuyere axis. The natural gas is supplied through the annular slot around this pipe. The combustion chamber before the nozzle serves for the partial preliminary mixing and burning of gas with oxygen, which prevents from flowing the melt into the tuyere and clogging of gas slot and oxygen nozzle by the droplets of metal and slag.

Water circulating in the closed circuit is chilled by industrial water with the temperature of approximately 20°C in the heat exchanger. To avoid the oversized heat exchanger due to the small temperature differential between the industrial water and the chilled water the average temperature of the latter is increased to approximately 40–50°C. Increase in the velocity and temperature of water in the circuit is not difficult since the circuit has its own water pump and is filled with the treated water, which does not precipitate deposits.

Calculation of Basic Parameters of Thermal Performance of the Tuyere

According to Table 12.2, $q_1 = 5000$ kW/m^2. The coefficient of heat flux concentration at the cell faces $K = 2$. The heat flux density to water $q_{\mathrm{WAT}} = K \times q_1 =$ 10,000 kW/m^2. The average density of heat flux across the face wall thickness of the tuyere ($\delta = 8$ mm): $q_{\mathrm{AVE}} = (5000 + 10{,}000)/2 = 7500$ kW/m^2. Let us find the surface temperature t_2 and t_1 in the critical zone of the tuyere: $t_2 = 40 + 10{,}000/150 = 107$°C; $t_1 = 107 + 7500 \times 0.008/0.379 = 265$°C.

The obtained by calculation values of t_1 and t_2 rule out the local boiling and allow to forecast high durability of the tuyere in spite of the ultra-high heat fluxes densities.

References

1. Markov BL (1975) Methods of open-hearth bath blowing. Metallurgia, Moskow
2. Fleisher AG, Kuzmin AL (1982) Izvestiya of higher education establishment. Ferrous Metallurgia 4:40–43
3. Neduzhyi IA, Alabovski AN (1978) Technical thermodynamics and heat transfer. University, Kiev
4. Idelchik IE (1972) Handbook for hydraulic resistances, 3-d issue. Engineering, Moskow
5. Bakhirev AI (1982) Collection of MFM of USSR, improvement of thermal performance and designs of metallurgical units. Metallurgia, Moskow

Chapter 13
Principles of Automation of Heat Control

13.1 Preliminary Considerations

Due to a sharp reduction in the duration of the heat, controlling the electric arc furnaces has become considerably more difficult. The existing methods of control fall behind the modern requirements. The objective of automation is a radical improvement of these methods. This quite complex task requires solving of many different problems from automation of control of numerous mechanisms of the EAF and calculations of charge to control of such complex physicochemical processes as heating and decarburization of liquid bath. In the present chapter, the main attention is paid to the so-called automated management, which, in contrast to automatic control, does not exclude participation of the operator.

Intensive research is being conducted and numerous innovations are being proposed in this field. The descriptions of new systems usually explain in detail what the automation system can do, however, it is never made clear how this is being done. The actual control algorithm is not revealed. These systems often remain a "black box" for the user, even after prolonged service [1]. Such situation makes the comparative analysis of different innovations and selecting the most effective ones much more difficult. Meanwhile the essence of any automated system can be revealed comparatively easily, if the methods for obtaining information used for controlling are taken as the basis for the analysis [2].

13.2 Automated Management Systems

13.2.1 Use of Accumulated Information: Static Control

Control of the heat process can be based on various types of information, which differ by their origin. First, it is based on the previously accumulated information, which summarizes lengthy experience of the furnace operation. Second, it is based on the operating data from the process gathered during the heat and reflecting its

Y. N. Toulouevski and I. Y. Zinurov, *Innovation in Electric Arc Furnaces*,
DOI: 10.1007/978-3-642-36273-6_13,

individual features. Obtaining sufficiently complete and reliable operating data from the process using sensors carries a high degree of difficulty which is being overcome just in recent years. Therefore, for decades, since steelmaking ceased to be an art of individual masters and started to rely upon objective quantitative criteria, and until now, the methods of EAF control have been based mostly on the accumulated information.

In the past, before the use of computers became widespread, the accumulated experience had been summarized through the instructional manuals on the energy mode and technological mode intended for the operators' usage. In these manuals, the entire heat was divided into separate operating stages, e.g., between charging of separate baskets, from charging of the last basket until complete melting of scrap and formation of flat liquid bath, etc. In the manuals, operating stages could be also determined by the amount of consumed electrical energy or by the time. During each of these stages, the operator was directed to operate at a certain step of the furnace transformer. The timing of turning of the burners on and off; the start and the end of the oxygen blowing of the bath; fuel and oxygen flow rates; and other process parameters were also prescribed. For the certain stages, the optional parameter values were provided depending on the conditions of the given heat. The operator was responsible for choosing the option.

Such manuals were used everywhere. It was assumed that maintaining of the given parameters by the operator ensured, on average, the best results for most of the heats. As the conditions of EAF operation changed, the manuals were updated. This prescriptive type of control, based mostly on the accumulated information, is passive with regard to the inevitable and frequently substantial deviations of real heat conditions from the nominal ones. This type of control can be called static control.

In practice, the operation manuals have never been followed precisely. The operators used to correct the prescribed parameters, when considering the specifics of the certain heats based on their personal experience. Such actions of the operators were usually deemed to be undesirable. On the other hand, however, the general understanding existed that following manuals too literally reduced the effectiveness of control.

13.2.2 Mathematical Simulation as Method of Control

Simultaneously with the computerization of steelmaking units, a new direction of automation has begun to develop, i.e., control with the aid of mathematical models. In connection with outstanding results obtained with the method of mathematical modeling in many areas of science and technology, it has been expected that this method would allow successful resolution of all general problems and, eventually, lead to complete control automation. However, these expectations have not realized due to the reasons common to both EAFs and oxygen converters. Let us note that a lot of efforts have been made toward development and implementation of control based on mathematical modeling, especially in oxygen-converter steelmaking. The experience gained from the converters is of great interest for the EAFs.

Various approaches toward development of mathematical models of steel melting processes are known. Regardless of a number of other features, the main advantages and drawbacks of the model are determined, first of all, by the information which it is based upon. Let us stress that the computer brings practically limitless capabilities of processing data input. Nevertheless, if certain necessary information regarding the process is not contained in these data, even in a hidden or non-evident form, lack of thereof is impossible to compensate either with amount of data involved in calculations or with somewhat special methods of its processing. In this regard, a statement is valid that the computer itself does not create any essentially new information.

By the type of information used in the models, it is common to classify them as statistic and determinate. The first ones are based on the equations obtained through the statistical analysis of data from the previous heats, i.e., the use of accumulated information. The control based on these models is static, just like the control based on the instruction manuals. Basically, the difference between them is that, when the mathematical modeling method of control is being used, the switching of the transformer steps and changing of other process parameters, in accordance with the stage of the heat, is done automatically, as opposed to manually. Proponents of minimal operator involvement regard this as a significant advantage, although a possibility of operator's interference into operation of automated system is always open. The adequacy of statistic models utilizing only the accumulated information to the real process does not meet the requirements of the practice. This concerns in particular the accuracy of prediction of current and final values of metal temperature t_{MET} and its carbon content C_{MET}. There have been attempts to improve the accuracy of prediction of these parameters with the help of the so-called determinate models.

Determinate models seem to describe processes strictly in terms of cause and effect. They are based on the equations of mass and heat balance of the heat, as well as on the kinetic equations describing processes of melting, heating, and decarburization of the bath in the course of the heat. The kinetic equations are based on the existing assumptions about the nature of these processes using well-known laws of physics and physical chemistry. Many (but not all) coefficients of the kinetic equations are the combinations of known physical chemistry parameters and constants. This creates an impression of determinate, or strictly cause-and-effect, description. However, the other not less important coefficients of the same equations are impossible to determine in theoretical way with acceptable accuracy. They can be calculated only statistically based on the data from the previous heats. The same fully applies to the coefficients of kinetic equations, which describe the rate of melting and heating of the bath, the rate of absorption of oxygen and carbon blown into the bath, the rate of oxidation of carbon, iron and its alloys, etc.

Strictly cause-and-effect mathematical description of the heat processes (without using statistical data) for the purpose of control of steelmaking units is impossible in principle. This can be explained by the fact that, under the real conditions of EAF or converter, the course of each heat is determined by a variety of specific features and inevitably has probabilistic, random nature. Therefore, using of accumulated information to determine a number of coefficients for both statistic and determinate mathematical model is unavoidable. The accuracy of calculation of these coefficients ultimately determines the adequacy of the models. Thus, as practical automation of electric arc furnaces and oxygen converters shows both statistic and so-called determinate models assure approximately the same results.

Insufficient effectiveness of static control of heating and decarburization of the bath, regardless the method of mathematical statistics used to calculate coefficients for the models based on accumulated data, is caused by statistic features of these processes. Under real production conditions, they are characterized, on the one hand, by the high degree of sporadic disturbances and abrupt, irregular variability of a number of determining parameters and, on the other hand, by quite strict requirements to the accuracy and reproducibility of final results of control. Many parameters, which significantly affect the heat processes, vary with time. These parameters are characteristics of scrap, hot metal and other charge materials, condition of the steelmaking unit and its operating conditions, position of blowing tuyeres relative to the level of the bath, composition and amount of slag, and numerous other factors. These variations are not only random fluctuations from heat to heat relative to a constant average level, but they also exhibit systematic drift over prolonged time periods.

Variableness and instability of the processes cause discrepancies, which are impossible to resolve within the framework of static control. Statistically reliable evaluation of the coefficients of the mathematical model, which have to be updated regularly, requires using the data from at least few hundred heats. It does not pose a problem, when computers are used. However, the more the heats are used to determine the model coefficients, the longer the time span, which separates the main mass of these heats from the current moment, and the higher the probability of accumulation of the significant changes in the operating conditions as a result of their systematic drift.

Starting with a certain optimum number (which is usually equal to only few tens of heats), the further increase in the number of the heats, with the purpose of improvement of the model, makes the model more accurate in case of averaged probabilistic description of the process, but less suitable for predicting the nearest future. Thus, the models based on a small number of heats, while being relevant to the current conditions, have insufficiently accurate coefficients, and the more accurate models based upon statistical analysis of the large number of heats become obsolete at the moment of their creation.

The shown discrepancy strictly restricts the possibilities of mathematical simulation as a method of control of the processes of heating and decarburization of the bath. The systems with automatic correction of the coefficients of mathematical models based on the results of the certain number of previous heats are frequently called "adapting," or even "self-learning." This hardly matches the reality, since the constant repeating of the same correction does not improve the above-mentioned limited possibilities of this procedure and does not increase the accuracy of determining the coefficients.

The required improvement of the quality of the static control can be ensured by increasing the level of stability of operating conditions of the steelmaking units. With regard to the heat in EAFs and converters, standardizing of the charge is of greatest importance. In case of completely stabilized process, the accumulated information becomes comprehensive and allows creating an ideal adequate mathematical model eliminating almost all of the control problems. However, such stabilization is practically unachievable, since it would require huge economically unjustifiable costs.

As mentioned above, the efficiencies of all systems using static control do not differ much, therefore, it is recommended to give preference to the simplest of them. It has to be clearly understood that mathematical models based on accumulated data, strictly speaking, simulate not the real processes in the furnace, but rather the instructional manuals on heat operating control.

13.2.3 Dynamic Control: Use of On-line Data

At the process operation rates achieved to date, the numerous conditions determining the operation mode of the EAF change practically continuously in the course of the heat. Strictly speaking, the power of the electric arcs, the flow rates of oxygen and various types of fuel, the intensity of gas evacuation from the furnace, and other process parameters should be continuously adjusted in accordance with these conditions. Such control can be called dynamic. It has to be based mainly on the on-line data representing the actual course of the processes at any given moment.

In principle, the dynamic control of such fast and unstable process as the heat in the modern EAF is the most efficient. However, realization of such control at the present time does not seem possible. First of all, the main obstacle is lack of technical means (sensors) required to obtain sufficiently complete on-line data on the heat processes over the entire duration of the heat. Inadequacy of on-line data is the fundamental problem of the automated dynamic control.

On-line data can be fed into the control system continuously or at intervals. Accordingly, the dynamic control can be either continuous or discrete. The discrete or non-repeating on-line data is, for instance, a signal indicating that the water temperature at the exit from some furnace element exceeded the acceptable level. As a reaction to such signal, the dynamic control system can use different control actions depending on its structure. For example, the system can turn off the burners located in that element for the entire time until its overheating is over. Such local systems of dynamic control protecting the furnace from hazardous situations and accidents make the operator's job much easier, and as a result they have become widespread.

The gas pressure under the furnace roof can serve as an example of continuous on-line data. The pressure value depends on the amount of infiltrated atmospheric air, the intensity of gas evolution from the bath, and the draft level of the exhauster. Using the on-line data on the gas pressure under the roof, it is possible to implement continuous dynamic control of the draft of the exhauster, which would maintain pressure during the heat at the required optimal level. In this case, on-line data on gas pressure as well as the signals indicating the occurrence of hazardous situations are unambiguously related to the control task. These on-line data can be called direct. The dynamic control using direct on-line data is the most efficient and, at the same time, the simplest means of process automation.

Unfortunately, as we mentioned before, in many important cases there are no sensors capable to detect direct continuous on-line data on heat processes. First of all, this refers to the task of dynamic control of temperature of liquid metal t_{MET} and its carbon content C_{MET}. The sensors for continuous control of t_{MET} and C_{MET} for the EAFs and oxygen converters have not been developed so far. It is interesting to note that, for open-hearth furnaces, the problem of continuous measurement of t_{MET} during the last few hours of the refining has been successfully resolved as far back as 30 years ago [2]. However, inertness and other features of the t_{MET} sensors of the open-hearth type prevented their use under conditions of short heat in the modern steelmaking units.

In oxygen converters, where the problem of dynamic control of the heating and decarburization of the bath was the key problem for years, the developers tried to compensate lack of direct on-line data on t_{MET} and C_{MET} with indirect data related, to a certain degree, to the changes in these parameters. Gas pressure under the hood, distinctive characteristics of noise and vibration accompanying oxygen blowing of the bath, and other process parameters changing in the course of the heat have been tried out as indirect on-line data. The variation of these parameters is indeed related to the changes in t_{MET} and C_{MET}, but their correlation is not just unambiguous, it is not even sufficiently close, since the changes in the above-mentioned parameters are affected significantly by numerous other factors. Therefore, all attempts to use such indirect data to improve the efficiency of t_{MET} and C_{MET} control using mathematical models did not yield the expected results.

The greatest efforts were spent on development and implementation of the method of determining C_{MET} by the composition of the converter off-gases. Although the composition of the converter off-gases relative to C_{MET} is indirect information, this method seemed to be rather promising. The idea of this method is as follows. By determining the content of carbon in the initial metal charge and that of leaving with of off-gases in the form of CO and $CO_{2,}$ it is possible to determine the amount of carbon remained in the bath at any given moment of the heat. However, this method based on material balance of carbon requires to continuously measure not only the composition, but also the amount of gases, as well as the mass of scrap, hot metal, liquid metal in the converter, and a number of other variables included in balance equations. Determining all of these values involves errors, which are quite significant compared to the required accuracy of measurement of C_{MET}. As a result, despite using the most expensive precision measuring instruments for gas analysis, the accuracy of determining of C_{MET} based on balance method turned to be insufficient. This called for development of new technical means which could directly and sufficiently accurately determine both C_{MET} and t_{MET} in the course of the converter heat without its interruption.

The result is that the measuring sondes have been developed which operate as follows [3]: 1.5–2.0 min prior to the end of oxygen blow, the sonde is very quickly inserted into the converter through its throat. The sonde tip is immersed into the liquid metal for a few seconds. Then the sonde is returned to the initial upper position. Before each immersion, the sensors for single measurement of C_{MET} and t_{MET} are installed in the sonde tip. Recharging the sensors into the sonde tips is done automatically.

Direct on-line data obtained in this way is provided to the operator and introduced into the control system. The predicting device calculates the additional amount of oxygen, which is necessary to blow into the bath in order to reduce carbon content to the final value C^0_{MET}, and the expected increase in t_{MET}. If the predicted final value of t^0_{MET} does not fall within the given limits, then the temperature t_{MET} is adjusted during the additional blow either with the help of calculated amount of cooling agent, or through the change of the tuyere position above the bath.

Dynamic heat control has significantly increased the converter productivity and recovered all expenditures for equipping them with new sensors. These devices are unique due to their size. The weight of the converter sonde with its mechanisms is about 60 ton. Sonde diameter reaches 0.5 m, and its length is such that it requires to increase the shop building height above the converter. This example shows how important and irreplaceable the direct on-line data on C_{MET} and t_{MET} is under present-day conditions, if such complex and expensive devices are being used to obtain these data.

The experience of the control automation accumulated in oxygen-converter steelmaking should be fully taken into consideration when analyzing different approaches to the same problem in the EAFs. For a long time, the dynamic control devices have been used in EAF solely to protect the furnace from hazardous situations and accidents related to unacceptable increase in the temperature of off-gases, water-cooled elements and bottom lining.

The new means of obtaining direct on-line data on gas pressure under the roof and gas composition at the exit from the freeboard have been implemented only recently [4]. With introduction of reliable pressure sensors, the implementation of the dynamic control of the exhauster draft has begun. This control in the course of the entire heat assures that the pressure is maintained at the constant optimal level, which yields significant economic effect [5].

The on-line data on gas composition is used successfully to prevent explosions in the furnace gas duct. If the content of CO and H_2 in the off-gases approaches the explosion limit, the natural gas supply to the burners is automatically reduced or stopped, and the intensity of oxygen blowing of the bath is reduced. This increases the efficiency of natural gas and oxygen use and leads to their reduced consumption. A sharp increase in hydrogen concentration can indicate the presence of water in the freeboard. This allows to timely detect the water leak and take the appropriate measures. As far as the potential use of on-line data on the off-gases composition for dynamic control of EAF bath decarburization is concerned, judging from the converter experience, this method cannot be considered sufficiently effective and promising.

For a long time, the attempts were being made to use various indirect on-line data such as spectral characteristics of noise and current harmonics of the electric arcs, their stability, position of the electrodes, and levels of heat fluxes directed toward on the water-cooled elements for the dynamic control of decarburization, as well as of heating of the bath in the arc furnaces. As in case with oxygen converters, this direction has not yielded good results.

The problem of obtaining direct on-line data on t_{MET} and C_{MET} at the final stage of the EAF's heat is successfully solved with the help of recently developed automated robotic devices for taking metal samples through the slag door [6]. These robots are similar to the converter sondes with regard to their functions, but are significantly simpler and cheaper. They are controlled by the operator and allow to carry out several measurements in the course of the heat. The t_{MET} and C_{MET} sensors located on the robot's sonde holder are replaced automatically or manually before each immersion into the melt. During this replacement, the operator stays at a safe distance away from the slag door.

More complete automation of dynamic control of the heat at the final stage before tapping can be achieved with the help of utilization of the predicting devices similar to those used in oxygen converters. Using on-line data input on t_{MET} and C_{MET}, such device predicts the tapping time, the required additional oxygen consumption, and other controlling actions assuring that the final values of t^0_{MET} and C^0_{MET} fall within the specified limits. Simple extrapolation formulae with constant coefficients obtained through statistical analysis of direct measurements of t_{MET} and C_{MET} during the previous heats are used for the prediction. Due to the fact that prediction is carried out for a short period of time, its accuracy is quite high despite using the elements of static information. This additional information allows operator to determine the tapping time with high accuracy. It must be understood that with tap-to-tap time of about 36 min, an unjustified delay for only 1.5 min reduces the furnace productivity by approximately 4 %.

The dynamic control of metal temperature and carbon content in the metal becomes the main task only during the relatively short period of the heat before tapping, when, after the full melting of the scrap remaining on the furnace bottom, the t_{MET} and C_{MET} sensors can be immersed into the bath without fear of their damage. At the rest of the time, the other complex problems are most important. These problems include control of heating and melting of scrap, oxygen blowing of the bath, carbon and fluxes injection, and slag foaming. The quality of the control of these processes determines in many respects the heat duration, electrical energy consumption, yield, and other basic technical and economic performance indices of the furnace.

These processes take place under continuously varying conditions. Therefore, in principle, all of them require dynamic control. Realization of their dynamic control is hindered not only by lack of technical means for obtaining the necessary on-line data. Until now, it is still unclear which process parameters, when measured, could serve as necessary information.

From this standpoint, let us review the process of heating the scrap by the burners. It is known that in the course of stationary sidewall burners' operation, the efficiency of natural gas utilization η_{NG} drops quite rapidly and reaches the level so low, that the burners have to be turned off. Obtaining direct on-line data on η_{NG} is practically not feasible. The sharp drop of this value occurs in case of significant fuel underburning caused by physical–chemical processes of interaction of the burner flames with the highly heated scrap, Chap. 5, Sect. 5.4.2.

Fuel underburning in the burners is accompanied by a significant increase in the temperature of off-gases and in their CO and H_2 content. However, the same phenomena are observed when CO evolution from the bath increases sharply, as well as during the combustion of flammable components of the charge and the evaporation of the moisture. Therefore, using the temperature and composition of the off-gases as indirect on-line data seems insufficiently effective. In practice, the fuel consumption and the burners' operation time are prescriptive and are set with the aid of automation which does not exclude the operator's interference. Thus, instead of the required dynamic control, the static instructional control is used.

In the modern furnaces, the sidewall burners are located along the entire perimeter of the freeboard. In case of charging with baskets, the amount and the size of scrap lumps as well as their placement in front of each burner vary from heat to heat. Therefore, the optimum duration of operation of different burners should not be the same. The burners should not be turned off simultaneously, but rather consecutively, some earlier, others later, depending on the actual positioning of the scrap in front of the burners.

The reliable indication that the particular burner has to be turned off is intensive melting and settling down of the scrap resulting in forming of a void in front of the burner and in loss of contact between the flame and the scrap lumps. The dynamic control of each burner using this visual indication would promote more uniform scrap melting along the perimeter of the bath, increased efficiency of natural gas and oxygen utilization, and, ultimately, shortening of tap-to-tap time.

The problem of dynamic control of the processes of oxygen and carbon injection into the bath is also related mostly to a lack of any reliable source of on-line data on these processes other than visual observation. The source of the direct objective on-line data can be pointed out for the process of slag foaming. It is known that the thickness of foamed slag should be maintained at such a minimum level, which allows the complete immersion of the electrical arcs into the slag. The optimum slag foam level can be controlled in automatic mode by the sensors of the radiant heat flux from the arcs onto the furnace walls. At the moment of arcs transition into the immersed state, the heat flux density drops by several times. The development and implementation of such sensors for EAFs do not impose any particular difficulties since the principles of their operation and construction are well known. For many years, they were successfully used to control heat fluxes on the lining of the open-hearth furnaces' roofs and of the electric arc furnace walls [2].

The problem is that in practice the flow rates of oxygen and carbon injected into the bath greatly exceed the values necessary to maintain the thickness of the foamed slag layer at the optimum level. Such regimes of blowing are used to increase the EAF productivity, which is considered preferable to other performance indices, Chap. 1, Sect. 1.3.2. Thus, the control of the slag foaming becomes a part of the overall task of dynamic control of flow rates of oxygen and carbon used for the bath blowing.

As these flow rates increase, the rate of oxygen and carbon useful absorption by the bath drops. Therefore, this increase is expedient only up to a certain limit. In actual practice, this limit is frequently exceeded, which does not provide a noticeable gain in productivity, but reduces performance indices of the furnace. Optimum flow rates of oxygen and carbon depend on design features of the blowing devices, as well as on many other factors determining the hydrodynamics of the bath. All these factors vary in the course of the blowing. It is unknown which process parameters should be measured in order to provide the on-line data required for the optimal dynamic control of the O_2 and C flow rates. Therefore, in practice, just as for control of the burners, these flow rates are limited prescriptively.

It seems that the only reliable and quite significant source of the on-line data on the course of the blowing is visual observation of the bath behavior. Such observation provides integrated multi-faceted information regarding the blowing processes, which cannot be formally evaluated. The boiling and circulation behavior of the bath, the distribution of the intensity of CO evolution over its surface area, splashing of the melt by the oxygen jets, and a number of other visual indications allow the experienced highly qualified operator to get the correct understanding of efficient oxygen and carbon flow rates of the individual blowing devices distributed along the bath perimeter. By adjusting these flow rates in accordance with his estimation of the course of the blowing, the operator can achieve much better results in comparison with the "blind" prescriptive control.

A multiyear experience with BSE manipulator in the numerous EAFs around the world, Chap. 11, Sect. 11.1.2, demonstrates the great possibilities of using the visual information regarding the bath behavior for improving the effectiveness of control of the oxygen and carbon blowing devices. Usually the control of the manipulator was assigned to the most experienced well-trained operators capable of taking into consideration the specifics of the heat. It was noticed that in those cases when the manipulator was controlled by a less experienced operator who unvaryingly carried out the operations as prescribed by the instruction acting in a robot-like manner, the furnace performance results worsened significantly. In case of controlling the mobile water-cooled blowing tuyeres, the same results were observed, Chap. 11, Sect. 11.1.3. The examples given, as well as other examples, prove that the dependence of the results of manual control in cases when it is actually required on the operator's qualification should not serve as an argument to substitute manual control with less effective static programmed control.

At the present time, the modes of furnace operation with the closed slag doors are used everywhere. This is caused by the need to completely eliminate infiltration of atmospheric air into the freeboard. However, as it is shown in the next chapter, the infiltrations can be successfully eliminated by another method, namely with the help of the jet air curtains. Therefore, there are enough reasons to think that denying the operators the opportunity to observe the heat process through the slag door at least periodically is not fully justified. A certain managerial pressure was necessary to accustom the operators to work with the furnace "blindly." In the integrated control system, which includes both the automation and the operator, visual observation of the furnace operation must be regarded as one of the most important sources of on-line data.

13.3 Rational Degree of Automation

The purpose of automation is to improve the control efficiency rather than to completely eliminate the operator's participation. There are no reasons to assume that these purposes coincide and that the automatic control is always more effective than the manual one. When the steel is produced in the EAFs, there is always a quite high probability that the normal course of the heat process is disrupted and that the critical or hazardous situation occurs which would result in producing the defective metal or in the breakdown of the steelmaking unit. If this happens, even if not often, the losses are so significant that they make the presence of a highly qualified operator responsible for the furnace operation absolutely necessary at every heat.

Under these conditions, it is most reasonable not to restrict operator's participation in the process control, but on the contrary, to fully utilize his capabilities. Such approach to the problem, with the appropriate training and upgrading of the operators qualification, makes it possible to develop the automated management systems, which meet the highest requirements. The complete automation for the sake of automation itself, as a stand-alone task, is meaningless.

On the one hand, the rational level of automation depends on operator's capabilities and on the other hand it depends on the automation capabilities. The relative level of these capabilities is determined by the on-line data available to the operator as compared to the on-line data which can be input into the control system. If the operator possesses more complete on-line data regarding the given process and is capable of assuring more efficient control, then it is impractical to limit his participation in the system operation. On the contrary, full automation is absolutely necessary in cases when only quick-operating automatic controllers can assure the required quite high speed of control. This is particularly relevant to the systems which maintain the optimum parameters of the electric arcs and gas pressure under the furnace roof.

As it has been already shown, the static control with the aid of mathematical models cannot compete in efficiency with the control with participation of the experienced operator. However, such automated systems when used as operator's guide are helpful beyond the doubt. They warn the operator that it is time to change the parameters of the furnace operational mode to start or to stop certain technological operations, e.g., to charge the next basket of scrap or start measuring t_{MET} and C_{MET}. These signals are not error-free with regard to time, yet they substantially ease and improve the control on condition that operator is capable of not only following the guidance of the automation, but also of making the necessary corrections.

There is an opinion that the operation of the static automated management systems in the guidance mode is only the first stage of their development and that these systems, as they are perfected, can completely replace the operator. In the reality, the operation in this mode is the limit of capabilities of such systems. The full control automation will be achieved only when the problem of the development of the sensors to obtain on-line data on the heat processes in the required scope is resolved.

References

1. Riedinger D, Pfaff H-P (2007) Optimization potentials of DC furnace regulating systems. MPT Int 5:68–70
2. Toulouevski YN, Nechaev EA (1978) Informational problems in intensification of steelmaking processes. Metallurgia, Moskow
3. Wallner F, Fritz E (2002) Fifty years of oxygen converter steelmaking. MPT Int 6:38–43
4. Cordova E, Khan M, Zuliani D et al (2007) Advanced dynamic control technology for EAF steelmaking and other combustion processes. MPT Int 2:28–32
5. Toulouevski Y, Preisman M (2007) A self-cleaning, water-cooled pressure probe implemented under the EAF roof. MPT Int 2:34–35
6. Eaddy DC, Moody VA (2002) DC Furnace Upgrades – Operating Results. 7th European Electric Steelmaking Conference, Venice

Chapter 14
Off-Gas Evacuation and Environmental Protection

14.1 Preliminary Considerations

Harmful effects of electric arc furnaces on the environment and human beings are high level of noise and emission of highly dust-laden toxic gases. The basic sources of noise are the electric arcs. The highest noise level (120–130 db) is observed at the high-power furnaces during the period of unstable arcing at the beginning of scrap melting. As the arcs are submerged into the foamed slag, the level of noise reduces approximately to 90 db.

Noise transmission beyond the boundary of the steel melting shop is prevented by soundproofing of its external walls and enclosures. The operating personnel are stationed in the soundproof air-conditioned rooms. This is applicable to all control panels located in the working platforms of the shop, as well as to the crane cabs. This book does not examine the problem of protection from noise in more detail. The methods and devices for removal of dust from the off-gases and the problems of dust processing and utilization are not examined either. As far as protection from the dust and gas emissions is concerned, we analyze only those aspects of this problem which are directly related to the furnace operation and have a significant effect on the performance indices and intensification methods.

14.2 Formation and Characteristics of Dust–Gas Emissions

14.2.1 Sources of Emissions

Dust and gases forming the emissions evolve both inside the freeboard and outside the furnace during the tapping. The basic sources of gas generation in the freeboard are carbon contained in the bath, which is oxidized during the oxygen blowing, and oxy-gas burners. For a short term, large amount of gases is generated during charging of contaminated scrap due to oil and plastics burning out as well

Y. N. Toulouevski and I. Y. Zinurov, *Innovation in Electric Arc Furnaces*, DOI: 10.1007/978-3-642-36273-6_14, © Springer-Verlag Berlin Heidelberg 2013

as due to evaporation of moisture. The carbon charged together with the scrap also partially burns out.

Dust is formed during the metal evaporation in the electric arcs zone with subsequent oxidation of the condensing particles, as well as due to the bath splashing. The thus-formed dust is fine dispersed; up to 70% of its particles smaller than 2 μm (1 μm = 10^{-6} m). In the modern EAFs operating with intensive oxygen blowing of the bath, both small and large particles and even the conglomerates of liquid metal and slag are ejected from the freeboard into the gas duct. The ejected materials form the caked build-ups in the gas ducts.

14.2.2 Primary and Secondary Emissions

It is common to distinguish the so-called primary and secondary emissions. The dust-laden gases formed in the freeboard and forcedly drawn out through the opening in the furnace roof are classified as the primary or "controlled" emissions. Total specific amount of these gases per heat M_{GAS} (kg/ton of steel) basically depends on the consumption of carbon and natural gas (kg/ton; m^3/ton). For the furnaces with approximately equal consumptions, the value M_{GAS} can be assumed to be relatively constant.

On the contrary, the intensity of primary gases formation per unit time, V_{GAS} (m^3/min), which must correspond to the intensity of sucking-off of the gases, can vary within quite wide limits. This value increases and is directly proportional to the mass of the heat and is inversely proportional to its duration. The intensity of primary emissions fluctuates sharply in the course of the heat. The maximum values $V_{GAS.MAX}$ must be considered for selecting the basic parameters of the direct evacuation system.

Two peaks of the value $V_{GAS.MAX}$ are observed during the period of processing of scrap contaminated with combustible components (oil, plastics, etc.) in the furnaces operating with intensive oxygen blowing of the bath and equipped with the large number of oxy-gas burners. The first of them, with the maximal intensity, occurs during the first several minutes after charging of baskets of scrap, when the combustible components of the scrap are burning out. The second peak is observed during the blowing of the bath, when all burners are working simultaneously, although with the sharply reduced power in the mode of protective pilot flame.

Dust-laden off-gases which are captured with the aid of the exhaust hoods outside of the furnace are called secondary, or "uncontrolled," emissions. These gases are ejected into the atmosphere of the shop from the ladle during tapping, through the electrode gaps in the furnace roof, and during scrap charging while the freeboard is not covered by the roof. The amount of secondary emissions directly at the locations of their ejection into the atmosphere of the shop is relatively small. It is equal to approximately 15–20% of the amount of primary emissions. However, the secondary dust-laden off-gases mix with the large volumes of air before they are captured by the hoods. As a result, both their total amount per heat, directed to baghouse, and intensity of formation per unit time at the locations of the exhaust hoods can dozens of times exceed the respective volumes of the primary dust-laden off-gases.

14.2.3 Composition, Temperature, and Heat Content of Off-Gases

The chemical composition of the secondary off-gas emissions in the zones of their capturing is close to the composition of air contaminated with CO_2, CO, and other impurities. In comparison with the primary emissions, the dust content in these off-gases is small, yet exceeding considerably the limit values permitted for the emissions into the atmosphere. As a result, the secondary emissions are directed for purification just as the primary, which requires to sharply increase the carrying capacity of the gas evacuation system and the power of the exhausters.

The chemical composition of primary off-gases leaving the freeboard can vary within wide limits depending on the charge materials, the intensity of oxygen and carbon injection into the bath, the power of the oxy-gas burners, the amount of air infiltrated into the furnace, and a number of other factors. The so-called "typical compositions" of primary off-gas emissions, which can be found in different publications, give an idea of the range of these variations. One of them is given below as an example (for the dry gases), %: 10–15 CO_2, 5–15 O_2, 5–25 CO, 5–10 H_2, and 50–70 N_2. However, these generalized data do not give an idea about the dynamics of fluctuations of composition of the primary emissions, whereas understanding the specifics of the dynamics of composition fluctuations is of great practical value.

The typical pattern of fluctuations of concentrations of CO and H_2 over the course of the heat in 100-ton EAF is shown in Fig. 14.1 [1]. The concentrations of CO_2 and O_2 change in the same manner. It is only relatively recently that obtaining such data on a regular basis has become possible, when many of the EAFs had been equipped with automatic high-speed gas analyzers. The fast pace of development of this direction is associated with the activity of H. D. Goodfellow and his TGI Company. These data confirmed the concept of considerable non-uniformity of the carbon oxidation process. The intensity of carbon monoxide evolution from the bath periodically increases sharply. Within 2–3 min the content of CO in the off-gas can double or even triple. This points the non-uniformity of oxygen absorption by the bath during intensive blowing. Periodically, oxygen is accumulated in the bath, and over-oxidation of metal sharply increases for a relatively short-time period. Then thus accumulated oxygen is rapidly consumed with intensity considerably exceeding the intensity of blowing. Then this process can be repeated.

Taking into account the response time of gas analyzers, it is possible to assume that in actuality the maximum CO concentrations for the short periods of time can reach even higher values than those shown in Fig. 14.1. As far as the peak concentrations of H_2 are concerned, they occur during the first several minutes after charging of each basket of scrap and are caused, basically, by impurities burning

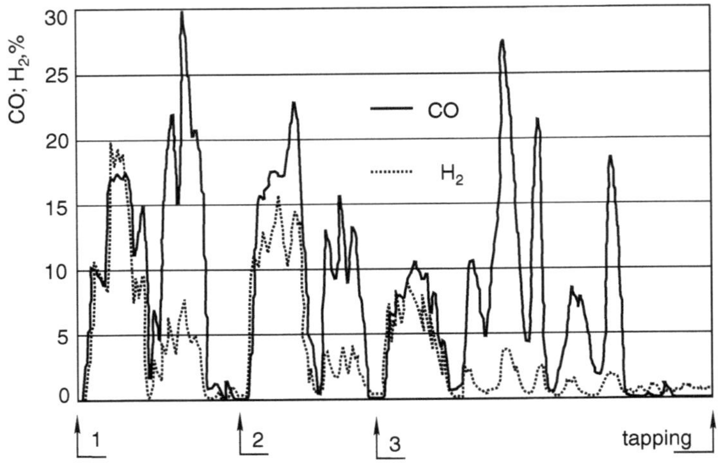

Fig. 14.1 CO and H_2 content in primary gases during the heat, % [1]. Figures at *arrows* are the numbers of baskets with scrap

out under oxygen deficiency, as well as by dissociation of water vapor in the arc zone. The content of H_2 in the primary off-gases drops sharply during the liquid bath stage and becomes insignificant by the end of the heat, Fig. 14.1. The examined data lead to the conclusion that some kind of average values of CO and H_2 should not be used for solving the problems related to gas evacuation. These problems (i.e., determining basic parameters of the gas evacuation system, selecting methods of pressure adjustment in the freeboard, cooling of off-gases, post-combustion of combustible components in the off-gases, and ensuring explosion safety) require a consideration of maximum possible concentrations of CO and H_2. For the short moments, they can exceed 30 and 20%, respectively [2].

The dust content in the primary off-gas emissions varies in the course of the heat from 15–30 to 50–60 g/m^3. The chemical composition of dust depends on the grade of steel and composition of the scrap. During production of carbon steels from the dormant steel scrap, the dust consists of up to more than two-thirds of iron oxides; the rest is Si, Mn, Mg, and Ca oxides [3]. If the metal charge contains large amounts of automobile galvanized sheet scrap, the content of zinc in the dust reaches 20%, which means that the dust becomes toxic. The content of iron oxides and other elements is reduced, respectively.

In the course of the heat, the temperatures of off-gases exiting the freeboard can vary within a wide range from 400–600°C at the beginning of the melting after charging of the first basket of scrap, to 1500–1600°C during the period of the intensive oxygen blowing of the bath. Even higher temperatures can occur in case of rather complete post-combustion of CO in the freeboard.

The total heat content of the primary dust and gas emissions escaping from the furnace per unit time has a great impact on the basic parameters of gas evacuation systems. This value, representing the heat power of the gas flow Q_{GAS}, MW, is the sum of two components: the so-called physical heat of gases $Q_{GAS.PH}$ and their chemical energy $Q_{GAS.CH}$. The physical heat of gases is determined by their flow rate, m^3/min, heat capacity, and temperature. The chemical energy is determined by the content of CO and H_2 in the gases. These combustible components must be burnt completely within the boundaries of the gas duct. In the modern furnaces, the chemical energy of primary emissions, in the majority of the cases, exceeds considerably their physical heat content.

The characteristic property of the EAFs is the significant fluctuations of the value Q_{GAS} in the course of the heat. When the total power of oxy-gas burners is high and with intensive oxygen and carbon injection into the bath, the maximum values $Q_{GAS.MAX}$ can be observed both at the beginning of the heat during the stage of scrap melting and during the stage of blowing of the bath. The heat power of primary dust–gas emissions at some of the moments of the heat can reach quite high values comparable with the power of the furnace transformer.

14.3 Capturing Emissions: Preparing Emissions for Cleaning in Bag Filters

14.3.1 General Description of the System

Two basic types of gas evacuation systems have gained acceptance. For the purpose of cleaning of primary and secondary emissions, the first system uses the general baghouse, whereas the second one uses two separate baghouses. To a large extent, the basic structural elements of these systems are the same. Let us examine the first of them, Fig. 14.2. The primary emissions are captured by the system of direct evacuation of the off-gases from the freeboard through the opening in the furnace roof (1). The secondary emissions are captured by the hood (2) located above the furnace, or by similar devices used instead. The direct evacuation system includes the following elements: the roof elbow (3), the stationary gas duct (4), the dust chamber (5), and the gas duct (6) leading from the chamber to the baghouse (11). The gap (8) with width *L* between the elbow (3) and the gas duct (4) allows to swing the roof for scrap charging. In certain cases, the initial section of the gas duct (4) is made movable, which makes it possible to adjust the gap width *L*. In front of the filters, the gas duct (6) connects the gas duct (9), which evacuates the secondary emissions from the hood (2). The gas ducts (4) and (9) are equipped with the valves (10), which make it possible to control the intensity of sucking-off of the primary emissions from the furnace and the secondary from the hood (2).

The preliminary preparation of primary emissions for the cleaning in the bag filters includes the following operations: post-combustion of CO and H_2, suppression of toxic dioxins, cooling of off-gases, and their rough cleaning from the largest fractions of dust in the chamber (5). Post-combustion of CO and H_2 is usually carried out with air infiltrated through the gap (8) or through the special

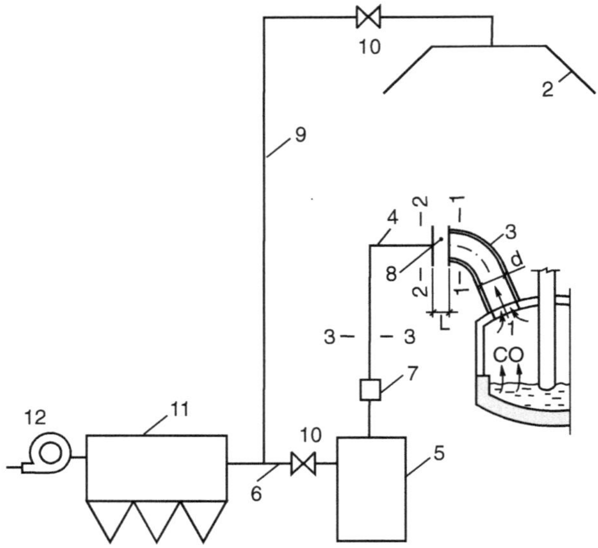

Fig. 14.2 Schematic diagram of evacuation and purification of gases from EAF, 12 – exhauster (the rest of designations are given in the text)

branch pipe equipped with control valve. The deep cooling of off-gases is required because of operating conditions of the bag filters. When the bags made of relatively low-cost materials are used, the temperature of off-gases at the entrance of the filters must not exceed 120–140°C. When the electric steel melting shops are reconstructed with an increase in the power of the EAF, in some cases it might be profitable to use the bags made of more expensive materials allowing to increase the temperature of off-gases to 200°C. The primary gas emissions are cooled by injecting into the flow of hot off-gases of finely atomized water with the aid of the device (7), as well as by mixing the off-gases with cold secondary emissions at the gas ducts (6) and (9) connection, Fig. 14.2. Cooling of off-gases by the outside air sucked through the special valve is also being used.

14.3.2 Problems of Toxic Emissions

In case of utilization of contaminated scrap, the varying compositions of highly toxic compounds of halogens with hydrocarbons, known under the general name of dioxins, are formed when plastics and oils are burning out. In the majority of the countries, the permissible content of these compounds in the emissions into the atmosphere is legislatively restricted to the quite low level of the order of 10^{-10} g/m^3. This makes the problem of cleaning of the off-gases from the electric furnaces difficult to solve.

It is known that in case of smelting of alloy and special steels, when clean scrap is used, there are practically no dioxins in the furnace off-gases. However, for mass production of electric steel, avoiding the use of cheap although contaminated scrap, as well as ensuring its thorough cleaning, does not seem possible due to economic reasons. Therefore, cleaning off-gases from the dioxins is, at the present time, an obligatory element of production process.

The following specifics of the behavior of dioxins during heating and cooling the off-gases are known. If gases containing dioxins are heated to the temperatures considerably exceeding 1250°C, then the latter decompose practically completely within approximately 1s [4]. However, during further relatively slow cooling of these gases, decomposed dioxins can synthesize once again. The secondary synthesis occurs in the range of temperatures from 600 to 200°C.

These specifics are the basis for the simplest and effective method of cleaning primary emissions from dioxins, which consists of the following. First, during CO and H_2 post-combustion, the off-gases are heated to high temperatures. The dioxins decompose in the process. Then, finely atomized water is injected into the flow of gases downstream. This water evaporates rapidly. As this takes place, the temperature of the gases drops sharply to 200°C and lower. The secondary synthesis of dioxins during this "quenching" of gases is completely eliminated.

It is necessary to understand that water injection also eliminates the possibility of further post-combustion of CO and other combustible gases. Therefore, the use of the described method of suppression of dioxins requires to meet the following additional condition. It is necessary to ensure such a complete post-combustion of CO within the duct before the point of water injection, which, considering further dilution of gases, will make it possible not to exceed the maximum permissible content of this component in the emissions into the atmosphere.

In order to ensure rapid burn-out of CO within a restricted length of the gas duct, it is necessary to intensify the mixing of the flow of the furnace gases exiting through the roof elbow with the air flow infiltrated through the gap (8), Fig. 14.2. Formerly, the sections of the elbow and of the duct adjacent to this gap were positioned, as a rule, horizontally and coaxially. In this case, the annular air flow was moving through the periphery of the duct enveloping the central flow of gases. Such aerodynamics of the flows is characterized by their relatively slow mixing.

A new design concept has gained acceptance in recent years, whereby the elbow (3) and the gas duct (4) are joined at an angle, both of them descending steeply after jointing: the elbow to the opening in the furnace roof and the duct to the chamber (5), Fig. 14.3. In case of such layout, the flows of furnace gases and infiltrated air intersect, which increases considerably the intensity of their mixing and the rate of CO burn-out. Thus, two problems are being solved simultaneously, i.e., suppression of dioxins by water injection and reduction of the CO content in the off-gas to the required level. Furthermore, the new layout considerably decreases the deposits on the walls of the elbow and the duct, which makes their cleaning easier. At the same time, capturing of large particles of dust by the chamber (5) improves.

Dioxins and heating of scrap by off-gases. The severization of standards for permissible content of dioxins in the EAF's gases has led to the fact that, in the majority of cases, preheating of contaminated scrap by off-gases has become technically impractical. This is applicable both to the shaft furnaces and to the Consteel furnaces. The temperatures of the gases entering the shaft or the conveyor preheaters are close to the freeboard temperatures. While heating the scrap, they are saturated with dioxins. After transferring the portion of their heat to the scrap, the gases leave the preheaters at relatively low temperatures of the order of 600–700°C. In order to decompose the dioxins, it is necessary to reheat these gases by the gas burners in the special chambers to the temperatures close enough to those at the entrance to the heaters.

The resulting energy effect of this process is that reduction of electrical energy consumption obtained due to scrap preheating is by approximately 75% cancelled out by the consumption of energy for high-temperature reheating of off-gases. The additional effect consists in shortening of tap-to-tap time in case of furnace operating with preheating of scrap. This effect does not always compensate the significant costs on construction and operation of complex devices such as shaft and conveyor scrap preheaters. As a result, preheating of scrap by off-gases is no longer considered in the design of the furnaces of the new series by the world's leading furnace manufacturers, Chap. 1, Sect. 1.4.2.

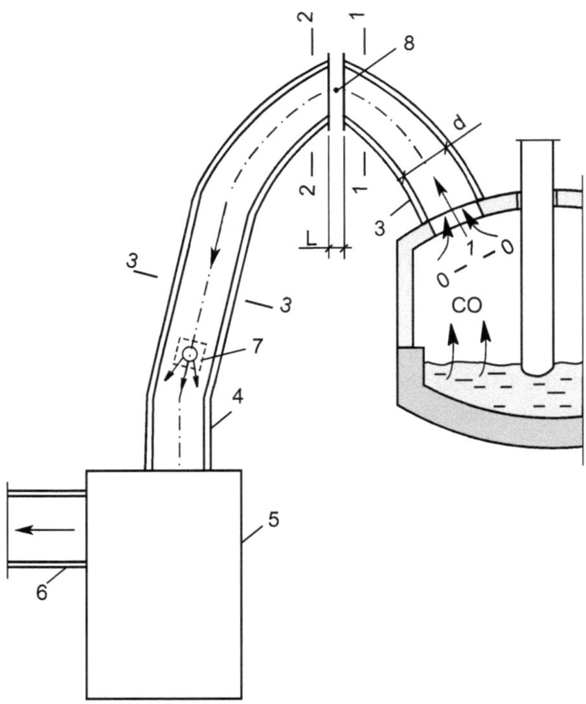

Fig. 14.3 Schematic diagram of direct gas evacuation (designations of elements are similar to those in Fig. 14.2; all designations are given in the text)

14.3.3 A Simplified Method of Gas Parameters' Calculation in the Direct Evacuation System

The analysis of operation of the direct evacuation systems and the innovations in this area requires relevant calculations. The effectiveness of the direct evacuation has the utmost effect on the economics of the entire system of gas evacuation in general. An insufficient carrying capacity of direct evacuation system is unacceptable since it sharply increases the secondary emissions and the cost of their capturing and cleaning. Besides, frequently, it is specifically the limited capacity of direct gas evacuation that hinders the increase in the furnace productivity. Therefore, particular attention is given further below to the calculations of the direct evacuation systems.

Pronounced instability of the processes of gas formation in the EAFs and the necessity to consider the momentary maximum values of gas evolution intensity create a significant uncertainty when estimating the initial data and limit substantially the potential accuracy of calculations. This makes unreasonable to consider the secondary details and allows using simplified calculation techniques on one mandatory condition, namely that they sufficiently fully take into consideration the most important features of the processes in the modern furnaces.

As an example, let us review the calculation of gas parameters in the direct gas evacuation system of 120-ton EAF during the period of oxygen blowing of the bath when the intensity of primary emissions reaches its maximum. The calculation technique used in the given example can be easily applied to the other periods of the heat, as well as to the furnaces of a different capacity with other regime parameters and equipment. Such calculations are simple but cumbersome, therefore their understanding is difficult. To make understanding of these calculations easier, the designations used in them are listed below with explanations. The given designations should be reviewed together with Fig. 14.3.

Calculations

List of Designations

V – amount of gases per unit time (flow rate), m^3
V_{O_2} – oxygen flow rate for bath blowing
$\mathcal{K}_{O_2}$ – coefficient of oxygen use for carbon oxidation
V_{CH_4} – flow rate of natural gas (methane CH_4) in burners
V_{GAS}^{BRN} – amount of combustion products (gases) produced by all burners
$V_{CO_2}^{BRN}$ – amount of CO_2 in combustion products of burners
$V_{H_2O}^{BRN}$ – amount of H_2O in combustion products of burners
V_{AIR}^{0-0} – amount of primary air infiltrated into freeboard, cross-section (0–0) (Fig. 14.3)

V_{CO}^{0-0} – amount of CO evolving from bath into freeboard
V_{-CO}^{0-1} – amount of CO burned to CO_2 in freeboard and roof elbow between cross-sections (0–0) and (1–1)
$V_{+CO_2}^{0-1}$ – amount of CO_2 evolved as a result of primary post-combustion of CO between cross-sections (0–0) and (1–1)
V_{CO}^{1-1} – amount of CO in roof elbow at cross-section (1–1) after primary post-combustion CO to CO_2

Amounts of any other gas component at other cross-sections are designated similarly, e.g.:

$V_{N_2}^{3-3}$ – amount of nitrogen at cross-section (3–3)
V_{GAS}^{1-1} – overall amount of gases in roof elbow, cross-section (1–1)

Overall amount of gases at other cross-sections is designated in the same way.

$V_{-O_2}^{2-3}$ – amount of oxygen consumed for secondary post-combustion of CO in roof elbow between cross-sections (2–2) and (3–3)
$V_{+CO_2}^{2-3}$ – amount of CO_2 evolved as a result of secondary post-combustion of CO
c_{GAS}^{1-1} – heat capacity of gases in cross-section (1–1), kJ/m^3 × °C
t_{GAS}^{1-1} – gas temperature at cross-section (1–1),°C
Q_{GAS}^{1-1} – heat content of gases at cross-section (1–1), kJ/m^3

c, t, and Q at other cross-sections are designated similarly.

p^{2-2} – maximum negative pressure at cross-section (2–2), Pa
$V_{GAS.t}^{1-1}$ – actual amount of gases per second at cross-section (1–1) at temperature t, m^3/s

Pressures p, velocities v, and density ρ of gases at other cross-sections are designated similarly. designated similarly.

$v_{GAS.t}^{1-1}$ – actual gases velocity at cross-section (1–1) at temperature t, m/s

Amount of CO Evolving from Bath V_{CO}^{0-0}, m^3/min

Oxygen blown into the bath is consumed to oxidize carbon dissolved in it in accordance with the reaction:

$$2\,C + O_2 = 2\,CO. \tag{14.1}$$

According to reaction (14.1), 2 m^3 of CO evolves per 1 m^{3}of O_2. In reality, oxygen is consumed to oxidize not only carbon but also iron and its alloys which are oxidized without forming the gaseous products. A portion of oxygen is not absorbed by the bath at all and evolves into the freeboard. In the calculations, this fact is taken into consideration by introducing of the so-called coefficient of using oxygen for carbon oxidizing $\mathcal{K}_{O_2} < 1$. If the overall amount of blown oxygen is V_{O_2}, then oxygen consumption for oxidation of carbon in the bath in accordance with reaction (14.1) is defined as $\mathcal{K}_{O_2} \times V_{O_2}$.

On the other hand, carbon is oxidized not only by oxygen blown into the bath, but also by air infiltrated into the furnace, as well as by iron oxides accumulated in the slag during the previous stages of the heat. It does not seem possible to estimate the effect of all of the above-mentioned factors with the acceptable accuracy. This justifies the calculation of intensity of CO evolution from the bath with the aid of a simplified technique using formula:

$$V_{co}^{0-0} = 2 \times K_{0_2} \times V_{0_2}, \tag{14.2}$$

which takes into consideration the effect of all of the additional factors by using the coefficient $\mathcal{K}_{O_2} < 1$. In the furnaces operating with the specific intensity of oxygen blowing close to 1 m^3/ton × min, it is recommended to assume $\mathcal{K}_{O_2} = 0.7$ [5]. The results of such calculations are well correlated with the practical data.

Assuming the specific intensity of oxygen blowing equal to 0.95 m^3/ton × min for the 120-ton EAF under consideration, we obtain $V_{O_2} = 0.95 \times 120 = 114$ m^3/min (6840 m^3/h); and $V_{CO}^{0-0} = 2 \times 0.7 \times 115 \cong 160$ m^3/min.

Amount of Combustion Products Generated by Burners V_{GAS}^{BRN} m^3/min

In the modern furnaces, during the period of intensive bath blowing, the oxy-gas burners operate at sharply reduced power in the protective pilot flame mode. The number of the burners in the furnace is equal to 8. Let us assume their total power in the pilot flame mode to be 8.4 MW, which corresponds to the flow rate $V_{CH_4} =$ 16 m^3/min when recalculated for methane (CH_4). Natural gas is substituted with methane in order to simplify calculation of burning. Such simplification introduces a small error only into the composition of the primary gases, which is insignificant.

In the burners, gas burns completely without underburning in accordance with the reaction:

$$CH_4 + 2\,O_2 = 2\,H_2O + CO_2. \tag{14.3}$$

According to (14.3), 1 m^3 of CO_2 and 2 m^3 of H_2O are formed per 1 m^3 of burned CH_4. Therefore, when 16 m^3/min of gas are burned, 16 m^3/min of CO_2 and 32 m^3/min of H_2O are formed. The total amount of combustion products is $V_{GAS}^{BRN} =$ 48 m^3/min.

Amount of Primary Air Infiltrated into Freeboard, $V_{AIR}^{\bullet-0}$ m^3/min

The amount of infiltrated air depends on the furnace roof height above the sill level H, area of the slag door opening, gas pressure under the roof p_{RF}, and density differences between the furnace gases and external air. If $p_{RF} = 0.0$ Pa, the value of V_{AIR}^{0-0} can be calculated using the simplified formula:

$$V_{AIR}^{0-0} = 180(\varepsilon \times h \times b) \times \sqrt{H - 0.5\varepsilon \times h},\ m^3/min, \tag{14.4}$$

h – overall height of opening of slag door, m
ε – portion of height h to which the doorplate is raised
b – door width, m

Formula (14.4) is obtained using the known dependence of gas flow through the aperture with "thick" walls under the following conditions: external air temperature is 20°C, density is 1.2 kg/m^3, temperature of the gases in the freeboard is 1650°C, and the density of gases is 0.18 kg/m^3. If p_{RF} values are not equal to zero, the following approximate relations can be used; if p_{RF} = – 5 Pa, the value of V_{AIR}^{0-0} calculated using formula (14.4) should be increased 1.2 times, and if p_{RF} = + 5 Pa, this value should be decreased by the same number of times.

In 120-ton furnace, $H = 2.6$, $h = 1.25$, and $b = 1.0$ m. Let us assume that the slag door is opened by a quarter-height ($\varepsilon = 0.25$). Inserting these data into formula (14.4), we obtain V_{AIR}^{0-0} = 87.4 m^3/min. In the mode of p_{RF} = −5 Pa, which guaranties that the gases do not escape from the freeboard, the maximum amount of the primary air is: $V_{AIR}^{0-0} = 87 \times 1.2 = 105$ m^3/min.

Maximum Amount of Primary Gases at the Outlet of the Roof Elbow, V_{AIR}^{0-0} m^3/min

Let us find the maximum amount of gases at cross-section (1–1):

$$V_{GAS}^{1-1} = V_{CO}^{0-0} + V_{GAS}^{BRN} + V_{AIR}^{0-0};\ V_{GAS}^{1-1} = 160 + 48 + 105$$
$$= 313\,m^3/min\,(18780\,m^3/h).$$

Composition of Primary Gases, %

When determining the composition of gases at the outlet of the roof elbow, it should be taken into consideration that a portion of CO burns in the freeboard and in the elbow in accordance with the reaction:

$$CO + 0.5\,O_2 = CO_2. \tag{14.5}$$

According to (14.5), when 1 m^3 of CO burns, 1 m^3 of CO_2 is formed and 0.5 m^3 of O_2 is consumed. Therefore, during post-combustion of CO to CO_2, the volume of gases reduces due to the oxygen consumption.

In the modern furnaces which operate most of the time with closed door, it can be assumed that the primary post-combustion of CO is carried out using oxygen delivered specifically for this purpose through the sidewall burners or special tuyeres. Such post-combustion does not affect the amount V_{GAS}^{1-1} of primary gases; only their composition changes. To determine this composition, let us assume that 35% of CO is burned in the freeboard. Under this condition, let us calculate the composition of humid and dry gases at the outlet of the roof elbow. The composition of humid gases includes water vapor. The composition of dry gases corresponds to the gas analyzer readings. The gases enter the analyzer at room temperature after water vapor condensation.

The amount of CO_2, formed between cross-sections (0–0) and (1–1) $V^{0-1}_{+CO_2}$ is equal to the amount of burned CO (14.5).

$$V^{0-1}_{+CO_2} = 0.35 V^{0-0}_{CO} \quad V^{0-1}_{+CO_2} = 0.35 \times 160 = 56\,\mathrm{m^3/min}.$$

The overall amount of CO_2 at cross-section (1–1) is equal to:

$$V^{1-1}_{+CO_2} = V^{0-1}_{CO_2} + V^{BRN}_{CO_2} \quad V^{0-1}_{+CO_2} = 56 + 16 = 72\,\mathrm{m^3/min}.$$

The amount of CO after the primary post-combustion is V^{1-1}_{CO} =160–56 = 104 m^3/min CO% = 104/313 = 33%.

Similarly, when $V^{BRN}_{H_2O}$ = 32 m^3/min and V^{0-0}_{AIR} = 105 m^3/min, we find H_2O% = 32 × 100/313 = 10%; O_2% = 0.21 × 105 × 100/313 = 7%; N_2% = 0.79 × 105 × 100/313 = 27%.

The composition of dry gases for four components, i.e., CO_2, CO, O_2, and N_2, is calculated the same way with consideration given to reduction of overall amount of gases V^{1-1}_{GAS} by a value of $V^{BRN}_{H_2O}$ = 32 m^3/min, i.e., from 313 m^3/min to 313–32 = 281 m^3/min. For the dry gases CO_2% = 72 × 100/281 = 26%, etc.

The Temperature of Primary Gases t^{1-1}_{GAS}, °C

The processes of heat exchange in the freeboard are so complex that the temperature of the gases at the exit from the freeboard cannot be calculated theoretically with acceptable accuracy. Therefore, with a good approximation to practical data, we can assume that the temperature t^{1-1}_{GAS} in the currently reviewed period of oxygen blowing of the bath exceeds the metal temperature by 50°C and equals to approximately 1650°C. We assume that the temperature value takes into consideration, in particular, both heat losses by the gases through radiation onto the water-cooled walls and the roof of the furnace and heat input from partial post-combustion of CO in the freeboard.

Physical Heat Content of Primary Gases $Q^{1-1}_{GAS.PH}$

In accordance with accepted simplifying assumption with regards to the temperature t^{1-1}_{GAS}, the heat content of the primary gases per second $Q^{1-1}_{GAS.PH}$, kW, is defined by the expression $Q^{1-1}_{GAS.PH} = c^{1-1} \times t^{1-1}_{GAS} \times V^{1-1}_{GAS}/60$, kJ/s or kW (1 J/s = 1 W).

In those equations where heat apacities are expressed in kJ/m^3 × °C, the values of V_{GAS} must be measured in m^3/s.

$$c^{1-1} = 1.70\,\mathrm{KJ/m^3} \times {}^\circ\mathrm{C} \;-\; \text{heat capacity of humid gases at } t^{1-1}_{GAS} = 1650^\circ\mathrm{C},$$

$$Q^{1-1}_{GAS.PH} = 1.7 \times 1650 \times 313/60 \cong 14630\,\mathrm{kW}.$$

Amount of Gases at the Inlet of the Stationary Gas Duct (Cross-Section 2–2) V_{GAS}^{2-2} Including Amount of Secondary Air Infiltrated Through the Gap (8), V_{AIR}^{2-2}, m³/min

The amount of primary gases after passing through the gap (8) between the roof elbow and the stationary gas duct increases to V_{GAS}^{2-2} due to infiltrated through the gap secondary air, Fig. 14.3. The value of V_{GAS}^{2-2} can be calculated using the approximate formula:

$$V_{GAS}^{2-2} = V_{GAS}^{1-1} \times [1 + L/d \times (17 + 9 \times 10^{-6} \times t^2)], \qquad (14.6)$$

L – gap width, m
d – hydraulic diameter of the roof elbow, m
t – temperature of gases in the roof elbow, t_{GAS}^{1-1}, °C

This formula was obtained as a result of studies with cold and fire models of gas evacuation devices, as well as from the measurements in the operating 100-ton and 200-ton furnaces [5]. In the range of regular ratios of the area of cross-section (2–2) to the area of (1–1), inaccuracy of formula (14.6) is no more than 20%. If V_{GAS}^{1-1} = 313 m³/min, L = 80, d = 1250 mm, and t_{GAS}^{1-1} = 1650°C, the calculation using formula (14.6) gives the maximum value V_{GAS}^{2-2} = 1139 m³/min (19 m³/s). The amount of secondary air infiltrated through the gap (8) is in this case equal to V_{GAS}^{2-2} = 1139–313 = 826 m³/min.

Temperature t_{GAS}^{2-2} and Physical Heat Content $Q_{GAS.PH}^{2-2}$ of Gas Mixture with Air at the Inlet of the Stationary Gas Duct

The temperature t_{GAS}^{2-2} is determined from the heat balance equations for gases at the cross-sections (1–1) and (2–2) on condition that at the cross-section (2–2) only the dilution of gases by the secondary air occurs without the post-combustion of carbon monoxide CO entered from the freeboard (104 m³/min). Under this condition, the heat content of gases at the cross-section (2–2) remains the same as at the cross-section (1–1).

$$V_{GAS}^{2-2} \times c^{2-2} \times t_{GAS}^{2-2} = 14,630 \text{ kW}, \qquad (14.7)$$

Hence $t_{GAS}^{2-2} = 14630/19.0 \times c^{2-2} = 770/c^{2-2}$

Since the heat capacity of the gases c^{2-2} depends on the unknown temperature t_{GAS}^{2-2}, Eq. (14.7) is solved by the method of consecutive approximations. In a first approximation, c_1^{2-2} is assumed in accordance with the expected temperature, e.g., 400°C. At 400°C, c_1^{2-2} = 1.33 kJ/m³ × °C (considering the composition of mixture of gases at the cross-section (2–2), determination of this composition is not shown.). By solving (14.7) at c_1^{2-2} = 1.33 kJ/m³ × °C, we find that in the first approximation t_{GAS}^{2-2} = 770/1.33 = 579°C. In the second approximation, the calculation is repeated using the value of heat capacity c_2^{2-2} = 1.36 kJ/m^3 × °C, which corresponds to the new chosen value t_{GAS}^{2-2} = 600°C. We find the new value t_{GAS}^{2-2} = 770/1.36 = 566°C. The rounded value $t_{GAS}^{2-2} \cong$ 570°C can be accepted as a final result, since further correction of the values of c^{2-2} and t_{GAS}^{2-2} is not required. If the first chosen temperature was much more different from the final value and the heat capacity changed with the temperature more sharply, a third approximation would be necessary. In practice, two approximations are usually enough.

Amount of Gases After Secondary Post-combustion in the Stationary Gas Duct, V_{GAS}^{3-3}, m^3/min

In the case under consideration, complete post-combustion of CO is realized due to oxygen from the air infiltrated through the gap (8) and oxygen contained in the primary gas. The post-combustion must be finished at the cross-section (3–3) before atomized water is injected into the gas duct, Figs. 14.2 and 14.3. The required value is defined by the expression $V_{GAS}^{3-3} = V_{GAS}^{2-2} - V_{-O_2}^{2-3}$, where $V_{-O_2}^{2-3}$ is the amount of oxygen consumed for secondary post-combustion of CO between the cross-sections (2–2) and (3–3). The amount of CO at the inlet to the stationary gas duct is V_{CO}^{1-1} = 104 m^3/min. According to reaction (14.5), the amount of CO_2 obtained from secondary post-combustion is the same: $V_{+CO_2}^{2-3}$ = 104 m^3/min and $V_{-O_2}^{2-3}$ = 52 m^3/min.

$$V_{GAS}^{3-3} = 1139 - 52 = 1087\,\text{m}^3/\text{min}\,(18.1\,\text{m}^3/\text{s}).$$

Composition of Humid Gases After Secondary Post-Combustion, %

This composition is calculated similarly to the composition of gases at the crosssection (1–1):

$V_{CO_2}^{3-3} = V_{CO_2}^{1-1} + V_{+\Delta CO_2}^{2-3}$; $V_{CO_2}^{3-3}$ = 72 + 104 = 176 m^3/min; CO_2 = 176 × 100/1087 = 16%.

$V_{H_2O}^{3-3} = V_{H_2O}^{1-1}$ = 32 m^3/min; H_2O% = 32 × 100/1087 = 3%.

The amount of oxygen in the cross-section (3–3) is a sum of amounts of oxygen in primary and secondary air except $V_{O_2}^{3-3}$:

$V_{O_2}^{3-3}$ = 0.21 × (105 + 826) – 52 = 144 m^3/min; O_2% = 144 × 100/1087 = 13%.

Similarly $V_{N_2}^{3-3}$ = 0.79 (105 + 826) = 735 m^3/min; N_2% = 735 × 100/1087 = 68%.

Total Heat Content and Temperature of Gases After Secondary Post-Combustion Q_{GAS}^{3-3},kW and t_{GAS}^{3-3}, °C

These values are determined from the equation of heat balance for gases at the cross-sections (2–2) and (3–3).

$$Q_{GAS}^{3-3} = Q_{GAS.PH}^{2-2} + V_{CO}^{2-2} \times Q_{CO}. \tag{14.8}$$

$Q_{GAS.PH}^{2-2}$ = 14630; $V_{CO}^{2-2} = V_{CO}^{1-1}$ = 104 m^3/min (1.73 m^3/s)

Q_{CO} = 12,700 kJ/m^3 CO–thermal effect of reaction (14.5)
Q^{3-3}_{GAS} = 14,630 + 104 × 12700/60 = 36,643 kW; $Q^{3-3}_{GAS} = c^{3-3} \times t^{3-3}_{GAS} \times V^{3-3}_{GAS}$.
Hence $t^{3-3}_{GAS} = 36,643/18.1 \times c^{3-3}$

Taking into consideration the composition of gases at the cross-section (3–3) and using the method of consecutive approximations we find c^{3-3}= 1.57 kJ/m^3 × °C and t^{3-3}_{GAS} ≅ 36,643/18.1 × 1.57 ≅ 1290°C.

Negative Pressure at the Inlet of the Stationary Gas Duct Necessary for Prevention of Uncontrolled Emissions Through the Electrode Ports

To directly evacuate the entire amount of primary gases at the moments of their most intensive formation, the gas exhausters must provide minimum required rarefaction (negative pressure) $-p^{2-2}_{MAX}$, Pa, at the cross-section (2–2). In the past, the typical error in the design of gas evacuation systems was using of gas exhausters with a quite sufficient productivity, but with a pressure not big enough to create such rarefaction. As a result, the uncontrolled emissions from the free-board could not be eliminated.

Hydraulic calculation of the portion of the gas duct between cross-sections (1–1) and (2–2) aimed to determine the value $-p^{2-2}_{MAX}$ has significant specifics, which were not encountered in the calculation of water-cooled elements, Chap. 12. The main difference is that the amount of primary gases increases downstream due to infiltration of air through the gap between the roof elbow and the gas duct. In the water-cooled elements, the water flow rate remained constant along the duct length.

Due to this specific of the portion of the duct under consideration, to determine the maximum value of negative pressure $-p^{2-2}$ (here and further the index "max" is dropped), we should use the Bernoulli equation. It expresses the law of conservation of energy of the incompressible gas flow. When it is applied to the problem under consideration, this equation can be rearranged as follows:

$$(p + \rho v^2/2)_{2-2} = (p + \rho v^2/2)_{1-1} - \zeta_8(\rho v^2/2)_{1-1}. \qquad (14.9)$$

The first two terms of Eq. (14.9) represent the energy of flow at the cross-sections (2–2) and (1–1) equal to the sum of potential energy of pressure p and kinetic energy $\rho v^2/2$. The indices at the parenthesis are applicable to all values inside the parenthesis.

The physical meaning of Eq. (14.9) is that the energy of the flow at the cross-section (2–2) is equal to the energy at the cross-section (1–1) less irreversible losses of energy, which is converted to heat when the flow of the primary gases collides with the flow of infiltrated air. The amount of losses is determined by the last term of Eq. (14.9), where ζ_8 is the coefficient of local resistance of the gap between the elbow and gas duct. This coefficient depends on geometrical parameters of the gap and can be calculated using a simplified formula:

$$\zeta_8 = 0.7 + 15L/d_{\text{h}}, \qquad (14.10)$$

L – width of gap, m
d_h – hydraulic diameter of roof elbow

Formula (14.10) was obtained as a result of the same studies which led to formula (14.6) for determining V^{2-2}_{GAS}.

By solving Eq. (14.9) for p^{2-2}, we obtain the following:

$$p^{2-2} = p^{1-1} + (\rho v^2/2)_{1-1} \times (1 - \zeta_8) - (\rho v^2/2)_{2-2}. \tag{14.11}$$

Let us assume that the pressure of the gases under the roof and their velocity in the furnace freeboard is equal to zero; the diameter of roof opening is equal to the diameter of the roof elbow, and quite insignificant hydraulic resistance due to friction in the elbow can be ignored. Under these conditions, using the corresponding Bernoulli's equation for the cross-sections (0–0) and (1–1) we can obtain the following expression:

$$p^{1-1} = -(\rho v^2/2)_{1-1} \times (1 + 0.5), \tag{14.12}$$

where 0.5 is the coefficient of local resistance for gases entering the roof elbow.

Let us find the actual velocity of gases at the temperature of 1650°C, $v^{1-1}_{GAS.t}$. The density of the gases at this temperature is $\rho^{1-1} = 0.18$ kg/m^3. The actual amount of gases per second is $v^{1-1}_{GAS.t}$ = 313 × (1650 + 273)/273 × 60 = 36.7 m^3/s. For the diameter of the roof elbow of d_h = 1.25 m and its cross-section area equal to F^{1-1} = 1.23 m^2: $v^{1-1}_{GAS.t}$ = 36.7/1.23 = 29.9 m/s.

Using expressions (14.12 and 14.10), we find P^{1-1} = – 0.18 × 29.9^2 × 1.5/2 ≅ −121 Pa; ζ_8 = 0.7 + 15 × 0.080/1.25 = 1.66.

Let us find the actual velocity of gases at the cross-section (2–2) $v^{2-2}_{GAS.t}$. The density of the gases at this cross-section at t^{2-2} = 570°C is ρ^{2-2} = 0.43 kg/m^3. The amount of gases per second recalculated for this temperature is as follows:

$$V^{2-2}_{GAS.t} = 19.0 \times (570 + 273)/273 = 59\,\text{m}^3/\text{s}.$$

For the cross-section area of the gas duct F^{2-2} equal to twice the area of the elbow F^{1-1}, the actual velocity of gases is $v^{2-2}_{GAS.t}$ = 59.0/1.23 × 2 = 24 m/s.

Then, using Eq. (14.11), let us find the unknown maximum rarefaction (negative pressure): p^{2-2} = – 121 + (0.18 × 29.9^2/2)(1–1.66)–0.43 × 24^2/2 = −300 Pa.

We must point out that as the diameters of the roof opening, the roof elbow, and the gas duct decrease, the required negative pressure p^{2-2} sharply increases.

14.3.4 Energy Problems

The most important feature of the modern systems of gas evacuation of the EAFs is a quite unusual for other steelmaking units ratio between the amount of process gases formed inside of the furnace and the amount of gases emitted into the atmosphere after the gas purification. As it was shown in Sect. 14.3.3, the maximum intensity of evolution of gases in the freeboard of 120-ton furnace, including the infiltrated air, is about 19,000 m^3/h. When these gases enter the bag filters for dust removal, their amount increases due to the dilution with air up to 1.4–1.5 million m^3/h, i.e., by 70–80 times. As a result, in the modern EAFs, about 10% of entire electrical energy used for steelmaking is consumed for evacuation and purifying of quite insignificant, in essence, amount of primary process emissions.

Such excessive energy consumptions for gas purification are caused mostly by a necessity to capture the secondary uncontrolled emissions by low-efficiency exhaust hoods located above the furnace. The modern environmental standards for control of air pollution with dust require complete absence of any visible dust emissions above the still mill, which is controlled by the sensitive optical instruments. To satisfy these requirements, huge amounts of dust-laden air must be exhausted through the hood.

Satisfactory operation of the exhaust hoods is hampered by the large distances vertically from the furnace to the hood, the presence of the crane, and the basket during the scrap charging between them, as well as the air flows along the shop building which entrain the dust-laden gases. To diminish the effect of these flows, the areas of the aperture in the shop walls for the rail carriers are minimized; the apertures are enclosed with air curtains; the air delivery uniformly distributed along the shop perimeter is provided by the systems of forced ventilation, and the hood configuration is improved. Nevertheless, the studies and the accumulated experience show that even under the most favorable conditions, to ensure the complete capturing of the secondary uncontrolled emissions, the velocity of suction of air–gas mixture through the open from below hood aperture must not be less than 2 m/s. In 120-ton furnaces, with the area of hood aperture equal about 200 m^2, such velocity of suction results in the volumes of gases passing through the hood, which are shown above ($200 \times 2 \times 3600 = 1{,}440{,}000$ m^3/h).

One of the trends in reduction of energy costs for gas evacuation is the installation of furnaces in the small-volume air-tight doghouses, which excludes the usage of hoods. In the 1980 s, such doghouses were widespread. They enclosed not only the furnace itself, but also the zone of tapping of metal into the ladle. Therefore, all the secondary uncontrolled emissions were concentrated in the doghouse. To provide passage for the charging baskets and the ladle for steel tapping, these doghouses were equipped with tight-closing doors of different design. Due to relatively small volume of doghouses, the amount of gases evacuated from them, assuring the complete capturing of the secondary emissions, reduced by a number of times compared to the hoods. At the same time, the doghouse protected the personnel from the furnace noise.

Despite the advantages noted above, with the further increase in furnace productivity, the use of the doghouses had to be ceased. This was caused by the following reasons. At every heat, massive doghouse overhead half-doors for the passing of the charging baskets and half-doors for passing of the ladle car must be moved repeatedly. As the number of heats per day was about 40, even insignificant delays of the technological process related to these additional operations caused unacceptable drop of the furnace productivity. Besides, the necessity to work inside the doghouse greatly complicates the maintenance of such furnace elements as oxy-gas burners and devices for oxygen, carbon, and lime injection into the bath. The doghouse prevents the use of manipulators at the working platform, particularly to control temperature and composition of the metal.

Recently, a number of effective innovations to reduce volumes of uncontrolled emissions and energy costs for gas evacuation were developed and implemented. Furnace operation modes with closed door became widespread. In the course of the heat, the door opens only when it is necessary and for a short time, for example to remove slag. When the reliable sensors for gas pressure under the roof became available, the automatic control of draft of gas exhausters to maintain the optimum pressure during the entire heat has been successfully implemented. The operation of the furnace with the closed door and with maintaining of the low negative pressure under the roof allows sharply reducing air infiltration into the furnace and completely eliminating the secondary uncontrolled emissions through the electrode ports and the annular gap between the roof and the shell of the furnace.

Amount of the secondary emissions at the tapping can be sharply reduced, if the exhaust hood capturing them is located close to the ladle. Despite the attempts to develop such design, the satisfactory solution has not been found so far. However, other possibilities exist. In the most modern furnaces designed for charging the entire amount of scrap with one basket, the duration of a single dust–gas emission does not exceed 3–4 min. The emission during the tapping has approximately the same duration. Short duration of these single emissions combined with total lack of the frequent escape of gases through the electrode ports enables a new approach to the solution of the problem of capturing of the secondary emissions under the shop's roof.

The new approach is to use the so-called deep accumulating hoods which are the air-tight pockets built-in the top part of the shop building. The volume of these pockets is approximately 10 times greater than the volume of the regular hoods. This allows accumulating the entire single emission of the dust-laden gases at the scrap charging, as well as at the tapping. In comparison to air, these gases have higher temperature and lower density, hence they can remain in the pockets long enough. Therefore, they can be exhausted gradually as opposed to the very short time period of emission as in the regular hoods.

The highest intensity of the secondary emissions is reached during the charging of scrap containing large amounts of oil and plastics. In these moments, the power of the exhausters has to be approximately doubled compared to the other periods of the heat. This significantly increases the costs of the gas purification, since it requires the respective increase in its carrying capacity. Substitution of the hoods with the accumulating pockets allows to stretch exhausting of a single emission in time and, therefore, to sharply reduce the required maximum power of the exhausters and energy costs.

Significant contribution into increase in the energy expenditures in the gas evacuation system comes from a simplest but quite cost-ineffective method of the cooling of the primary emissions by diluting them with cold air. As calculations show, the amount of air at the temperature of 25°C, which has to be mixed with 1 m^3 of hot gases to cool them down from 1600 to 150°C, is approximately 15 m^3. Unfortunately, incomparably more effective method of cooling of the primary emissions by atomized water is not used in all cases, since it requires strict control of the amount of injected water and atomization quality in the course of the heat. Water must evaporate completely. Otherwise, unacceptable moisturizing and plugging of the bag filters occur.

If the potential of the reviewed innovations is fully utilized, they allow to significantly reduce the ultimate volumes of emissions and costs for their capturing and purification. Nonetheless, the energy problem of gas evacuation in the EAFs is quite far away from being satisfactorily resolved. As the capacity and productivity of the furnaces increase and the environmental protection standards become stricter, even further aggravation of this problem can be expected. This will require new, more radical approaches to its resolution.

14.4 Use of Air Curtains

The EAF operation mode with closed door, which became widespread, eliminates the air infiltration into the freeboard. However, this mode has significant disadvantages. It is necessary to eliminate quite efficient mobile door burners located at the manipulators. In the absence of such burners, the complete melting of the bath near the sill is somewhat delayed. Accordingly, the moment when it becomes possible to control the temperature and the composition of the bath, is pushed back as well. This, in turn, delays the tapping. The caked slag–metal build-up forms on the sill near the closed water-cooled door. The removal of this build-up is associated with certain difficulties. Pushing the slag–metal build-up into the bath even further delays its melting near the sill. Besides, the operation with closed door complicates the heat control as it deprives the operator of important visual information regarding the condition of the freeboard, Chap. 13, Sect. 13.3.

The alternative to the closed door mode is using of air curtains, which is free from the disadvantages mentioned above. However, to successfully substitute the door with the air curtain, the elimination of infiltration of the external air must be assured on condition that the jets of the curtain itself do not enter the freeboard. Such mode can be called the mode of complete aerodynamic locking of the door. Besides, the curtain must not worsen the environmental situation at the working platform of the furnace.

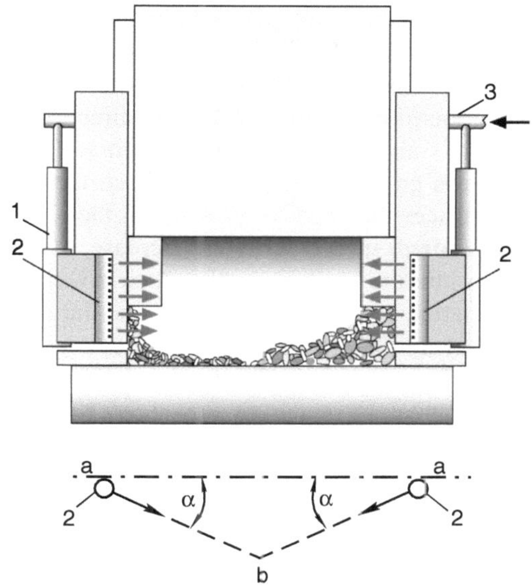

Fig. 14.4 Air curtain for 60-ton EAF, 3 – delivery of compressed air (the rest of designations are given in the text)

The development of efficient air curtains for EAFs was hindered by lack of sound scientific concepts regarding the very mechanism of their operation in the mode of complete aerodynamic locking and, consequently, by lack of calculation methods of such curtains. Due to this reason, the purely empirical attempts to develop efficient curtain designs for the EAFs for long time did not yield positive results.

The studies conducted by the authors on the cold and hot models of various air curtains, as well as on the EAFs of 100- and 200-ton capacity, has shown that the mechanism of air curtain operation in the complete aerodynamic locking of the door has nothing in common with the air curtains, which are installed in the building doors to prevent penetration of cold external air [5–7]. The obtained data allowed offering a simple calculation method and efficient design of the air curtains for the EAFs. A double-sided air curtain at the slag door of the 60-ton EAF at the plant in Akko (Israel) is an example of such design, Fig. 14.4.

The required air pressure in front of the curtain nozzles does not exceed 0.3 bar, which corresponds to the initial jet velocity of approximately 190 m/s. However, to reduce the overall dimensions of air ducts in the door zone, compressed air under pressure of 4–6 bar rather than fan air is supplied to the curtain. To use the excessive compressed air pressure effectively and to reduce its flow rate, the injectors (1) are built-in into the curtain design. In the injectors, the compressed air is mixed with the atmospheric air. Due to this fact, the flow rate of the compressed air does not exceed 300 m^3/h (5 m^3/ton), which is only about 60% of the entire air used in the curtain.

The low pressure (about 0.3 bar) air obtained in the injectors enters the collectors (2) with the nozzles spread over their height. Air jets flowing from the nozzles converge and form flat flows *a–b*, directed at the angle α to the plane of the door aperture *a–a*, Fig. 14.4. Such direction of the jets eliminates the possibility of curtain's air penetration into the furnace freeboard.

Laws governing the jets in this curtain are different in principle from the free jets. Under the negative pressure in the open slag door, they behave as the jets in the limited space. With an increase in the distance from the nozzle the jet momentum reduces, as well as their kinetic energy which is expended to form the circulation zones on both sides of the flat flows *a–b*. As a result, the static pressure field of the gases in the door zone changes substantially.

At the certain value of the ratio of the initial momentum of curtain jets I_{CUR} to the momentum of the air flow I_{AIR}, which would be infiltrated into the furnace without the curtain, the mode of complete aerodynamic locking the door arises. In this mode, the static pressure differential between the circulation zones divided by the flat air flows of the current *a–b* becomes equal to zero, and flowing of air through the boundary *a–b–a*, Fig. 14.4, stops. Under these conditions, the curtain behaves as a partition practically impenetrable to the air flows. Under this mode, the air exchange between the circulation zones is insignificant and occurs only through the turbulent pulsations.

The studies conducted has shown that the value of the ratio

$$I_{CUR}/I_{AIR} = K; (K > 1), \tag{14.13}$$

which assures the complete aerodynamic locking of the door does not depend on the initial velocity of jets, the area of the open door, and the negative pressure in the door. The criterion $\mathcal{K}$ is determined only by the basic geometrical relations between the door and the curtain such as angle α, Fig. 14.4. In the calculations of the door curtains for EAF it can be assumed $\mathcal{K} \cong 3$.

The calculations of the air curtains are performed using the following method. Using formula (14.4), for the given height of the EAF roof over the sill level, the dimensions of the slag door, and the degree of its opening, let us find the amount of air infiltrated into the furnace freeboard without the curtain, V_{AIR}, m^3/s. Since in this case we find the amount of air not per minute but rather per second, the coefficient equal 3 must be introduced into formula (14.4) instead of the coefficient equal 180. By dividing V_{AIR} by the area of the open aperture $F = \varepsilon \times h \times b$, m^2, let us find the flow velocity of the air infiltrated through the aperture v_{AIR}, m/s and its momentum $I_{AIR} = \rho \times V_{AIR} \times v_{AIR}$, where $\rho = 1.2$ kg/m^{3}is air density outside the furnace. Then, for $\mathcal{K} = 3$, let us find the required momentum of the curtain $I_{CUR} = 3$ I_{AIR}. Assuming the initial jet velocity to be equal v_{CUR}, let us find the air flow rate into the curtain: $V_{CUR}= I_{CUR}/\rho \times v_{CUR}$. For the given number of the nozzles in the collectors, we find the overall area and the nozzle diameter using the values V_{CUR} and v_{CUR}.

The curtain momentum I_{CUR} increases directly proportionally to v_{CUR}. Therefore, in order to reduce the air flow rate into the curtain, the velocity of the air jets should be increased by reducing the nozzle diameter. However, the increase in v_{CUR} makes sense only up to 160–190 m/s. At the higher velocities, due to pressure increase in the collectors, the injection coefficient in the injectors drops and a substantial reduction of compressed air flow rate cannot be achieved. Besides, the noise from the curtain rises to undesired level.

If the curtain momentum differs from the minimum required momentum for the given area of the open door aperture and from the average value of the negative pressure in it, then the mode of the complete aerodynamic locking of the aperture is disrupted. At the lower momentum, the complete elimination of air infiltration into the furnace cannot be achieved. If the momentum is significantly higher, the curtain jets begin to entrain the dust-laden gases from the furnace onto the working platform. However, if the momentum fluctuates within ±20% from the required level, these phenomena are practically not detected. The best results are assured when the curtains are used simultaneously with the automatic control of the gas pressure under the furnace roof.

Under the conditions of 60-ton EAF at USM, the efficient operation of the curtain could be observed visually. When the curtain was turned off, the gases flow rising from the slag ladle is sucked into the open door at high velocity. When the curtain is turned on, this flow, which is clearly visible due to its high dust content, passes by the door without being deflected toward it. Simultaneously, the vortices of red fume appear at the door aperture; they pulsate very close to its exit, but do not escape outside, Fig. 14.5a, b.

Due to elimination of air infiltration into the furnace, the use of the curtain increased the yield by 0.8% on average. This result was confirmed with quite high accuracy by conducting the special experiments. In these experiments, the small series of heats with and without the curtain were conducted alternating randomly numerous times. The calculations show that the decrease in heat input from iron oxidation due to the observed increase in the yield compensates to a considerable extent the reduction of the heat losses due to elimination of cold air infiltration into

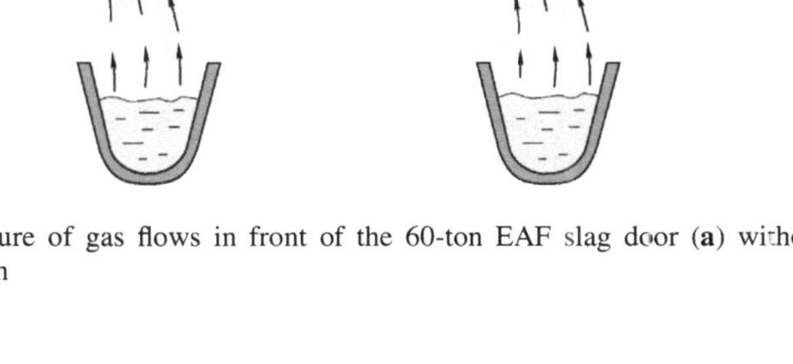

Fig. 14.5 Picture of gas flows in front of the 60-ton EAF slag door (**a**) without curtain and (**b**) with curtain

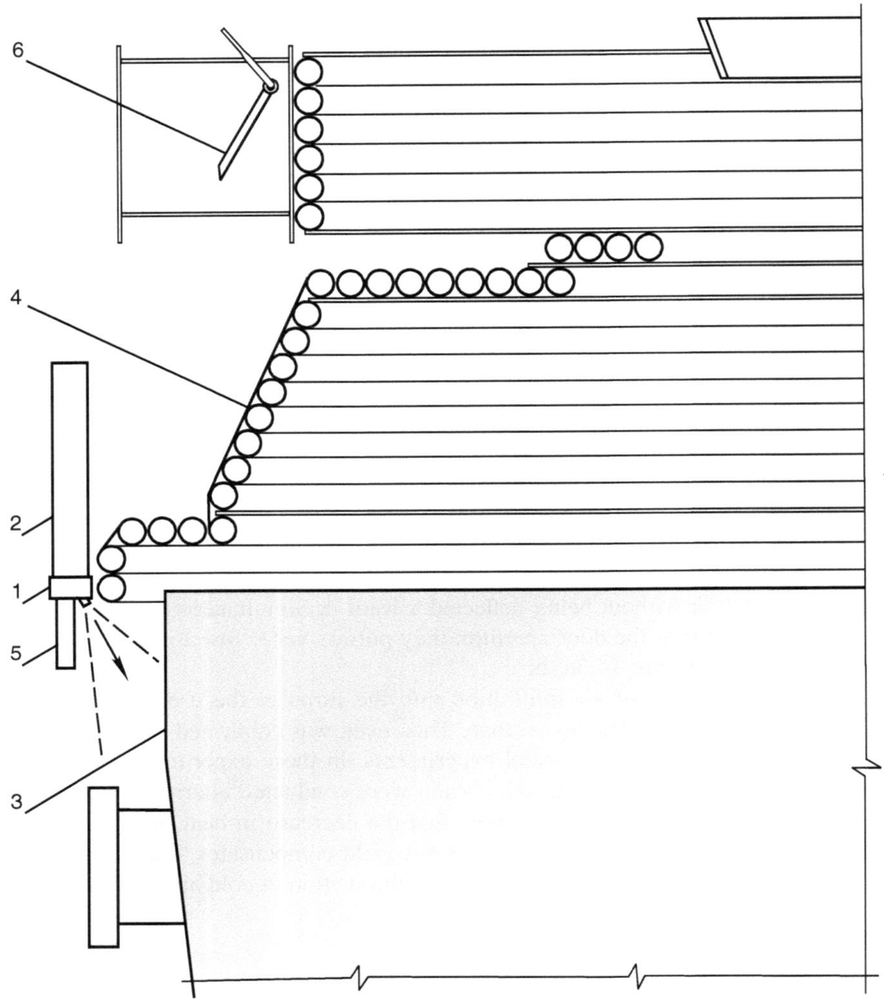

Fig. 14.6 Air curtain for a ladle furnace (designation are given in the text)

the furnace. Therefore, when the curtain was used, the reduction of electrical energy consumption was insignificant. The design of the curtain at USM turned out to be quite reliable. During a few years of continuous operation it practically did not require any maintenance or repair [8].

The highest effectiveness of curtain operation is provided by a combination with automatic adjustment of gas pressure under the EAF's roof. As is known, such an adjustment has a number of important advantages, particularly, allows improving operation of the gas evacuation system. Required pressure probes being lacked before are developed at present. The probe developed by the Preisman Technology Company is one of most reliable probes [9].

In melt shops, it is reasonable to use air curtains not only for EAFs but also for ladle furnaces. Air infiltration under a ladle lid hinders production of certain high-quality grades of low-nitrogen steel. The air is infiltrated through an annular gap between a ladle lid and ladle top. Because of slag build-ups on the ladle the width of such a gap is generally no less than 150–200 mm since a regular build-ups removal from ladles is associated with considerable difficulties.

Application of the air curtain is the optimal solution of this problem. Schematic diagram of the Preisman Technology's curtain is shown in Fig.14.6. Compressed air is delivered to the curtain. In order to reduce the pressure and save compressed air the injectors are used as it takes place in the slag door curtain, Fig.14.4.

An annular air collector (1) with nozzles is divided into two or four separate parts for the convenience of assembly. At each of parts the injector (2) is installed. Such a design provides uniform air distribution for all the multiple nozzles of the curtain. Annular water-cooled shield (5) is connected with the collector (1) in order to reduce consumption of the air which ensures the complete aerodynamic locking of the gap between the ladle top (3) and lid (4). The curtain jets are directed angularly downward tangentially to both the shield and ladle sidewall.

A minimal compressed air consumption can provide maintaining the gas pressure under an upper part of the lid at a level of +5 Pa by means of a butterfly damper (6) regulating the draft. At such a regulation the compressed air consumption for the curtain of a 200-t ladle furnace amounts to about 100 m^3/h (at a pressure of 6 bar abs.).

References

1. Amoldo L, Khan M, Evenson E et al (2003) Results of Goodfellow EFSOP at Hylsa. Planta Norte, ISS Mexico
2. Cordova E, Khan M, Zuliani D et al (2007) Advanced dynamic control technology for EAF steelmaking and other combustion processes. MPT Int 2:28–32
3. Stark CB (1990) Gas cleaning apparatus and units for metallurgical production. Metallurgia, Moscow
4. Shults LA, Kochnov YM, Kochnov MY (2006) Current state and development of systems for evacuation, utilization and purification of gases at large capacity high power electric arc furnaces. Ferrous Metals, pp 18–28
5. Kiselyov AD, Toulouevski YN, Zinurov IY (1992) Increase in efficiency of gas evacuation from electric arc furnaces. Metallurgia, Moscow
6. Idelchik IE (1972) Handbook for hydraulic resistances, 3-d issue. Engineering, Moscow
7. Toulouevski YN (1999) Air curtain for 60-t EAF at United Steel Mills Ltd. In: Proceedings of 6th European Electric Steelmaking Conference, Düsseldorf
8. Toulouevski YN (2001) Air curtain for EAF slag doors. MPT Int 4:128–129
9. Toulouevski Y, Preisman M, Coe L (2007) A self-cleaning water-cooled pressure probe implemented under the EAF roof. MPT Int 2:34–35

Index

Y. N. Toulouevski and I. Y. Zinurov, *Innovation in Electric Arc Furnaces*,
DOI: 10.1007/978-3-642-36273-6, © Springer-Verlag Berlin Heidelberg 2013

MIX
Papier aus verantwortungsvollen Quellen
Paper from responsible sources
FSC® C105338

If you have any concerns about our products,
you can contact us on
ProductSafety@springernature.com

In case Publisher is established outside the EU,
the EU authorized representative is:
Springer Nature Customer Service Center GmbH
Europaplatz 3, 69115 Heidelberg, Germany

Printed by Libri Plureos GmbH
in Hamburg, Germany